The Professional Knowledge Base of Science Teaching

Deborah Corrigan • Justin Dillon
Richard Gunstone
Editors

The Professional Knowledge Base of Science Teaching

Springer

Editors
Assoc. Prof. Deborah Corrigan
Monash University
Fac. Education
Centre for Science, Mathematics
& Technology Education
Clayton, Victoria
Australia
debbie.corrigan@monash.edu

Em. Prof. Richard Gunstone
Monash University
Fac. Education
Wellington Road
3800 Clayton, Victoria
Australia
richard.gunstone@monash.edu

Prof. Justin Dillon
King's College London
Dept. of Education & Professional Studies
Franklin-Wilkins Bldg.
150 Stamford Street
London SE1 9NH
United Kingdom
justin.dillon@kcl.ac.uk

ISBN 978-90-481-3926-2 e-ISBN 978-90-481-3927-9
DOI 10.1007/978-90-481-3927-9
Springer Dordrecht Heidelberg London New York

Library of Congress Control Number: 2011922162

Printed on acid-free paper

Springer is part of Springer Science+Business Media (www.springer.com)

Preface

This is the second book in the series from the Monash University–King's College London International Centre for the Study of Science and Mathematics Curriculum. This centre was established in 2002 with initial support from the Monash University Research Fund (new areas), and in the context of the signing of an agreement between Monash and King's, two years earlier, that led to the establishment of the then Monash University London Centre.

The first book in the series, *The Re-Emergence of Values in Science Education* (D. Corrigan, J. Dillon & R. Gunstone [Eds.], Rotterdam: Sense Publishers, 2007), considered the state of science education in the twenty-first century through a lens of values. The book presented a 'big picture' of what science education might be like if values once again become central in science education. However, overwhelmingly the experiences of those who teach science have been in an environment which has seen the de-emphasizing of values in both science and science education. So there is a disparity between the evolutionary process that science is undertaking and that undertaken by science education (and school science education in particular). In this book, *The Professional Knowledge Base of Science Teachers,* the focus is on exploring what expert science education knowledge and practices may look like in the emerging 'bigger picture' of the re-emergence of values.

We used the same approach to the creation of this book as we did with the previous book focussed on values in science education. In order to attempt both the creation of a cohesive contribution to the literature and having authors able to assert their own voices without restrictive briefs from us as editors, we again organised a workshop involving the authors and ourselves to enable a more interactive and formative writing process. Authors completed a first draft of their chapters in time to distribute them to all workshop participants before we met. The workshop then involved discussions of individual chapters and feedback to authors, and considerations of the overall structure and cohesion of the volume. Authors then rewrote their chapters in the light of these forms of feedback. As with the values book, the workshop was scheduled around the European Science Education Research Association (ESERA) conference, but on this occasion the workshop took place at the Monash University Centre in Prato (Italy) rather than in the same city as ESERA.

As well as for the values book, this procedure had previously been used very successfully in the production of two other books in which the editors had variously been involved *The Content of Science: A Constructivist Approach to its Teaching and Learning* (P. Fensham, R. Gunstone & R. White [Eds.], London: Falmer Press, 2000); *Improving Science Education: The Contributions of Research*, (R. Millar, J. Leach & J. Osborne [Eds.], Milton Keynes: Open University Press, 1994). We believe that this process significantly improves the quality of the final product and provides an opportunity for what is, sadly, a very rare form of professional development—considered and formative and collaborative (and totally open) discussions of one's work by one's peers.

We gratefully acknowledge the funding of the workshop through contributions to the Monash-King's College London International Centre for the Study of Science and Mathematics Curriculum from the Monash University Research Fund and from King's College London.

May 2010

Richard Gunstone
Deborah Corrigan
Justin Dillon

A very sad postscript
Late in August, as this book was in its final stages of production with Springer, we received the tragic news that Sandi Abell had lost her battle with cancer. In 2009 this illness meant Sandi had to return home to USA from the ESERA conference, and so could not attend our workshop for this volume in Prato. Even so, as we remember with both affection and sadness, her desire to maintain engagement with our workshop meant we had a wonderful discussion of her chapter via Skype, with her in her home and all the rest of us at our workshop. We are grateful for her contributions to this book. Much more importantly we acknowledge her major contributions to science education research, and through that to the thinking of many researchers around the world including the three of us.

August 2010

Richard Gunstone
Deborah Corrigan
Justin Dillon

Contents

Contributors

Sandra K. Abell University of Missouri, Columbia, USA

Glen Aikenhead University of Saskatchewan, Saskatoon, Canada

Jacinta Bartlett Monash University (Peninsula campus), Frankston, VIC, Australia

Amanda Berry Monash University (Clayton campus), Clayton, VIC, Australia

Anthony Clark University of British Columbia, Vancouver, Canada

Rebecca Cooper Monash University (Clayton campus), Clayton, VIC, Australia

Deborah Corrigan Centre for Science, Mathematics & Technology Education, Monash University (Clayton campus), Clayton, VIC, Australia

Bronwen Cowie University of Waikato, Hamilton, New Zealand

Justin Dillon Dept. Education & Professional Studies, King's College London, Franklin-Wilkins Bldg., 150 Stamford Street, London, SE1 9NH, UK

Gaalen Erickson University of British Columbia, Vancouver, Canada

Maria Evagorou University of Nicosia, Nicosia, Cyprus

Peter J. Fensham Monash University & Queensland University of Technology, Brisbane, QLD, Australia

Helmut Fischler Freie Universität Berlin, Berlin, Germany

Richard Gunstone Centre for Science, Mathematics and Technology Education, Monash University (Clayton campus), Clayton, VIC, Australia

Beverley Jane Monash University (Peninsula campus), Frankston, VIC, Australia

Alister Jones University of Waikato, Knighton Road, Hamilton 3240, New Zealand

Stephen Keast Monash University (Clayton campus), Clayton, VIC, Australia

Heather King King's College London, London, United Kingdom

John Loughran Monash University (Clayton campus), Clayton, VIC, Australia

Hongming Ma Monash University (Gippsland campus), Churchill, VIC, Australia

Sandy Martinuk University of British Columbia, Vancouver, Canada

Masakata Ogawa Tokyo University of Science, Tokyo, Japan

Debra Panizzon Flinders Centre for Science Education in the 21st Century, Flinders University, Adelaide, SA, Australia

Léonie J. Rennie Curtin University of Technology, Perth, WA, Australia

Jill Robbins Monash University (Peninsula campus), Frankston, VIC, Australia

Marcelle A. Siegel University of Missouri, Columbia, USA

Lynn Uyen Tran Lawrence Hall of Science, University of California at Berkeley, Berkeley, CA, USA

About the Authors

Sandra K. Abell As noted by the editors in their foreword, Sandi Abell passed away during the production of this volume. During her outstanding academic career, Sandi was a remarkable and highly influential contributor to science education research and development. She served as a president of the National Association for Research in Science Teaching and was recognized as a Curators' Professor of Science Education at the University of Missouri. There, she directed the MU Science Education Center where she inspired learning, curiosity, and/or laughter on a regular basis. Much of her research was concerned with understanding the process of becoming a teacher of science, from pre-service preparation throughout the teaching career. Among her many publications were co-editing the *Handbook of Research on Science Education* (2007) and co-authoring *Seamless Assessment in Science: A Guidebook for Elementary and Middle School Teachers*. These, and much else, will continue to influence science education and teacher education for many years to come.

Glen Aikenhead is professor emeritus, Aboriginal Education Research Centre, University of Saskatchewan, Canada. His research and development projects over the years have emphasized relevance of school science for students' everyday lives. His current work in cross-cultural Indigenous science education involves science curricula, textbooks and teacher professional development that recognize the importance of Indigenous knowledge to understanding the physical world for all students. This has led to a co-authored book *Bridging Cultures: Indigenous and Scientific Ways of Knowing Nature* (2011).

Jacinta Bartlett is an early childhood educator with a passion for researching design and technology education. She completed a BEd (hons) (H1) at Monash University in 2010. Jacinta has co-lectured at Monash University in early childhood technology, science and mathematics education and has published and presented at several national and international conferences. She is currently working with young designers in a preschool setting as well as assisting with research into early childhood educators' understandings of technology with her colleagues at Monash University.

Amanda Berry is an associate professor in science education in the Faculty of Education, Monash University. Her research interests have grown from her experiences as a high school science teacher seeking to better understand and improve learning within her science classes. Her current research focuses on the development of science teacher knowledge, the professional learning of teacher educators and the development of a pedagogy of teacher education that responds to contemporary educational issues. Recent books include *Tensions in Teaching about Teaching: Understanding practice as a teacher Educator* (2008) and *Understanding and Developing Science Teachers' Pedagogical Content Knowledge* (2006) (with John Loughran and Pamela Mulhall).

Anthony Clark is a professor in the Faculty of Education at the University of British Columbia, and co-director of UBC's well known Centre for the Study of Teacher Education. Before coming to Canada he was a classroom teacher in Australia for a number of years. His research interests include the practicum in teacher education and teaching and learning in higher education.

Rebecca Cooper is a research fellow in science education in the Faculty of Education, Monash University. Whilst working as a Physics, Science and Mathematics teacher, she has also been involved in pre-service science teacher education and delivering professional development for science teachers. Her research interests include considering how science teachers develop pedagogical content knowledge, improving the quality of science teaching to increase student engagement and finding ways of making school science more relevant to students.

Deborah Corrigan is an associate professor in the Education Faculty at Monash University. Currently she heads the faculty's Science Education Research Group in the Centre for Science, Mathematics and Technology Education, and has previously been director of undergraduate and pre-service programs and associate dean (teaching). She has extensive research and development experience in the areas of science education, particularly chemistry education, and teacher education/teacher development/teacher learning (both pre- and in-service). She is currently involved in Victorian science teacher professional learning programs for both the Victorian Government and the Catholic Education Office, and the Science Learning Hub for the New Zealand Ministry of Research, Science and Technology. Her most recent book is the precursor to this volume, *The Re-emergence of Values in Science Education* (2007), which she also edited with Justin Dillon and Richard Gunstone.

Bronwen Cowie is currently director of the Wilf Malcolm Institute of Educational Research, Faculty of Education, and the former director of the Centre for Science and Technology Education Research at the University of Waikato, New Zealand. She has been a teacher of science in secondary schools. She has directed a number of large national research and development projects and has extensive experience in classroom-based research. Her research interests include the nature of assessment for learning interactions in science and technology classrooms, student voice, culturally responsive pedagogy and the role of ICTs in learning science.

Justin Dillon is professor of science and environmental education and head of the Science and Technology Education Group at King's College London. Justin joined King's in 1989 after teaching science in London schools for ten years. He is an editor of the *International Journal of Science Education* and president (2007–2011) of the European Science Education Research Association. His research interests include teaching outside the classroom, public engagement with science, environmental education and teacher development.

Gaalen Erickson is a professor in the Department of Curriculum and Pedagogy at the University of British Columbia. He has had a long-standing research interest in examining the methods and theories used to identify and interpret student and teacher learning in the context of science education. As a former director of the Centre for the Study of Teacher Education at UBC, he also developed with colleagues and graduate students a research programme on models of teacher professional development examining the relationships between research and practice as they are enacted in projects documenting the nature of practitioner inquiry and professional knowledge.

Maria Evagorou is a lecturer in science education at the University of Nicosia, Cyprus. Her research interests focus on exploring young students' argumentation and system thinking skills within the context of science. More specifically, the emphasis of her work is on young students' talk when they engage in the discussion of socio-scientific issues, and how the use of online technologies can enhance students' argumentation skills and their understanding of a system. Maria has received awards for her research in argumentation at the AERA Science SIG and EdMedia 2008 Conference, and has published in international journals.

Peter J. Fensham is an emeritus professor at Monash University where he was associated for many years with the development of its internationally acclaimed research in science education. His book *Defining an Identity* discusses how science education has evolved as an international field of research. His membership of the Science Expert Group for the OECD PISA project enabled him to contribute to novel ideas for assessing scientific literacy. His current research interests are in policy making with respect to science education and how control of the curriculum for school science is changing. At present he is an adjunct professor at the Queensland University of Technology.

Helmut Fischler is professor of physics education at the Freie Universität Berlin. His research activities are focused on students' learning and teachers' professional development. Investigations on students' learning refer to their problems to understand conceptions of modern physics. His research on (student) teachers analyses the relationship between their subject matter and pedagogical content knowledge on the one hand and their decision making in classrooms on the other. Currently he is guest professor of physics education at the Linnaeus University, Kalmar/Växjö, Sweden.

Richard Gunstone is now emeritus professor of science and technology education at Monash University where he continues to engage with research on science learning, teaching and curriculum, and the mentoring of younger academics. He initially joined Monash in 1974 after a career as a high school physics, science and mathematics teacher. Among his current projects is being editor-in-chief of Springer's *Encyclopedia of Science Education*, due to be published in 2012.

Beverley Jane was a science educator at Monash University where she linked theory and practice by teaching science curriculum units in local primary schools. Her popular books *Science for Children* and *Technology for Children* (with Marilyn Fleer) continue to influence the pedagogical development of pre-service teachers across Australia. Now an ordained minister, she plans workshops for grandparents on how to foster their grandchildren's understanding of science.

Alister Jones obtained his master's degree and PhD from the University of Waikato, New Zealand. He is currently dean of the Faculty of Education at the University of Waikato, and is the past director of the Wilf Malcolm Institute of Educational Research and of the Centre for Science and Technology Education Research. He has been a teacher of science in secondary schools and has been involved in research in science and technology education in both England and New Zealand. His research interests involve aspects of science and technology education, including teacher development in science and technology education, teaching and learning of physics and curriculum development, and assessment in technology education.

Stephen Keast is a lecturer in science education in the Faculty of Education, Monash University. He had 13 years teaching experience as secondary science and mathematics teacher before making the transition to tertiary education. His main research interests are in the professional learning of teachers, documenting teacher professional wisdom, articulating teacher educator pedagogy and investigating the developing pedagogy of pre-service science teachers. He works extensively with teachers in professional learning programs.

Heather King is a postdoctoral research associate working at King's College London on the European Union funded project, Towards Women in Science and Technology (TWIST). She is also an independent consultant specializing in museum education. Heather has a PhD in science education from King's, and was previously the director for the Center for Informal Learning and Schools. Her research interests focus on the role of museum explainers and their support of natural history inquiry.

John Loughran is the foundation chair in curriculum & pedagogy in the Faculty of Education, Monash University. John was a science teacher for ten years before moving into teacher education. His research has spanned both science education and the related fields of professional knowledge, reflective practice and teacher research. John is the co-editor of *Studying Teacher Education* and his recent books include: *What Expert Teachers Do* (2010); *Developing a Pedagogy of Teacher Education: Understanding teaching and learning about teaching* (2006); *Understanding and Developing Science Teachers' Pedagogical Content Knowledge* (2006) (with Amanda Berry and Pamela Mulhall); and lead editor of *The International Handbook of Self-Study of Teaching and Teacher Education Practices* (2004).

Hongming Ma is currently a lecturer in the Faculty of Education at Monash University, Australia. She studied engineering at Southwest Jiaotong University and used to work as a secondary vocational school teacher in China. Her research interests are related to cross cultural understanding of issues in science education. Specifically, the issues involve the nature of science/technology and its role in school science/technology curriculum at different levels, students' conceptual change in learning science/technology (including cognitive and socio-cultural influences) and the pedagogy of teaching and learning of science/technology and their relationship with other curricular domains.

Sandy Martinuk is a PhD candidate at the Department of Physics and Astronomy, University of British Columbia. He plans to be the first student to graduate from that department based on research conducted in physics education. He has been involved in tutoring and teaching introductory university-level physics for ten years. He co-founded the UBC Physics Teaching Assistant Training Program in 2007, and received the UBC Science Graduate Teaching Award in 2008. His current research explores students' perception of the relevance of classroom physics to the everyday world, and how their epistemological framing of classroom activities regulates their learning.

Masakata Ogawa is professor of science education at the Graduate School of Mathematics and Science Education at Tokyo University of Science. After getting his doctoral degree in plant physiology at Kyoto University in 1981, he taught science education classes at Ibaraki University, Hiroshima University and Kobe University (professor emeritus) before joining in Tokyo University of Science in 2009. His research interests lie mainly around cultural contexts of science education and science and technology education policy issues. He was the founding president (2007–2009) of the East-Asian Association for Science Education and the president (2004–2008) of the Japan Society for Science Education.

Debra Panizzon is associate professor and deputy director for the Flinders Centre for Science Education in the 21st Century, which is located at Flinders University. Prior to commencement in academia she taught junior science and senior biology and physics in secondary schools. Debra is an experienced science education academic having worked with both primary and secondary pre-service teachers. Her research interests lie in the areas of cognition, student acquisition of scientific concepts, rural and regional education, and assessment. Importantly, much of this research involves partnerships with science and mathematics teachers so that theory and practice inextricably linked.

Léonie J. Rennie is research professor at Curtin University, Perth, Western Australia. Her background is in science teaching and curriculum, and, currently, her research is focused on integrated curriculum in science, mathematics and technology, and school-community projects that aim to enhance scientific literacy. She has published over 160 books and monographs, book chapters and refereed journal articles. She has delivered keynote addresses to audiences in Australia, Brazil, India, South

Africa, Sweden, the United States and the Netherlands on her research relating to gender, learning and assessment in science and technology, both in school and out.

Jill Robbins is a senior lecturer in early childhood education in the Faculty of Education, Monash University. Her research interests focus on young children's thinking (especially in relation to natural phenomena); early childhood science and technology; early childhood teachers' understandings of technology and technology education; the role of grandparents in supporting young children's learning in science and technology, and sociocultural theory and its application to research and teaching. Among her teaching commitments are early childhood science and technology in on-campus and off-campus programs, including in Singapore.

Marcelle A. Siegel is an assistant professor of science education at the University of Missouri, Columbia. Her research focuses on classroom assessments for learning, equitable assessment and biology learning and teaching. Marcelle is currently principal investigator of a National Science Foundation grant on assessment reform.

Lynn Uyen Tran is an educational researcher with the Lawrence Hall of Science. She has a PhD in science education from North Carolina State University and has done her post-doctoral work with the Center for Informal Learning and Schools at King's College London. She conducts research and professional development on the knowledge, identity and work of science educators in museums.

Chapter 1
Approaches to Considering the Professional Knowledge Base of Science Teachers

Deborah Corrigan, Richard Gunstone and Justin Dillon

Introduction

Over the past 20 years much has been written about the knowledge bases claimed to be needed for teaching science. The most prominent and influential has been Shulman (1987), who proposed seven knowledge domains necessary for teaching [science or any other component of the curriculum]: content knowledge, pedagogical knowledge, school knowledge, knowledge of pupils, curriculum knowledge, pedagogical content knowledge (PCK) and knowledge of educational ends, purposes and values. Tamir (1989) proposed six somewhat similar domains of knowledge, specifically for science teachers; these focused on subject matter, pedagogy, subject matter specific pedagogy, general liberal education, personal performance and foundations of the teaching profession. Among the many examples of arguments supporting the importance of such knowledge bases for successful teaching of science have been those of Gess-Newsome (1999), for whom the development of teachers (including pre-service teachers) includes the integration of knowledge ba'ses, and Morine-Dershimer and Kent (1999), who suggest that the development of "context specific pedagogical knowledge that helps to guide teachers' decisions and actions" (p. 23).

Whatever structure one proposes for the knowledge base of (science) teaching, clearly aspects of this knowledge base change over time. The most obvious example is change forced on teachers by shifts in curriculum thinking. Iconic movements such as Science, Technology and Society (STS) (Aikenhead 1994), Science for All (Fensham 1985), Scientific Literacy (Roberts 2007), Science for Public Understanding (Osborne et al. 2003), Humanistic Science (Aikenhead 2006), and, indeed, evolving notions of the nature of science itself (Duschl 2000; Lederman 2006), have all impacted on the knowledge base required of science teachers. Further, these particular changes have influenced how science teachers are expected to, and actually

D. Corrigan (✉)
Centre for Science, Mathematics & Technology Education, Monash University
(Clayton campus), Clayton, VIC, Australia
e-mail: debbie.corrigan@monash.edu

D. Corrigan et al. (eds.), *The Professional Knowledge Base of Science Teaching*,
DOI 10.1007/978-90-481-3927-9_1, © Springer Science+Business Media B.V. 2011

do, view both science and science education. Aikenhead (2006) explores some of these notions in his book *Science Education for Everyday Life*, through his discussions of teacher orientations to a Humanistic Science Education.

There is a sharply increasing significance for the knowledge bases for science teaching in current trends in school science education. With the development of a standards-based approach to the quality of science teaching becoming increasingly common in the Western world, and phrases such as "evidence-based practice" becoming the common catch-cry of attempts to "measure" such quality, it is timely to look at what constitutes evidence of quality science teaching and to ask on what basis can such evidence be judged and how does such evidence reflect the knowledge basis of the modern day professional science teacher?

The contested nature of many of the phrases used in science education, such as "scientific literacy" and "pedagogical content knowledge", has become a central issue for consideration, as while these phrases are widely used in the literature it is increasing obvious that there is little consensus about their meaning. Within this volume, for example, there are instances of multiple interpretations across chapters of many of these phrases.

For these and other reasons, we believe the time is ripe for a collection of writings that considers the knowledge bases seen to be required for science teaching. The book brings together a number of researchers, who have worked with science teachers and science teaching in a number of ways and in a range of social and cultural contexts, in an attempt to make more explicit what can constitute valid evidence for making judgements about what is quality science teaching and what represents a valid knowledge base for professional science teachers.

The Profession of Science Teaching

While there may still be some debate about the extent to which teaching is a profession, an occupation, a vocation, or a job, the growing groundswell of support for teaching as a profession has heralded some significant changes. As Freidson (1994) has pointed out, significant elements of a profession are that it is self-regulated, has the capacity to organize its own work, selects recruits who have sufficient training and competence and ethical performance and "are capable of controlling themselves by cooperative, collective means and that in the case of complex work, those who perform it are in the best position to make sure that it gets well done" (p. 176).

Tran and King (Chap. 16) have used Abbott's (1988) definition of a profession as "a well-defined body of knowledge and skills necessary to provide a service to society that is not offered elsewhere". In the context of education, this view of a profession, which includes the notion of education as a public service, is a little different to that advanced by others (such as Freidson, above). This is particularly important to Tran and King as they consider the specific context of the profession of educators engaged in education/teaching outside the usual classroom, such as in museums and interactive centres. The criteria outlined by Freidson above become more problematic for this group of professionals, because this group has a more limited ability than teachers in usual classrooms to self-regulate and engage in the necessary complex work in a cooperative and collaborative manner. This limitation essentially derives from the more disparate nature of the work of this group of informal educators. Tran and King see this group as a profession still emerging, as there is currently a lack of articulation of a body of knowledge for science educators working in museums and science centres. They also note that the knowledge base required by these informal educators is more interdisciplinary in nature than that of many other professions, and includes areas such as context, choice and motivation, objects, content, learning and talk (as a mechanism for communication). Associated with this knowledge base then are components of knowledge (or a form of pedagogical knowledge) that include an educator's orientation towards facilitation, knowledge about affordances of objects, knowledge of the very different range of forms of learning involved when one is a visitor to such centres, knowledge of facilitation strategies, and an understanding of science.

While then there are variations in the ways one might conceptualize professions and professionalism of differing forms and in differing contexts, there is a ubiquitous underlying assumption about the work of professionals that it is usually complex. Clearly teaching (of all forms) is no exception. As Hargreaves (1998) points out, teachers' work is complex, difficult and demanding, and, he argues, requires teachers to engage in "intellectual" and "emotional" work as well as work organization (the capacity to organize and control their own work). This can provide a form of useful broad frame (or lens) through which one can analyse what it means to be a professional science teacher in a variety of contexts and from a variety of perspectives.

Given the considerable debates we have already alluded to regarding the profession of teaching and the knowledge base of teaching, conceptual frameworks for the consideration of the nature of teacher professional work, knowledge and so on are less common that might be expected. Because the Hargreaves (1998) framework has such explicit focus on teacher professional work, it is of interest to us in our considerations of the knowledge base of science teachers. This interest is enhanced by two further observations. Firstly, it is, in our experience, an analysis with which teachers often identify; this is a broad framework that teachers find they can apply in considering their own work. Secondly, the broad framework provides a powerful way of considering aspects of the common tensions between researchers and teachers—in general, researchers have for a long time been more focused on the "intellectual" work of teachers, while teachers themselves tend to be more concerned with what Hargreaves describes as the "emotional" and "organizational" forms of teachers' work.

Perspectives on the Knowledge Base of Science Teaching

The first and last of the remaining chapters in this volume (Chap. 2; Chap. 17) take two different very broad perspectives on the professional knowledge base of science teaching, and hence on the nature of the work of professional science teachers.

Rennie discusses the ways in which concerns for developing a science education with which more students choose to become engaged lead to reconsideration of the science curriculum in terms of a greater focus on the world outside the school. This in turn leads to the need to consider the complexity of a range of consequences such as the need for the curriculum to address real-world issues and problems, the then necessary interactions between science and other disciplines (an "integrated curriculum") and science, society, culture, politics, etc. Given that these developments necessarily place substantially different demands on the knowledge base of science teaching, Rennie then also gives us a valuable overview of the nature of teaching knowledge in the early stages of the twenty-first century. In the final chapter of the book, Fensham also considers ways in which the nature of the knowledge base needed for the teaching of science in the twenty-first century is changing.

Other than the final chapter by Fensham, we have used a loose adaptation of the Hargreaves (1998) framework we noted above in sequencing the remaining chapters of this book. We begin with those chapters that tend to be more focused on intellectual aspects of science teachers' work, and conclude with those chapters with greater focus on work organization. Between these two we have a group that has a greater focus on aspects of the emotional elements of teachers' work—those aspects concerned with social and cultural forms of science teachers' work. We also note here an extremely important qualifier to this logic: this is an *ex post facto* logic, not one the authors were given at any point in their creation and editing and rewriting. Hence, as one would expect, none of the chapters is in any way exclusively focused on any one of these three aspects of teachers' work.

The Intellectual Work of Science Teachers

In more recent times, the intellectual work of teachers has been equated with Shulman's (1986) concept of PCK as "the most powerful analogies, illustrations, examples, explanations and demonstrations—in a word, the ways of representing and formulating the subject that makes it comprehensible for others" (p. 9). The idea of PCK has been taken up by many of the authors in this volume (for example, see Rennie, Jones & Cowie, Corrigan & Gunstone, Evagorou & Dillon, Abell & Siegel, Robbins Jane & Bartlett). However, the construct of PCK, while useful to many educational researchers, has not become widely used by the teaching profession. This fact may be in part be due to the lack of consensus about the meaning of PCK and a lack of belief in its ability to encompass other aspects of the intellectual work of teachers, such as values, classroom reality, teachers' thinking, students' thinking and the numerous contexts that shape the teaching and learning of particular content in particular ways. Such considerations are well explored in many chapters in this volume.

The first such consideration, the chapter by Fischler (Chap. 3) is the most radical such discussion of different ways of considering the intellectual work of teachers. Fischler's career of teaching and scholarship in physics education has been in the

German Didaktik tradition, a tradition that has among its central differences with the Anglo-Saxon traditions of England, North America, Australia, etc., a quite different beginning point regarding the relationship between *teacher* and *curriculum*. Fischler details the implications for teacher professional knowledge in this fundamentally different way of considering how to transform knowledge-of-worth into appropriate learning experiences. He includes in this analysis the specific impact of that component of the German conceptualizing of Didaktik known as Bildung (the central notion describing both the process of personal development and the consequence of this development).

Jones and Cowie (Chap. 4) note that science teachers are under pressure to demonstrate the effectiveness of their work through student achievement while at the same time increasing student engagement with science. In primary schools many teachers do not have a broad and deep education in science, a fact which adds to the pressures that they face. Through their work on assessment for learning, Jones and Cowie have been able to enhance teachers' knowledge-as-action and show an impact on their students' learning. The authors argue teachers' knowledge-as-action is crucial because it encompasses teachers' understanding and enactment of how to help students to understand specific subject matter using individual and sequenced activities, representations and social groupings. In a related (but different) account, Loughran and Berry (Chap. 5) discuss the professional development of science teachers (primary and secondary) who took part in a one-year Science Teaching and Learning programme. A novel aspect of the experience was a writing day in which participants constructed a case intended to illustrate an aspect of their professional learning as a science teacher. Using these cases, Loughran and Berry explore how the teachers shifted their orientations towards teaching science in ways that began to bridge the gap between school science and the science in the worlds of their students. The cases illustrate these teachers' efforts to better understand the nature of science teaching and learning and how the relationship between teaching and learning might be addressed through better understanding of the details of what happens in their classrooms.

Corrigan and Gunstone (Chap. 6) write about their own learning, their own "intellectual work", in the processes of development of an on-line resource designed to support more learner-centered approaches to science teaching by both primary and secondary teachers. Through analysis of records of discussion about and evolving drafts of entries for this web resource, they describe their evolving understandings of science (and a range of research-practice-personal experiences linkages that inform their pedagogies). This analysis leads them to then argue, among other things, the need to see the components of any form of model of teacher knowledge as both dynamic and interactive, and to see forms of causal links between such components as bi-directional rather than the more common view of unidirectional.

As already noted, we do not see any of the chapters in the book as being exclusively located in one of our three (very loose) categories. For example, every chapter in the book has a contribution of substance to make to considerations of the intellectual work of science teachers, but we do not outline any further chapters in this subsection because of the prime emphasis in each.

Socio-Cultural Contexts for the Work of Science Teachers

In Hargreaves' (1998) descriptions of his three broad forms of teachers' work, he describes "emotional" work in these terms:

> Teaching is an emotional practice that also involves heavy investments of emotional labour. It cannot be reduced to technical competencies of clinical standards alone. The emotions of teaching are, in this sense not just a sentimental adornment to the more fundamental parts of the work. They are fundamental in and of themselves. They are deeply intertwined with the purposes of teaching, the political dynamics of educational policy and school life, the relationships which make up teaching and the senses of self which teachers invest in their work. (p. 368)

As noted above, there are a number of chapters in this volume that have a strong focus on aspects of teachers' knowledge (and thus work) that are socio-culturally embedded and determined. An obvious one is that by Fischler (Chap. 3) in his considerations of Didaktik and Bildung, and ways in which these broad constructs frame so much of a very different and very influential form of conceptualizing relationships between teachers, curriculum, and educational purposes.

Aikenhead (Chap. 7) advances a view of quality science teaching from a cultural perspective. He considers appropriate indicators and evidence of quality science teaching from two points of view. Based on documented failures of traditional school science, he first articulates what quality science teaching *is not*. Second, he develops indicators of school science culture that avoid such failures, thereby clarifying what quality science teaching is. He argues a strong case for specific features of "relevant school science content."

The next two chapters, by Ogawa (Chap. 8) and Ma (Chap. 9), each considers issues of forms of the science knowledge of science teachers in a particular non-Western cultural context (Japan for Ogawa, China for Ma). That might be seen to imply similarity between these two chapters; this is not the case. In the Japanese context that Ogawa investigates, the elementary (primary) curriculum has a component called Rika. Rika has two components that are in some ways incompatible: the first is essentially "western modern science", and the second "Shizen", a Japanese traditional cosmology that has a specific values system embedded within it. So Ogawa provides an analysis of a context where the formal curriculum includes a form of science that is explicitly culturally linked. Ma, by contrast, investigates the impact of traditional Chinese knowledge (which she terms "Chinese Native Knowledge" in recognition that not all indigenous knowledge can be categorized as "science") on Chinese science teachers who are teaching the western modern science that is the school curriculum. These strong contextual differences also mean that the two studies had different purposes, and thus interesting (and most appropriate) methodological differences. Among many fascinating conclusions from the studies are some somewhat common threads relating to the ways teachers perceive their teaching of and student learning of systems they see as not compatible (western modern science and the culturally linked way of understanding and predicting the natural world). Ogawa also notes some linkages that had previously been argued between Rika and Bildung (see Chap. 3).

Panizzon (Chap. 10) explores the impact on science teachers of a very different form of socio-cultural difference—that of teaching in what in Australia is termed a "rural and remote school". While there are many of these at the primary level in Australia (most being primary schools with six or seven grade levels and only one teacher), there are many at the secondary level as well. Panizzon focuses on science teachers in secondary schools in her analysis. These secondary contexts are characterized by small whole school student populations, very small senior level classes, sometimes multi-grade teaching, local contexts that are often quite unfamiliar to the teacher, and, most importantly, the absence of a significant number of science teaching colleagues—the absence of a professional learning community. Panizzon explores the particular challenges for teacher knowledge that these contexts provide, including ways of responding to needs for a professional learning community.

Work Organization Aspects of the Work of Science Teachers

A significant hallmark of professionalism is the capacity of its practitioners to organize and control their own work (Freidson 1994). While many could say that organization of work is managed for teachers since they are not as autonomous in their work as perhaps general practitioners in a medical surgery, nevertheless, as they engage in their work, particularly in the classroom, teachers have no more constraints placed on them than do other professional in surgeries or offices. In addition, Acker (1999) reminds us that communities of teachers in schools are adult working groups. "We are so obsessed with schools as places for (students) that we forget they are workplaces for adults" (p. 196). As teachers are professionals engaged in an enterprise of public service, they carry with them a great social responsibility. Sachs (2000, 2003) argues that such social responsibility is best manifested as activist professionalism in communities of practice characterized by collegiality, negotiation and reformation with an aim to "improve all aspects of the educational enterprise at the macro level and student learning outcomes and teacher's status in the eyes of the community at the micro level" (2000, p. 77). So in organizing their own work, professional teachers need to attend to the needs of the market (or workplace) while simultaneously being subjected to more scrutiny and obliged to adhere to centralized policies with respect to curriculum and assessment and to teaching standards. As noted by Ballet et al. (2006), these changing conditions in teachers' work have resulted in an intensification of work.

In relation to the work of professional science teachers, Rennie (Chap. 2) has highlighted how the intensification has been manifested by the growth of more centralized curriculum policies. Rennie highlights the disconnect not only between science in and out of school but also the age-old tension between science as a single discipline versus science playing a role in other human affairs. She stresses that science concepts in school are idealized and simplified while science out of school is not so simple, is interdisciplinary, messy and with many uncontrolled variables. Science-related issues are complex and not easily understood, and there are competing social, cultural, economic and political values providing conflicting interpretations of how to use science knowledge for the benefit of society. Using science to understand scientific issues is a necessary experience for students and involves learning how to cope with uncertainty and risk, an experience that not many students have.

Evagorou and Dillon (Chap. 11) provide an example of a shift in demand for approaches to teaching science through their consideration of teaching argumentation. Argumentation can be seen as a means of rethinking the explanations of phenomena/theory/discovery, as part of the process of knowledge construction, as an important thinking skill, and as a fundamental discourse of science (particularly scientific enterprise as part of scientific literacy).

Abell and Siegel (Chap. 12) highlight the intensification of science teachers' work through their consideration of assessment literacy. They provide a framework for science teacher assessment literacy (and while it is not always obvious how this framework is specific for science teachers, given the ways in which purposes and strategies for assessment tend to be applicable across the curriculum, this is not surprising). Abell and Siegel examine assessment from a socio-cultural perspective and note that the values and principles that teachers hold become important drivers of assessment. Their proposed framework includes consideration of the purpose of assessment, what to assess, how to assess and assessment strategies and reported actions.

Robbins, Jane, and Bartlett (Chap. 13) set out to develop pedagogical approaches to technology education that would enhance early childhood teachers' understandings of how children's creative thinking in the context of block play. They describe how their work in partnership with teachers led to identification of new pedagogies, including increasing use of the words *technology* and *design* by the teachers during interactions with children and other ways in which increased knowledge about technology among the early childhood teachers had positive impact on children's learning, and the need for children to have time to interact and to work together on design activities.

Martinuk, Clark, and Erickson (Chap. 14), in working with first-year undergraduate physics learners and teachers, take another approach to the study of the relationship between pedagogy and learning. The authors report on a bold move undertaken by physics academics faced with a realization of the inadequacy of their existing curriculum offerings. As Martinuk et al. note, the changes were made not only to improve student attainment and understanding but also to offer a meaningful and relevant physics education and thus to perhaps increase student engagement with physics and motivation to continue to study physics. The curriculum reform involved a team effort and a need to listen to the voices of the students. The resulting changes in curriculum and pedagogy led to a more meaningful experience for the students which, in turn, led to greater levels of satisfaction among the academics.

Keast and Cooper (Chap. 15) worked with an even older group—final (fourth)-year pre-service secondary science teacher education students whom they were teaching. They discuss the teaching use of a novel resource, Slowmation, and how it allows student teachers and their lecturers to focus on the complexity of teaching and impact on student learning. By watching short animated movies illustrating a science concept and created by school students in classes taught by the teacher education students, the pre-service teachers are able to reconstruct their pedagogical knowledge within the framework of Morine-Dershimer and Kent (1999)—a framework that their lecturers had been assisting them to deconstruct. Keast and Cooper identify the dismay of their student teachers who, while watching the animated movies in the company of their peers, realise that telling children the "right answer" does not change the children's conceptions. The multiple gazes of their colleagues enable the student teachers to see their own teaching and their students' learning from many perspectives.

We have already made reference to the contribution by Tran and King (Chap. 16), in the context of our initial considerations of ways of conceptualizing professions in education. While this chapter has clear links with work organization perspectives, we have placed their chapter at this point in the sequence of the book because it starts to expand our visions again. Tran and King take Rennie's arguments (Chap. 2) that, in terms of teaching outside the classroom, a different interpretation of content knowledge and pedagogical practice is required, and explore the nature of knowledge required by educators. A significant component of the need for this different interpretation derives from out-of-school settings, and the opportunities they afford, being inherently interdisciplinary. Through this process of examining the nature of out-of-school educators' knowledge Tran and King seek to develop a framework for understanding, guiding and enhancing the practice of these educators.

Future Aspects of the Work of Science Teachers

We began this section by noting that Rennie, in the first of the remaining chapters in this volume (Chap. 2), raises important issues about ways in which the nature of the knowledge base of science teaching is changing today. In the final chapter in this volume on the professional knowledge base of science teachers, Fensham (Chap. 17) also considers a bigger picture and argues there are challenges to teachers' professional knowledge that have recently emerged, from both beyond science (complexity and uncertainty in the new world of work, the "knowledge society") and within science (complexity and uncertainty in science per se). Fensham highlights that both these forms of challenge stem from the increasing complexity of society and notes that the way people learn has to be able to equip them to deal with such complexity. Fensham calls for radically different priorities for learning that recognize:

- new conceptions of knowledge content that incorporate notions of risk and trust,
- new pedagogies, and
- new assessment approaches.

Increasing complexity, both within science and in society in the broadest sense, means that the simple examples and stories, applications and implicit certainties represented by traditional school science approaches are no longer adequate. Moves towards using science examples from daily life, concerns, and decisions mean more complex and complicated and less certain contexts and cases. This will involve moving to more complicated cases, as represented by many of the STS approaches to science education, and extending to more complex cases now beginning to be represented by "Grand Challenges" and "Socio-Scientific Issues (SSIs)" in science education. Examples of such complex cases include climate variability, hydrological forecasting, and chemistry for sustainability. Teachers will need help to gain the knowledges needed for these new emphases.

References

Abbott, A. (1988). *The system of professions: An essay on the division of expert labor.* Chicago: The University of Chicago Press.

Acker, S. (1999). *The realities of teachers' work: Never a dull moment.* London: Cassell.

Aikenhead, G. S. (1994). What is STS teaching? In J. Solomon & G. Aikenhead (Eds.), *STS education: International perspectives on reform* (pp. 47–59). New York: Teachers College Press.

Aikenhead, G. S. (2006). *Science for everyday life: Evidence-based practice.* New York: Teachers College Press.

Ballet, K., Keltchtermans, G., & Loughran, J. (2006). Beyond intensification towards a scholarship of practice: Analyzing changes in teachers; work lives. *Teachers and Teaching: Theory and Practice, 12,* 209–229.

Duschl, R. (2000). Making the nature of science explicit. In R. Millar, J. Leech, & J. Osborne (Eds.), *Improving science education: The contribution of research* (pp. 187–206). Philadelphia: Open University Press.

Fensham, P. J. (1985). Science for all: A reflective essay. *Journal of Curriculum Studies, 17,* 415–435.

Freidson, E. (1994). *Professionalism reborn: Theory, prophecy, and policy.* Cambridge: Polity Press.

Gess-Newsome, J. (1999). Introduction. In J. Gess-Newsome & N. Lederman (Eds.), *Examining pedagogical content knowledge* (pp. 3–20). Dordrecht: Kluwer.

Hargreaves, A. (1998). The emotional politics of teaching and teacher development. *International Journal of Leadership in Education: Theory and Practice,* 1, 315–336.

Lederman, N. G. (2006). Syntax of nature of science within inquiry and science instruction. In L. B. Flick & N. G. Lederman (Eds.), *Scientific inquiry and the nature of science: Implications for teaching* (pp. 301–317). Dordrecht: Springer.

Morine-Dershimer, G., & Kent, T. (1999). The complex nature and sources of teachers' pedagogical knowledge. In J. Gess-Newsome & N. Lederman (Eds.), *Examining pedagogical content knowledge* (pp. 19–32). Dordrecht: Kluwer.

Osborne, J. F., Ratcliffe, M., Collins, S., Millar, R., & Duschl, R. (2003). What 'ideas about-science' should be taught in school science? A Delphi Study of the 'Expert'Community. *Journal of Research in Science Teaching, 40,* 692–720.

Roberts, D. (2007). Scientific literacy/science literacy. In S. K. Abell & N. G. Lederman (Eds.), *Handbook of research on science education* (pp. 729–780). Mahwah: Lawrence Erlbaum Associates.

Sachs, J. (2000). The activist professional. *Journal of Educational Change, 1,* 77–95.

Sachs, J. (2003). *The activist teaching profession.* Buckingham: Open University Press.

Shulman, L. (1986). Those who understand: Knowledge growth in teaching. *Educational Researcher, 15*(2), 4–14.

Shulman, L. (1987). Knowledge and teaching: Foundation of new reform. *Harvard Educucational Review, 15*(2), 4–14.

Tamir, P. (1989, April). *Subject matter and related pedagogical knowledge in teacher education.* Paper presented at the annual meeting of the American Educational Research Association, Washington.

Chapter 2
Blurring the Boundary Between the Classroom and the Community: Challenges for Teachers' Professional Knowledge

Léonie J. Rennie

"Teachers, students, parents and Dr C. We hope you enjoy our performance. We would like to tell you why we wrote the Potato Rap. We were given this challenge by Dr C. to see if we could inform you that this is the International Year of the Potato. Why would anybody want to make an International Year of the Potato, after all, it's just a potato, right? But once we started to research potatoes, we found out some interesting facts…"
Soon the spokesperson for the Year 4 class introduced Dr C., the school's visiting scientist (a potato pathologist), to the school assembly, who explained,
"I work with a lot of sick potatoes, and just like a doctor, we have to find out which potatoes are sick, why they are sick, and see if we can find a cure. The work that we do as scientists helps you to have a constant supply of potatoes."
He thanked the class for their performance and some rap music began to throb.
"Hey!"
shouted 25 potatoes in unison—the entire class was on the stage, dressed in coloured tights stuffed with crumpled paper—brown potatoes, yellow potatoes, purple potatoes, white potatoes, pink potatoes.
"What do you know?
All the world eats"
and 25 potatoes threw their arms in the air, shouting
"Potatoes!"
The rap music continued:
"How do you know if a potato is sick?"
(A large brown potato at the front of the group collapsed dramatically, but gracefully, to the stage.)
"Find a 'tato pathologist really quick!"
(A white potato comes to the rescue.)
"They are like a doctor, calling around"
(The brown potato is rapidly cured.)
"They'll keep potatoes healthy and brown!
Potatoes!"
Caught up in the class's enthusiasm, the audience of delighted parents and other students at the assembly raised their hands and joined the shout before the next verse of the Potato Rap.

L. J. Rennie (✉)
Curtin University of Technology, Perth, WA, Australia

D. Corrigan et al. (eds.), *The Professional Knowledge Base of Science Teaching*,
DOI 10.1007/978-90-481-3927-9_2,

This story illustrates one outcome of the *Scientists in Schools* project in Australia, which aimed to establish sustained and ongoing partnerships between scientists and school communities.[1] Evaluation of the project demonstrated benefits to scientists, teachers and students (Rennie and Howitt 2009), not the least of which was renewed enjoyment of and enthusiasm for science amongst all participants. Further, there was a strong dose of professional learning and increased confidence for teachers, especially those in primary schools who professed to know "not much" about science. One of these teachers was Mrs R., whose strong pedagogical knowledge and creative enthusiasm enabled very positive outcomes from her Year 4 class's research on potatoes with their scientist, Dr C.

Scientists in Schools is a project that blurs the boundaries between schools and communities. Scientists and teachers work together, often on a regular basis, in ways that enable students to experience "real-world" science and teachers to stay in touch with contemporary science outside of school. Projects that promote science and scientists to schools and their communities can make a positive difference to students' engagement and interest in science education. This chapter is about the place and promise of programs, like *Scientists in Schools*, that blur the school–community boundary, and it begins with some background to describe that boundary in terms of the present state of science education in the Western world.

Background: The Quality of Science Education

Scientists in Schools was launched in July, 2007, the same month that the World Conference on Science and Technology Education was held in Perth, Western Australia. Building on a policy options paper he had drafted for the World Conference, Fensham (2008) identified 11 emerging issues for policy makers, noting that "the quality of school education in science and technology has never before been of such critical importance to governments" (p. 4). One of those issues is interest in and about science, and he recommended that personal and societal interest about science should be the reference point for curriculum decision-making.[2] Two other issues concerned the quality of learning and the need for an effective assessment system. Similarly, Osborne and Dillon's (2008) reflections on science education in Europe identified a lack of perceived relevance, a pedagogy that lacked variety, and an assessment system that encouraged rote, rather than mastery learning,

[1] The *Scientists in Schools* project is an initiative of the Australian Government's Department of Education, Employment, and Workplace Relations, and is managed by the Commonwealth Scientific and Industrial Research Organisation.

[2] The focus here is on Western countries (from where national reports are more readily obtained), but projects like the Relevance of Science Education (ROSE) indicate that in terms of interest in, and commitment to, science, the Western countries are those where the situation is most dire (Schreiner and Sjøberg 2007). In developing countries, a science career is much more likely to be a passport to well-paid employment and so the value of science education remains high.

amongst the reasons for students' lack of engagement in science. The European Commission's High Level Group on Science Education called for action to increase students' interest in science through a renewed, inquiry-based pedagogy (European Commission 2007). The Group also argued for increased opportunities for cooperation between the formal and informal arenas, an argument endorsed and extended by Stocklmayer et al. (2010).

The search for a science education that students find engaging has turned attention towards a curriculum that places more focus on the world outside of school, and less on the discipline-specific kind of curriculum traditionally offered to our students. This refocus is based on the reasonable view that if students are to operate as informed citizens in a world that is becoming increasingly global, then the science curriculum they experience at school must be sufficiently meaningful and relevant for them to perceive links with what they experience outside the school doors. Unfortunately, creating and delivering a curriculum that has such a focus turns out to be a very large challenge to how science education usually works in schools, as Osborne et al. (2002) discovered in their evaluation of the pilot phase of the *AS Science for Public Understanding* course. Although students enjoyed it, Osborne et al. found that achievement of its broader aims was limited by the difficulties teachers experienced in changing the culture of their pedagogical practice. "Changing the curriculum is one thing", Osborne (2007) pointed out, "Asking teachers to change their pedagogy to meet the demands of such a curriculum is another" (p. 181).

In this chapter, some of the problems inherent in refocusing the science curriculum to include more links with the world outside of school are identified, and the conflicts that arise for teachers are explored. In particular, the underlying tensions between teaching for disciplinary knowledge and teaching for understanding of real-world, interdisciplinary problems are examined and an argument is made for a more balanced view of science curriculum that can serve the need for students to become scientifically literate citizens. Such a curriculum creates particular challenges for teachers, and the professional knowledge required and how it may be developed, are also explored.

Science Curriculum and Scientific Literacy

Increasingly, school science curricula have endorsed scientific literacy as a key outcome. The *National Science Education Standards* (National Science Council 1996) and *Twenty-First Century Science* (http://www.21stcenturyscience.org) are examples of curricula that state this emphasis clearly. However, scientific literacy is a contested concept, variously defined in various contexts. Further, how scientific literacy is envisaged in the documents describing the intended curriculum, can be quite different to how scientific literacy is portrayed in the curriculum implemented in the classroom.

In his analysis of scientific literacy/science literacy, Roberts (2007) sorted out some of the confusion in its meaning by referring back to

> a continuing political and intellectual tension that has always been inherent in science education itself. …two legitimate but potentially conflicting curriculum sources: science subject matter itself and situations in which science can legitimately be seen to play a role in other human affairs. (p. 729)

Roberts (2007) proposed, as a heuristic device, two visions of scientific literacy that reflect the extremes of these two sources: "Vision I gives meaning to SL [scientific literacy/science literacy] by looking inward to the canon of orthodox natural science, that is the products and processes of science itself" (p. 730), whilst "Vision II derives its meaning from the character of situations with a scientific component, situations that students are likely to encounter as citizens" (p. 730). Roberts gave examples of curricula and the Visions from which they were derived. Most traditional, discipline-based curricula draw from Vision I but *Twenty-First Century Science* draws primarily from Vision II.

Gardner (1975) illustrated how the curricula developed for science education during the 1950s and 1960s reflected the views of influential educationists of the time: "school subjects should serve as faithful and valid introductions to the academic disciplines whose names they bear" (pp. 1–2). These curricula exemplify a Vision I perspective of scientific literacy, focused on the key concepts of science independent of the real-world context in which those concepts might be applied. Most current curricula are traditional in the sense that they focus on the disciplinary knowledge of science, but some look beyond this, to the science-related issues students experience in their world.

Recently, Duschl (2008) documented "an important change in focus for science education, one that embraces a shift from teaching about *what* to teaching about *how* and *why*" (p. 270, original emphasis). Duschl noted "a connectedness in the practices of science that [is] not typically found in school classroom environments" (p. 272), and "the blurring of the boundaries between science and technology, and between different branches of the sciences themselves" (p. 274). The curricula that provide a connectedness between the discipline of science and the science students experience outside of the classroom draw from Roberts' Vision II. However, by making connections with science outside of school, these Vision II curricula do not ignore the discipline knowledge of science. The Visions are not mutually exclusive, as Roberts pointed out: "Vision II subsumes Vision I but the converse is not necessarily so" (p. 768).

Science in Everyday Situations

Aikenhead (2006) comprehensively reviewed research on the outcomes of traditional science curriculum and drew a conclusion that captured the central aspect of the boundary between science in school and science in the community. He stated that "research has produced one clear and consistent finding: *Most often canonical science content is not directly useable in science-related everyday situations*" (p. 29, original emphasis). Why is this so?

There are several reasons why it is difficult for students to use the canonical science content of traditional curriculum to help them make sense of the science in their everyday lives. Three of them are explored here. The first is that the science that happens outside of school differs from the science that is learned in school. This occurs because the science concepts learned in school are idealised and simplified, stripped of all the associated and confounding variables that operate in the world outside of the textbook. For example, Newtonian physics and its associated laws, such as F=ma, are useful approximations but cannot easily be applied outside of textbook physics problems. The process of simplification designed to enable students to "understand" the concepts, unfortunately also works as a barrier to using those concepts in the real world. Because there are so many uncontrollable variables, it is difficult to tease out how the school science concepts can have practical relevance in real-world situations. Even when using them well, abstract explanations and imperfect predictions are usually the best outcome. Further, often the concepts needed to understand science-related issues derive not only from science, but from other disciplines, such as mathematics and geography. Science issues in the real world, such as climate change, genetic modification and dealing with epidemics, are interdisciplinary.

A second and related reason is that the significant science-related issues in our daily lives are often complex and not completely understood. As Ryder (2001) found in his analysis of science understanding for functional scientific literacy, "the science knowledge featuring in everyday contexts is characterised by uncertainty and dispute amongst scientists" (p. 37). Arguments about global warming, greenhouse gases and carbon emissions are consistently in the news and provide ready examples of disputes and disagreements amongst people with different interpretations of similar but incomplete scientific evidence. These disagreements illustrate a third reason for difficulty in using disciplinary science to understand scientific issues. Almost always, there are competing social, cultural, economic and political values that provide conflicting interpretations of how to use science knowledge to take actions for the benefit of society (Corrigan et al. 2007). Further, conflicting interpretations and incomplete knowledge mean that making decisions is a risky business. Learning to cope with uncertainty and risk is an important part of becoming scientifically literate in the Vision II sense, but it has rarely featured in science curricula, except for those based on Science, Technology, Society and Environment.

Thus we see that the disciplinary science that students experience in the classroom is not immediately discernible in the issues and problems in which it resides outside the classroom, because it is melded immutably with knowledge and understanding in a range of other subjects, including mathematics, geography and economics, and also is imbued with social, cultural and political values. In short, science in the world is interdisciplinary and value-laden. Major problems facing our increasingly global world need to be tackled by interdisciplinary teams. How can our students be prepared to face this interdisciplinary world? What do they need to know?

Knowledge and the Science Curriculum

In terms of the teaching and communication of science, Duschl (2008) asked, "What is most worth knowing? Is it what we know? Or is it how we know and why we believe in it even in the face of plausible competing alternatives?" (p. 278). Vision I and Vision II offer very different views about the purposes of knowledge and education. This is not surprising. In Vision I, knowledge is treated as separate from experience and separate from its political and economic uses. Disciplinary science knowledge is valued for itself. In Vision II, the focus is on learning and knowing, rather than on knowledge. Knowledge is valued because it can be used to make sense of experience. The pedagogical approaches are more concerned about why and how to teach science than about what to teach (Duschl 2008).

What is significant and what makes the contrast between Vision I and Vision II important for this chapter, is that schooling itself evidences a mix of both perspectives. Schools are social institutions and, historically, have a major role in knowledge transmission. Traditionally, the nature of science knowledge to be transmitted is more like the canonical concepts described by Aikenhead (2006). But increasingly, schools are expected to ready their students for life in the outside world, much of which does not require an extensive disciplinary knowledge of science. These are different roles for the school curriculum, reminiscent of Fensham's (1985) point that the science curriculum has traditionally catered for the minority of students who wish to pursue further studies of science, and served less well the large majority who simply need enough science for citizenship. These two roles are conflicting rather than complementary, and for the most part, the conflict remains unresolved (Fensham 2008). Consequently, there will continue to be, at least for the time being, conflict between discipline-based curricula that provide orthodox, canonical science knowledge and integrated, interdisciplinary curricula that allow more flexibility in catering to students' needs and the interests of the local school communities. Teachers are in the middle of this conflict.

The heart of this conflict is that schooling, particularly secondary schooling, is not shaped to reflect the interdisciplinarity of real-world issues. Instead, school curricula are usually arranged in disciplinary areas. Most curricula have a section identified as science, even though, as Jenkins (2007) pointed out, school science is a term that covers a variety of sciences with major conceptual and philosophical differences. Further, Fensham (2009) noted that the Anglo-American tradition of teaching discrete subjects in a vertical fashion (that is, the content each year builds on the previous) promotes the inward-looking Vision I of scientific literacy. However, students arrive at school each day informed by their experiences in the community which are more closely related to the outward-looking Vision II, but generally are expected to set aside knowledge from those experiences and, while at school, work with school-based disciplinary science knowledge and understandings that often seem quite narrow and disparate to their own experience. As a result we see the creation of a boundary between the disciplinary science knowledge needed in school and the functional science knowledge used in the community.

Can this boundary be blurred? Teaching a science curriculum that includes interaction with significant science-related issues beyond the classroom, demands that teachers work in interdisciplinary ways and integrate at least some parts of the curriculum. However, curriculum integration is neither well-understood nor well-accepted in science education. Venville and her colleagues (Venville et al. in press; Venville et al. 2002) explored the reasons for this. They found problems of definition, disagreement about the reasons for integration, difficulties for teachers implementing integrated curricula and arguments about the quality of learning that resulted.

Drawing on the theoretical perspectives of Bernstein (1971), Venville et al. (2002) drew attention to the challenge posed by curriculum integration to the status and power of academic, disciplinary knowledge, arguing that integration was at odds with the traditional hierarchies, customs and culture of schooling, which are closely tied to disciplinary-based learning and its assessment. In contrast, functional knowledge from an integrated curriculum was perceived to be more "everyday" and less academic. It was perceived to be of lower status and hence as less worthwhile. In synthesising their research, Venville et al. found evidence that the status assigned to disciplinary knowledge (upon which the important tertiary entrance examinations were based) was a persuasive deterrent to the introduction of integrated curriculum, particularly in secondary schools. As a way forward, Venville et al. (2002) suggested that rather than try to work with two apparently competing curriculum paradigms based on the nature of knowledge, a pragmatic approach to curriculum integration was needed, an approach that did not ignore the established disciplines, but positioned them within a more holistic view of knowledge. Such an approach to integrated curriculum would recognise students' knowledge as grounded in their experiences and contexts, and attempt to meet the needs of students, the school and the local community. In such an approach, "the disciplines are there, but they are omnipresent rather than omnipotent" (Venville et al. 2002, p. 70). Venville et al. suggested that school science should provide students with opportunities to develop a scientific literacy that includes knowledge of the disciplines but also knowledge of the more interdisciplinary science-related issues students meet outside of school, in other words, a balanced curriculum that blurs the boundaries between disciplinary science in school and functional science outside of school (see also Rennie et al. in press).

Scientific Literacy in a Balanced Curriculum

The balanced curriculum described by Venville et al. (2002) has a meaning for scientific literacy consistent with Roberts' (2007) Vision II. This is the kind of scientific literacy that Goodrum et al. (2001) proposed should be an outcome of science education. It is forward-looking and concerns citizenship. Scientifically literate people are considered to be those who are interested in and understand the world around them; engage in the discourses of and about science; are able to identify questions, collect data and draw evidence-based conclusions; are sceptical and questioning of claims made by others about scientific matters and make informed decisions about the environment and their own health and well-being (Goodrum et al. 2001). Such a definition requires that people have certain skills and abilities that enable them to cope in life both within and beyond the classroom, and some of those skills and abilities are identified in Table 2.1.

Inspection of Table 2.1 reveals that scientific knowledge is needed, but it must be the kind of science knowledge that can be applied in new situations. The relevant knowledge is more likely to be functional science knowledge. The listed skills also strongly support the development of social responsibility, providing a better chance of harmonising the conceptual, epistemic and social learning goals, as argued by Duschl (2008). If students are to develop the skills and abilities listed in Table 2.1, then their school science curriculum needs to include significant interaction with the world outside of school.

There is evidence that the kinds of skills and abilities listed in Table 2.1 can be developed when there are effective school–community links. Further, the learning outcomes for students can be both powerful and worthwhile. In recent case study research, Venville et al. (2008) discovered that the outcomes of an integrated curriculum in which middle-school students learned about the social, economic and scientific issues related to the health of a local lake were very powerful for the learners. Even though the content followed the interests of students and their teachers and was certainly context-dependent, the curriculum approach and the

Table 2.1 Components of scientific literacy and underlying skills and abilities. (Based on Rennie 2006)

Scientifically literate people	Underlying skills and abilities
Are interested in and understand the world around them	Select and apply relevant science knowledge and skills in daily life Seek information to explain new phenomena or solve problems
Engage in the discourses of and about science	Feel comfortable to listen to, and to read, write and talk about science in everyday situations
Are able to identify questions, investigate and draw evidence-based conclusions	Analyse issues and identify, obtain and use needed information Understand how scientists go about finding answers to questions Construct and defend an argument
Are sceptical and questioning of claims made by others about scientific matters	Assess the trustworthiness of claims and sources of evidence
Make informed decisions about the environment and their own health and well-being	Recognise and cope with risk and uncertainty in decision making Choose to act responsibly and ethically

learning it engendered "moved" students well beyond their local and particular knowledge. Venville et al. (in press) concluded that what the integrated curriculum taught and what was learned during the case study provided students with usable scientific knowledge as well as values in social and civic responsibility. They were able not only to think in ways appropriate to the problems and issues that faced their community, but were able to communicate and debate these issues, and suggest ways of addressing those problems and issues.

An important afterword to the case study which gave rise to these findings is that following the introduction of state-wide achievement testing in three subject areas, including science, the integrated approach to curriculum in the middle school was abandoned in favour of a return to Vision I disciplinary-based approaches with the aim of enhancing performance on the tests. This move effectively reinstated the boundary that had been blurred, even bridged, by the local lake contribution to the curriculum.

Changing Curriculum, Changing Teaching

The case study concluding the preceding discussion illustrates how the curriculum can be opened up to a stronger focus on science-related issues outside of the classroom and work, in an interdisciplinary way, to build upon the students' own interests and concerns. The afterword also illustrates just how difficult it is to maintain that focus. There is no doubt that the kinds of skills and abilities these students were developing were those described in Table 2.1, and that these students were given opportunities to become scientifically literate in the Vision II sense. The argument presented in this chapter is that more students can be given such opportunities if their experiences of science in school and science in the community are brought much closer together by using community resources to explore science-related issues that have local relevance, thus blurring the boundary between school and community. The kinds of resources available are almost boundless. Rennie (2006) drew attention to families and friends, institutions such as museums with an educational role, community and government organisations and the media, as readily available resources that provide almost continuous opportunities for students to learn about science, both explicitly or implicitly, outside of school.[3]

However, making effective use of community resources requires considerable investment of time and effort. Already we have seen that science in the real world is complicated: We cannot control all of the relevant variables; much current scientific knowledge is uncertain and incomplete, leading to disputes and disagreements; and there are competing values and risks in making decisions about how knowledge is best used. There are significant pedagogical consequences of this "messiness" of

[3] Further information is available in recent reviews of research in out-of-school learning (Bell et al. 2009; Rennie 2007; Stocklmayer et al. 2010) and guidance for teachers in using a range of community resources (Braund and Reiss 2004).

science in the real world. Teaching a science curriculum that involves interaction with controversial science-related issues, or with new kinds of resources from beyond the classroom, requires of teachers an enhanced knowledge base and a suite of pedagogical skills that differ from those needed to teach a discipline-based, concept-oriented curriculum. The remainder of the chapter will address these pedagogical consequences.

Teachers' Content Knowledge

Teaching is a busy, full-time activity, and while teachers work in their classrooms, knowledge in the world outside is changing. Many teachers will need to broaden their content knowledge to enable them to bring contemporary science into the classroom. Teachers who read about, and keep up-to-date with, knowledge advances in their particular field will be well-placed to do this, but teachers of general science, particularly teachers in primary schools who are responsible for teaching more subjects than science, face a daunting challenge. Assistance for these teachers may need to come from the community itself. When students explore issues using community resources, their teachers have opportunities to learn as well as their students. Consequently, it is important that teachers are willing to learn from community members and resources and even from the students themselves. Mrs R., whose class's performance opened this chapter, is a good example. In working with scientist Dr C., she learned a great deal about potatoes and the nematodes attacking them, and also about science and how it works in the community.

Teaching About Community Issues in the Classroom

Not only content knowledge, but pedagogical knowledge is required for teachers to incorporate authentic, community issues into the classroom, or to move students outside of the classroom to work with issues in the community. Of course, teachers have always had excursions or field trips, but most research indicates that they are not well-integrated into the school curriculum (Rennie 2007). Excursions are often expensive and there are organisational and administrative hurdles to overcome. Further, it takes considerable effort for teachers to ensure that they are used effectively. Not all teachers know how to do this, although research suggests that teachers with good content knowledge are better able to integrate learning from excursions and field trips into their curriculum (Rennie 2007).

A particularly difficult area for teachers is dealing with socio-scientific issues that are controversial (Ratcliffe and Grace 2003). Good content knowledge is required, but also an ability to feel comfortable in dealing with the risk, uncertainty and ambiguity that reside in such issues. In other words, teachers themselves need to be scientifically literate in the Vision II sense so that they can be comfortable dealing with the everyday situations that arise in the community, including through the media. Recent work in New Zealand by Saunders (2010) revealed that although most teachers believed that such issues had a place in the classroom, they also believed they needed help in teaching them. Saunders developed and field-tested a professional learning model for teachers, and those who used it not only found it rewarding but were astounded at the interest evidenced by students and the high quality of work they produced. However, not all teachers see their role in this way. Levinson and Turner (2001) reported that a majority of science teachers they interviewed in the United Kingdom believed that teaching science should be about facts and explanations, and that dealing with associated social and ethical issues were not part of their role. It would seem that the students of these teachers would be limited to developing a Vision I perspective of scientific literacy.

Helping students to learn about and use science in everyday contexts requires a high level of pedagogical content knowledge, because using knowledge in different contexts often requires considerable reworking of that knowledge so that it can be used in new situations. Some years ago, Layton et al. (1993) explored how four groups of adults in different situations sought out and made use of science knowledge they needed to deal with particular issues in their lives. These researchers found that in order to make use of that knowledge, people had to rework it into a form that made sense to them. Layton et al. described this process of deconstructing and then reconstructing the information, as transforming knowledge, or constructing "knowledge for practical action in [their] specific situations" (p. 128). Teachers need to keep in mind that this process is very difficult and therefore take opportunities to assist students to develop the skills of using knowledge in new situations.

In one of their studies, Venville et al. (2004) found that students attempting to use content knowledge from their school lessons in science, mathematics and technology to build a solar-powered boat, frequently abandoned that source of knowledge as unhelpful, and drew on other sources, such as observing other students, and asking friends and family members for advice. This is consistent with an important conclusion from Aikenhead's (2006) review, that "when the science curriculum does not include the difficult process of transforming abstract canonical content into content for taking action, canonical science remains unusable outside of school for most students" (p. 30).

Students are continuously learning from sources in the world outside of school, and often that learning is not consistent with the disciplinary science knowledge presented in the classroom. Teachers need to be aware of what students have already "learned" from these external sources, not only to harness its potential to engage students' interests, but also to help them rework that knowledge so that it is meaningful in school science. Assisting students to transform knowledge into a form that can be used where it is needed requires considerable pedagogical content knowledge to determine what students do understand (and misunderstand), what they need to understand and then how to shift their understanding. Years of conceptual change research indicates that this is not easy to do. Students will resist if they see no reason to change their commonsensical, quite workable, but possibly misconceived, ideas. Jenkins (2007) put it well: "for most everyday practical purposes, common-sense, as distinct from scientific, thinking, is perfectly adequate" (p. 277).

Teachers' Professional Learning

The challenges to teachers' professional knowledge are significant for most teachers. Changing the curriculum to bring the classroom and the community closer together, means changing their teaching, and teacher change is rarely easy. Understanding the kinds of things that need to change can be a first step in assisting teachers to progress. Bartholomew et al. (2004) provided a framework that can be useful in this regard. Their research with a group of teachers asked to use the "ideas-about-science" approach to teaching the nature of science resulted in the identification of five dimensions of practice that recognised the salient, but not independent, components of effective teaching. These dimensions were first, the degree of teachers' confidence in their own knowledge and understanding of the nature of science; second, teachers' conception of their own role, as either a dispenser of knowledge or a facilitator of learning; third, teachers' use of discourse, as either closed and authoritative or open and dialogic; fourth, teachers' conception of learning goals as either limited to knowledge gains or inclusive of the development of reasoning skills and fifth, the nature of classroom activities, in terms of whether they were contrived and inauthentic, or by students and therefore were authentic (Table 3, p. 664). Bartholomew et al. found that

> effective teaching of "ideas-about-science" requires establishing a context in which it is possible for students to engage in reflexive epistemic dialogue.... For many teachers, enculturated in the habitus of traditional science teaching, this would require a shift in conception of their own role from dispenser of knowledge to facilitator of learning; a change in their classroom discourse to one which is more open and dialogic; a shift in their conception of the learning goals of science lessons to one which incorporates the development of reasoning and an understanding of the epistemic basis of belief in science as well as the acquisition of knowledge; and the development of activities that link content and process in tasks whose point and value is transparent to their students. (p. 678)

The parallels between teaching ideas-about-science and teaching about science-related issues are instructive. Bartholomew et al. (2004) noted that, as teachers' confidence grew and they became more used to dealing with the content, they began to offer a curriculum more like that described in the excerpt above. However, it was also clear that simply increasing teachers' knowledge and understanding about what is intended to be taught is insufficient to result in a change in practice. There must be considerably more opportunities for professional learning and resources to support such a change. There also needs to be change in the way students' learning is assessed. Many of the skills in the right hand column of Table 2.1 cannot easily be assessed using pencil and paper tests. Ways must be found to devise valid measures for the skills associated with a Vision II kind of scientific literacy. Abell's (Chap. 12) discussion of what counts as evidence for learning is central to this issue. Teachers will not change their practice without a concomitant change in assessment that demonstrably values the intended outcomes of the new curriculum to be implemented.

Changing Teachers' Mindsets: Ways Forward

This chapter presents an argument that students' interest in science and their perceptions of its relevance for them can be enhanced by bringing science in the community into science in the classroom. Some of the difficulties of doing this have been explored. The messy nature of real-world science, compared with the comparatively clear-cut, traditional canonical science concepts that typically compose the school science curriculum, was explored as one barrier to be overcome. The difficulties teachers often face in dealing with interdisciplinary, integrated science of the kind that exists in the community were also explored, as was the need to overcome the perception, particularly in secondary schools, that this was a move towards lower status, less powerful science knowledge. Because of pressures of time and the need to cover the curriculum, teachers are frequently caught in the conflict between teaching disciplinary-based science (that is promoting a Vision I perspective) and broadening the science they teach to the kinds of experiences students have outside the classroom (that is, moving the science curriculum closer to Roberts' Vision II). Blurring the boundary between school and the community requires that teachers believe that this is worthwhile. Aikenhead, in the Introduction to his chapter in this volume, points out that many teachers have a belief system that "seems to revere the memorisation of facts, abstractions, and algorithms". Moving teachers to implement a curriculum more aligned with Vision II than Vision I requires changing this belief system, or mindset. This requires four key changes in teachers' beliefs.

First, teachers must believe that allowing students to experience functional science in the real world and see scientists in their work place is important, and that understanding and using science in context is important. It gives students opportunities to see the relevance of disciplinary science concepts and learn how to transfer knowledge from in class to science experiences out of class. Science, as it is practised, is messy, uncertain and conflicted with values in the real world. Students need opportunities to find this out and learn to deal with the inherent ambiguities and risks.

Second, teachers need to believe that some (but of course not all) science concepts enshrined in current traditional curricula can be sacrificed to provide time and space for students to learn by devising and investigating their own questions about matters that are important to them. Teachers need to believe that the outcomes of this approach are worthwhile.

Third, teachers need excellent pedagogical knowledge to help their students develop the abilities and the skills described in Table 2.1. Teachers need to be able to "let go" of, or at least slacken their grip on, the learning reins, and allow students to take more control of their own learning. Of course this requires that students have opportunities to learn how to ask "good" questions that can be investigated by collecting and evaluating data to arrive at answers supported by evidence. In this way, students learn the relevant concepts and how they can be used.

Fourth, teachers need to believe that there are many valid ways to assess learning. The summative written tests so firmly entrenched in current assessment methods have variable validity in measuring learning outcomes. There are other assessment methods that enable students to demonstrate what they know and can do. Taking the science-related situation, rather than the science concepts, as the starting point for assessment allows for more creative ways of gathering evidence about students' learning. This is an important message (Fensham 2008), and curricula will not change unless the prevailing assessment methods change.

Teachers Learning by Doing

In making the changes that lead to students developing the kind of scientific literacy advocated in this chapter, teachers need support because, at least initially, most of them will be swimming against a strong tide of traditional pedagogical practice. How can they gain that support? The phrase "learning by doing" is often used for students, but it also works for teachers. In the evaluation of several federally funded Australian projects, strong support has been found for teacher change and development by being involved in school–community projects. The first of these projects was the Science Awareness Raising Program (Rennie and ASTA 2003), followed by the School–Community Industry partnerships in science projects (SCIps) (ASTA 2005), both led by the Australian Science Teachers Association (ASTA). In these projects, teachers in one or more schools worked with community members on a science-related issue that was important to the community. Successful projects at the primary, middle and secondary levels blurred the school–community boundary. They consumed considerable time and effort, but were found to be rewarding for teachers, students and parents, who were often key contributors from the community. Invariably, the students were very engaged and produced evidence demonstrating considerable learning, and often that learning was shared with the community.

Similar findings emerged from the larger scale, Australian Schools Innovation in Science, Technology and Mathematics (ASISTM) project (Tytler et al. 2008), in which schools worked with outside experts on innovative projects in their communities. Renewed self-confidence, content knowledge and confidence with science processes were gained, indicating that for many teachers participation in their ASISTM project was "a very potent and successful form of professional learning for teachers" (p. 39).

Rennie (2006, p. 9) argued that there are several important guidelines for effective school–community projects. Successful projects

- are based on some issue/stimulus which comes from within the community and is not imposed.
- require the input of community members to provide local knowledge to contextualise the issue.

- are educative, because they focus on science as a way of knowing, thinking and acting, and model science inquiry (working scientifically).
- are integrated into the school science curriculum and thus legitimise participation by students and teachers.
- involve negotiation and decision-making with the community in regard to social, political and economic factors, differing perspectives from different groups, and information collected (both local and science-related).
- have a tangible outcome to indicate when the project is complete and demonstrate that it has achieved something worthwhile.

In sum, such programs demonstrate that when the school–community boundaries are blurred, there is enhanced engagement and interest from students, and considerable professional learning for teachers. But this learning comes at a price. Working over boundaries is time-consuming and requires effort and commitment by the teachers and community members involved. Given this, such programs must be allocated a real place in the science curriculum, a place made possible by lessening the science disciplinary content by judicious selection of what is most meaningful for the students involved.

Final Word

Mrs R. joined the *Scientists in Schools* project when she was given the role of science coordinator in her primary school. She was resolved to learn more about teaching science, and wanted to promote science in her school. She believes that her partnership with Dr C. has improved her own pedagogical content knowledge as well as providing her class with exciting and challenging activities (some of which involved growing 15 different kinds of potatoes!) they otherwise could not have experienced. Allowing Dr C. into her classroom and working with him in a flexible and respectful way, reaped considerable benefits for Mrs R., Dr C. and her class of Year 4 students (Rennie and Howitt 2009). In terms of the raising students' interest and perceiving relevance in science, the *Scientists in Schools* project has worked well. It is appropriate to give the final words of this chapter to one of the 9-year-old girls in the class who, as part of the project's evaluation, wrote:

> Dr C has changed my life, the way I think about scientists. I thought science was boring, but I was wrong. If you think about it, if you put your mind to it, science is quite cool. As I said before, science has changed my life!

References

Aikenhead, G. (2006). *Science education for everyday life.* New York: Teachers' College Press.

Australian Science Teachers Association. (2005). *SCIps school industry partnerships in science: Final report.* Canberra: Department of Education, Science and Training. (Available at http://www.asta.edu.au/scips/gui/files/final_report.pdf)

Bartholomew, H., Osborne, J., & Ratcliffe, M. (2004). Teaching students "ideas-about-science": Five dimensions of effective practice. *Science Education, 88*, 655–682.

Bell, P., Lewenstein, B., Shouse, A. W., & Feder M. A. (Eds.). (2009). *Learning science in informal environments: People, places, and pursuits.* Washington: The National Academies Press.

Bernstein, B. (1971). On classification and framing of educational knowledge. In M. F. D. Young (Ed.), *Knowledge and control: New directions for the sociology of education.* London: The Open University.

Braund, M., & Reiss, M. (Eds.). (2004). *Learning science outside the classroom.* London: RoutledgeFalmer.

Corrigan, D., Dillon, J., & Gunstone, R. (Eds.). (2007). *The re-emergence of values in science education.* Rotterdam: Sense.

Duschl, R. (2008). Science education in three-part harmony: Balancing conceptual, epistemic, and social learning goals. In G. J. Kelly, A. Luke, & J. Green (Eds.), *What counts as knowledge in educational settings, disciplinary knowledge, assessment, and curriculum. Review of Research in Education, 32*, 268–291.

European Commission. (2007). *Science education now: A renewed pedagogy for the future of Europe.* Brussels: European Commission, Directorate-General for Research.

Fensham, P. J. (1985). Science for all: A reflective essay. *Journal of Curriculum Studies, 17*, 415–435.

Fensham, P. J. (2008). Science education policy-making: Eleven emerging issues. A report commissioned by UNESCO, Section for Science, Technical and Vocational Education. http://unesdoc.unesco.org/images/0015/001567/156700e.pdf. Accessed 10 Oct 2008.

Fensham, P. J. (2009). The link between policy and practice in science education: The role of research. *Science Education, 93*, 1076–1095.

Gardner, P. L. (Ed.). (1975). Science and the structure of knowledge. In P. L. Gardner (Ed.), *The structure of science education* (pp. 1–40). Hawthorn: Longman.

Goodrum, D., Hackling, M., & Rennie, L. (2001). *The status and quality of teaching and learning of science in Australian schools: A research report.* Canberra: Department of Education, Training and Youth Affairs.

Jenkins, E. (2007). School science: A questionable construct? *Journal of Curriculum Studies, 39*(3), 265–282.

Layton, D., Jenkins, E., Macgill, S., & Davey, A. (1993). *Inarticulate science? Perspectives on the public understanding of science and some implications for science education.* Nafferton: Studies in Education Ltd.

Levinson, R., & Turner, S. (2001). *Valuable lessons: Engaging with the social context of science in schools.* London: The Wellcome Trust.

National Science Council. (1996). *National science education standards.* Washington: National Academy Press.

Osborne, J. (2007). Science education for the twenty first century. *Eurasia Journal of Mathematics, Science & Technology Education, 3*(3), 173–184.

Osborne, J., & Dillon, J. (2008). *Science education in Europe: Critical reflections.* London: The Nuffield Foundation.

Osborne, J., Duschl, R., & Fairbrother, R. (2002). *Breaking the mould? Teaching science for public understanding* (a report commissioned by the Nuffield Foundation). London: The Nuffield Foundation. (Available at www.kcl.ac.uk).

Ratcliffe, M., & Grace, M. (2003). *Science education for citizenship.* Maidenhead: Open University Press.

Rennie, L. J. (2006, August). The community's contribution to science learning: Making it count. Plenary address to the ACER Research Conference 2006, "Boosting science learning—What will it take," Canberra.http://www.acer.edu.au/documents/RC2006_Rennie.pdf. Accessed 13 March 2007.

Rennie, L. J. (2007). Learning science outside of school. In S. K. Abell & N. G. Lederman (Eds.), *Handbook of research on science education* (pp. 125–167). Mahwah: Lawrence Erlbaum Associates.

Rennie, L. J., & Howitt, C. (2009). "Science has changed my life!" Evaluation of the scientists in schools project. A report of its evaluation. http://www.scientistsinschools.edu.au/evaluation.htm. Accessed 15 April 2009.

Rennie, L. J., & The Australian Science Teachers Association. (2003). *The ASTA science awareness raising model: An evaluation report prepared for the Department of Education Science and Training.* Canberra: ASTA. (Available at http://www.dest.gov.au/sectors/school_education/publications_resources/profiles/science_awareness_raising_model_evaluation.htm).

Rennie, L. J., Venville, G., & Wallace, J. (2010). Learning science in an integrated classroom: Finding balance through theoretical triangulation. *Journal of Curriculum Studies,* 1–24. First published on: 02 December 2010 (iFirst).

Roberts, D. (2007). Scientific literacy/science literacy. In S. K. Abell & N. G. Lederman (Eds.), *Handbook of research on science education* (pp. 729–780). Mahwah: Lawrence Erlbaum Associates.

Ryder, J. (2001). Identifying science understanding for functional scientific literacy. *Studies in Science Education, 36,* 1–42.

Saunders, K. (2010). Engaging with bioethics: A professional learning programme for science teachers. In A. Jones, A. McKim, & M. Reiss, (Eds.), *Ethics in the science and technology classroom: A new approach to teaching and learning* (pp. 103–128). Rotterdam: Sense.

Schreiner, C., & Sjøberg, S. (2007). Science education and youth's identity construction—two incompatible projects. In D. Corrigan, J. Dillon, & R. Gunstone (Eds.), *The re-emergence of values in science education* (pp. 231–247). Rotterdam: Sense.

Stocklmayer, S. M., Rennie, L. J., & Gilbert, J. K. (2010). The roles of the formal and informal sectors in the provision of effective science education. *Studies in Science Education, 46,* 1–44.

Tytler, R., Symington, D., Smith, C., & Rodrigues, S. (2008). An innovation framework based on best practice exemplars from the Australian School Innovation in Science, Technology and Mathematics (ASISTM) Project. Canberra, Australia.http://www.dest.gov.au/sectors/school_education/publications_resources/profiles/innovative_framework_asistm.htm. Accessed 1 June 2009.

Venville, G., Rennie, L., & Wallace, J. (2004). Decision making and sources of knowledge: How students tackle integrated tasks in science, technology and mathematics. *Research in Science Education, 31*(2), 115–135.

Venville, G., Rennie, L. J., & Wallace, J. (in press). Curriculum integration: Challenging the assumption of school science as powerful knowledge. In B. J. Fraser, K. Tobin, & C. McRobbie (Eds.), *Second international handbook of science education.* Dordrecht: Springer.

Venville, G., Sheffield, R., Rennie, L., & Wallace, J. (2008). The writing on the classroom wall: The effect of school context on learning in integrated, community-based science projects. *Journal of Research in Science Teaching, 45*(8), 857–880.

Venville, G., Wallace, J., Rennie, L. J., & Malone, J. (2002). Curriculum integration: Eroding the high ground of science as a school subject. *Studies in Science Education, 37,* 43–84.

Chapter 3
Didaktik—An Appropriate Framework for the Professional Work of Science Teachers?

Helmut Fischler

Didaktik and Current Developments in Science Education

Three factors can be identified that contribute to a marginalization of content, particularly in the perception of teachers but also in the activities of researchers in science education. Each of the three comes from a different aspect of science education:

The first factor derives from the Anglo-American curriculum tradition. In this curriculum tradition a division of labour takes place in which curriculum experts formulate content standards independently of the practitioners responsible for teaching and learning the content. Teachers are to implement effectively a curriculum of content "as an agency for the institutionalized teaching of a 'content', seen unproblematically in terms of this or that view of and selection from a subject matter" (Westbury 1998, p. 62).

The second factor contributing to a marginalization of content comes from developments in science education research—the shift to give priority to essential ideas of cognitive psychology in research on teaching and learning has reinforced this marginalization process. The role of subject matter became more and more underestimated in empirical studies on teaching and learning. "Such neglect is surprising given the needs to be specific about issues of knowledge when we address the curriculum of 'knowledge societies': *What should we teach* is subsequently pushed into the background" (Klette 2008, p. 4; emphasis in original).

The third factor is the currently very strong presence of large-scale assessment studies such as TIMSS and PISA. These have supported a process of standardization in the many countries participating in these studies (see Waddington et al. 2007, for an overview). The development of knowledge tasks for these assessments is restricted to small groups of determiners, and takes place in the absence of broad discussions about fundamental aspects of general education and about subject-related instructional goals. Educational policy makers, school administrators, teachers, and

---

H. Fischler (✉)
Freie Universität Berlin, Berlin, Germany

D. Corrigan et al. (eds.), *The Professional Knowledge Base of Science Teaching*,
DOI 10.1007/978-90-481-3927-9_3, © Springer Science+Business Media B.V. 2011

students in the participating countries are only very minor players in the design processes that lead to a standardization of the knowledge content of these tests.

In this situation Didaktik can be a corrective, bridging content-related issues on the one hand and pedagogical aspects on the other. Didaktik provides a teacher with a language and intellectual scaffolding with which he/she becomes able to scrutinize the content topics of the curriculum mandated by the state in terms of their contributions to a value-oriented education of students. The teacher as a professional practitioner has to embed the topics into an educational context. Didaktik "seeks to model forms of teacher thinking that might direct the teacher to systematic hermeneutic reflection about the ways in which classroom environments might support a personal subjective encounter, or relationship, with the educative 'content' represented in the curriculum, the ultimate forms of social life, and the like" (Westbury 1998, p. 57).

The spelling of "Didaktik" is deliberately distinguished from *didactic* because of the very different connotation of these different words. The latter, *didactic,* describes a methodological conception that has pejorative vibes. "Someone who is *didactic* tells people things rather than letting them find things out or discussing things" (Collins COBUILD English Dictionary 1995). *Didaktik*, on the other hand, has no linkage with this description of a particular form of teacher/teaching.

Didaktik offers a response to many critics who claim that improvement of science teaching and learning is not only a matter of teaching methods but also an issue of science content. Fensham (2001) reminded the science education community "that the disciplinary knowledge of the sciences is not automatically appropriate for school science" (p. 38). Criteria are needed for the processes of selection of topics, their elementarization and their construction for instruction, including as a later part of this process the students' cognitive and affective preconditions (as an example see Duit et al. 1997, also Duit et al. 2007).

It is not suggested that the continental European concept of Didaktik is to replace the Anglo-American tradition of curriculum. But it can supplement the curriculum tradition by emphasizing content-related aspects at the level of daily teaching. Thus, Shirley (2008) saw opportunities to combine positive results within curriculum-oriented developments on the one hand and Didaktik-oriented principles on the other. The standards and accountability movements, together with the tendency to embed investigations on teaching and learning oriented to cognitive psychology and disregarding issues of content into research designs, have generated several different consequences. They have caused discussions among teachers on how to meet these standards and therefore contributed to a deeper awareness of learning efficiency problems; in some cases this has resulted in positive practitioner collaboration. On the other hand the "division of labour" mentioned above has been strengthened. Shirley (2008) calls for a "post-standardization" phase in which these aspects are both taken into account: "The challenge in a new era of post-standardization, then, will be to sustain the momentum that reformers have made enabling teachers to collaborate and to innovate, but to do so in such a way that befits the full human dignity of learners who aspire towards autonomy (*Mündigkeit*) and self-activity (*Selbsttätigkeit*) as free and sovereign beings" (Shirley 2008, p. 38).

Bildung as an Essential Element of Didaktik

Bildung and Didaktik

In the German-speaking countries, and to some extent also in the Northern European countries, "Bildung" is the central notion describing the process of personal development and the result of this development process. Bildung is more than education; therefore no English term denotes the concept of Bildung appropriately. Some scholars translate Bildung as "formation", covering the forming of a personality and the product of this formation. It may be helpful for readers who come from the Anglo-American curriculum tradition to read what an American educational researcher proposes as a valid description after having struggled with numerous attempts to clarify the meaning of Bildung:

> *Bildung* is a noun meaning something like "being educated, educatedness". It also carries the connotations of the word *bilden*, "to form, to shape". *Bildung* is thus best translated as "formation", implying both the forming of the personality into a unity as well as the product of this formation and the particular "formedness" that is represented *by* the person. (Westbury 2000, p. 24; see also the descriptions by Nordenbo 2002 and Wimmer 2003, Wimmer is quoted by Ogawa, in Chap. 8)

Even in the German language it is not possible to find a clear and brief definition of Bildung. Among other reasons this is due to the fact that the concept of Bildung has undergone various changes of its meaning over recent history. Wolfgang Klafki, the most prominent exponent of a modern conception of Bildung, drafted the most significant indicators of this development in some decades around 1800, by absorbing stimuli from the European Enlightenment, "a few fundamental points in common emerged, not least the idea of the self-responsible, cosmopolitan person, contributing to his own destiny and capable of knowing, feeling and acting" (Klafki 1998, p. 313).

For Klafki, the terms "*self-determination, freedom, emancipation, autonomy, responsibility, reason,* and *independence*" are crucial notions denoting Bildung (Klafki 2000a, p. 87). This set of concepts describing qualities individuals should strive for could be misinterpreted as a portrayal of Bildung as an individualistic conception, but Klafki goes on to say: "…the basic concept of subject- or self-determination is anything but subjective!" (Klafki 2000a, p. 88). Bildung is also characterized by a second group of determinants: "*humanity, humankind* and *humaneness, world, objectivity, the general*" (Klafki 2000a, p. 88). Bildung, therefore, develops in the interplay between individual attributes, achievements and expectations on the one hand and the conditions a person has to cope with on the other. These conditions are results of societal processes and comprise different kinds of social life as well as systems of norms and beliefs that pertain to the fields of politics, arts, science and other domains.

Although Bildung refers to an individual's community, Klafki perceived a lack of an in-depth analysis of an individual's environment: "…the economic, social, and political conditions needed for the realization of this general demand for Bildung" was not examined consistently by those who strove for a widely accepted conception of Bildung (Klafki 2000a, p. 89). He proposed a further development that takes account of contemporary approaches to "a more differentiated and critical determination of the relationship between Bildung and society" (Klafki 1998, p. 313). Three abilities were, in this way, to be promoted by Bildung (Klafki 1998, p. 314):

- Self-determination
- Co-determination (all people are invited to take part in the development of the society)
- Solidarity (with those "whose opportunities for self-determination and co-determination are limited")

Bildung and Scientific Literacy

The generally accepted understanding of Bildung becomes more clear when compared with and contrasted to the way scientific literacy has often been used in the last two decades. For example, in the context of the OECD PISA project, scientific literacy stresses the application of knowledge and therefore has a more functional connotation than Bildung has. The cognitive aspects of students' scientific literacy "include students' knowledge and their capacity to use this knowledge effectively…" (OECD 2006, p. 22, see also Fensham 2007). Another characteristic feature of the PISA program is its claim to test whether or not students are well prepared for the demands imposed on them during their whole life: "PISA 2006 covers the domains of *reading, mathematical* and *scientific literacy* not so much in terms of mastery of the school curriculum, but in terms of important knowledge and skills needed in adult life" (OECD 2006, p. 8). Bildung also claims to help students withstand the challenges of their future life but by a *general* preparedness that is not simply acquired knowledge and skills.

The dominant position of the term "scientific competency" in the description of the PISA program (OECD 2006) signals additional differences between Bildung and scientific literacy. The focus on functionalist aspects of students' knowledge (competencies and skills) contrasts with the concept and *process* of Bildung; this concept and process are not primarily aimed at gaining specific qualifications that result in substantial benefits, but at helping a learner to acquire a characteristic individuality that allows one to successfully approach the above mentioned attributes of a person with Bildung. Therefore, a phrase such as "We teach children to be competent in a special domain" is not in line with this perception of Bildung. Knowledge is a part of Bildung, but the knowledge is embedded into a holistic view of the personality of an individual. Within this view both aspects of education—to help students to achieve a considerable state of Bildung as well as to prepare them to meet the requirements of private and vocational life—are two sides of the same coin. One of the most distinguished contemporary German pedagogues, Hartmut von Hentig, well known as an author of fundamental reflections on Bildung and as a school and university teacher, has used a pictorial metaphor to illustrate this situation. Bildung describes the tension or the bridge between ideals passed on and current needs of competence, between philosophical self-assurance and practice-oriented self-preservation of the society. According to Plato's great Cave Allegory: Bildung is both, the rise towards sunlight and the descent towards the cave. The one side without the other is senseless (v. Hentig 1996, p. 58).

Teachers Within the Concept of Didaktik

More than 50 years ago Klafki presented reflections on a possible transformation of a subject matter into an educational content. A series of five questions was proposed as a guidance for a teacher's reflections when preparing lessons, reflections leading to designing "one or several opportunities for children to make fruitful encounters with certain contents of education (Bildungsinhalte)" (Klafki 2000b, p. 143). This early version of a content analysis in Didaktik ("Didaktik analysis") was based on Klafki's first approach to a connection between the classical conception of Bildung and its significance for teachers' daily work. Under the perspective of the more modern interpretation of Bildung, Klafki expanded his comments on the five main questions towards the integration of social conditions and the processes of interaction. The starting question for the Didaktik analysis refers to a teacher's situation at the beginning of his/her lesson planning: "What questions, therefore, should a teacher ask in the preliminary phase of instructional preparation….?" (Klafki 2000b, p. 151). The five questions mirror the wide range of reflections teachers are requested to make:

I. What wider or general sense or reality does this content exemplify and open up to the learner? What basic phenomenon or fundamental principal, what law, criterion, problem, method, technique or attitude can be grasped by dealing with this content as an "example"?
II. What significance does the content in question, or the experience, knowledge, ability, or skill to be acquired through this topic already possess in the minds of the children in my class? What significance should it have from a pedagogical point of view?
III. What constitutes the topic's significance for the children's future?
IV. How is the content structured (which has been placed in a specifically pedagogical perspective by Questions I, II and III)?
V. What are the special cases, phenomena, situations, experiments, persons, elements of aesthetic experience, and so forth, in terms of which the structure of the content in question can become interesting, stimulating, approachable, conceivable, or vivid for children of the stage of development of this class? (Klafki 2000b, pp. 151–155).

In Germany and some other countries, generations of teacher students were introduced to the procedure of Didaktik analysis which helps teachers to reflect on the school contents' contributions to develop students' Bildung and to make content-related decisions about their teaching grounded on this analysis (Hopmann 2000). Student teachers learn that reflections on these questions do not deliver definite responses, but they open a discourse—preferably with colleagues—in which aims of instruction, students' cognitive, social, and affective perspectives, and the scientific structure of a topic under question are linked to each other, so that at the end of an iterative process an appropriate content structure for instruction becomes visible ("educational reconstruction", Duit et al. 1997, p. 602, Kattmann et al. 1995). In many cases a consensus on broader domains of content is achieved quite easily, but it is basically more difficult to scrutinize details. There is no doubt that the principles of quantum physics are a significant example of modern physics. The photo-electric effect and the Franck-Hertz-Experiment are widely accepted as parts of a syllabus at the upper secondary level and most teachers agree that these effects can be learned by students without serious learning problems. But there would be less agreement about the Compton Effect. How "fundamental" is this effect for understanding the principles of quantum physics?

Some aspects of Klafki's questions have been taken on and further developed by educationalists who, from various perspectives, have contributed to efforts to improve science education. For example, Klafki's II comprises students' prior knowledge and conceptions, but also includes their emotions connected with a topic. In a proposal that received wide attention Klafki suggested a way to achieving general education (Allgemeinbildung) by orientation at "key issues" (Schlüsselprobleme) that are to be defined as typical and topical for a given time period. For our cultural existence, topics such as peace, environment, impact of technology on the society, human rights, and others are to be considered. The "science-technology-society" (STS) movement in science education can be interpreted as a part of this idea. The attempt to derive concrete themes from these overarching frames necessarily fails, taking into account Klafki's criteria as a whole. Klafki's Didaktik analysis does not offer a means for a detailed determination of topics in science education, but it helps teachers to reflect on criteria that are oriented at students' cognitive and emotional preconditions, as well as at the significance of topics for students' current and future lives, and at requirements demanded by the society.

Referring to Didaktik analysis Shirley (2009) complains about the absence of a theoretical basis like Didaktik in the American tradition: "The loss of a living link to the Didaktik tradition is especially unfortunate because the moral values at the center of Didaktik are unavailable to contemporary American educators—at least through this venue" (p. 199).

Bildung Within Natural Sciences

Among other scholars, Martin Wagenschein (2000a) has been particularly prominent. He has written numerous basic articles, and with many and varied examples described how students' Bildung in natural sciences can be achieved. His central ideas are known by nearly all science teachers in the German-speaking countries and many teachers have read at least one of his publications.

For Wagenschein, the main goal of science education is to help students understand phenomena of the natural world. To "understand" means to have gained insight into the essence of scientific relationships, it does not mean just to know the formula or to be able to apply it to a concrete problem. According to Wagenschein, there are three characteristic teaching–learning situations in which Bildung in this sense is developing:

- *Exemplary teaching:* In order to gain a deep understanding of a piece of content it is necessary to invest a sufficient amount of time. Therefore, "we need the courage to leave gaps, in other words to be thorough and to deal intensively with selected examples" (Wagenschein 2000a, p. 116).
- *Genetic teaching:* If the knowledge is to become an integral part of a student's Bildung it is important that he/she has the opportunity to search productively for the solution of a problem, to find it, and to check it critically. With this position Wagenschein, already at the beginning of the 1950s, of the last century, introduced elements of an idea that later, in its cognitive dimension, was portrayed as the constructivist view of learning. Wagenschein emphasized the development of knowledge much more than the result of the process of acquiring knowledge.
- *Socratic teaching:* A teaching–learning process which focuses on the development of knowledge is best arranged in a Socratic conversation. The teacher has to talk with his/her students not in a lecturing and dogmatic way but, like Socrates in his dialogues, focussing on their ideas and moderating their learning processes.

According to Wagenschein, teaching environments with this triad of principles are particularly suitable (and often necessary!) for learning phases in which a basic understanding of central notions and processes in natural sciences are to be acquired. This is especially the case in the upper grades of elementary school and lower grades of secondary school. However, Wagenschein's triad is meant to be effective at all levels, since the process of Bildung does not come to an end. But weightings shift priorities: at higher levels the preparation for vocational or academic studies is dominant.

In order to substantiate the idea of an exemplary, genetic, and Socratic way of teaching, an example described by Wagenschein and translated into English (Wagenschein 2000b) is now given to clarify this conception:

The starting point for this example is Wagenschein's observation that after having been a student at school most people remember the term $g/2 \times t^2$ when asked for the characteristics of a free fall, even though they are not able to describe what this term really means. In his example Wagenschein pleads for teachers to ask the students to begin a series of investigations starting with Galileo's inclined plane experiments that give them the chance to refine their measurements from very simple methods, i.e. weighing the amount of water that flows into a bucket while the rolling ball covers a definite distance (as Galileo describes it), using a ruler and a metronome or a stopwatch, up to the application of electronic devices. At the end of such an extended investigation of reflections, deliberations, and experimental improvements the sequence of the numbers 1, 3, 5, 7,... may result. This denotes the distances between the points passed by a constantly accelerated (e.g. a falling) body in fixed time intervals. Wagenschein argues that if the teaching goal is not simply to be able to apply a formula, but to understand the characteristic feature of the free fall, the odd-numbered sequence is a much more appropriate description than the term mentioned above.

A comparison of this example with Matthews' (2000) proposal for teaching pendulum motion reveals both similarities and fundamental differences. The similarities are related to the historical and philosophical references that are emphasized by both authors as important parts of a science curriculum. A basic difference is revealed by the authors' conceptions about learning physics. For Wagenschein, the often strenuous and sometimes long-lasting work of students who follow their own suggestions to find approaches to solving problems could lead to a concept or a theory in the final stage of their work. Such a process contributes to students' Bildung even in those cases where their endeavours do not lead to a result they are comfortable with. In Matthews' transmission view these activities of students would be a waste of time: "...at the heart of science are concepts, and these need to be understood first" (Matthews 2000, p. 280). He goes on to argue that teachers have to provide their students with the correct scientific view before any observation begins: "...the theoretical structure that precedes observation is something that students need to receive from teachers" (p. 279).

A chemical example is now given to demonstrate how students can approach basic ideas in chemistry on their own. Under the perspective of Bildung, a central appeal to science teaching emphasizes the significance of phenomena which should have the priority over their explanation by means of models, at least in a first phase of a course. In chemistry teaching, chemical reactions are often described too early by chemical equations that mirror an interpretation which is not easily understood by students: The symbols in a chemical equation reflect the existence of atoms which remain unchanged in a reaction. In the view of many teachers, the idea of "conservation" matters a great deal, and atoms are appropriate entities to meet this principle. However, students cannot perceive conservation but they do observe changes and transformations in chemical reactions.

Buck and Mackensen (2006) describe a chemistry-related teaching–learning example that is, as they state, inspired by Wagenschein. They report on ideas of de Vos (2002) who proposed presenting a chemical reaction to students which dispenses with all effects that could students distract from the main point, namely from the conversion of one substance into another one (therefore a chemical; reaction with no fire, no detonation, no "fizzling", no electricity, etc.). A simple and beautiful reaction happens when solutions of lead nitrate and potassium iodide are mixed. A magnificent yellow precipitate is formed which slowly sediments from the solution. Chemists regard the chemical equation as the optimal form for describing the process:

$$Pb(NO_3)_2 + 2KI \rightarrow PbI_2 + 2KNO_3$$

Many teachers try to reach this equation quite early with their students without asking whether the students have understood the basic assumptions connected with this equation. A genetic approach aims at just such an understanding of assumptions.

The series of experiments begins with mixing the two substances without any solvent but using a pestle and a mortar. Rather quickly a bright yellow colour appears. The colour becomes visible during the process of rubbing and is restricted to this area. The teacher does not need to ask the students, this phenomenon raises its own questions. Many students believe in the conservation of substances, at least in the conservation of their characteristics in any process. Therefore, the following statement is one consistently given by students: "The yellow substance has already been in the grains (like yolk of an egg), the rubbing has opened the grains and freed the yellow colour." The rubbing of the pure substances and of the mixture can contribute to test this hypothesis. The influence of the rubbing can be qualified by putting the two substances together so that they have an area of contacting each other. A weak yellow line becomes visible. The pestle is only a mechanical instrument to intensify the contact.

Another characteristic of students' questions and statements is their refusal to speak of a yellow substance; they mostly mention a yellow colour. It is hard for students to accept that, in a reaction, substances disappear and new ones are created. Therefore, subsequent investigations are used to reinforce this aspect. In a Petri dish a layer of distilled water covers the base and small portions of the two substances are placed into two sectors of the dish opposite to each other. Both substances dissolve and after a while a thin yellow line emerges that grows in length and breadth: a dune of gold. The separation of manipulation and reaction is a central feature of this process; dissolution, transport and chemical reaction take place in different areas of the Petri dish at different times and each phenomenon can be thus observed separately.

The discussions about these de-accelerated phenomena of "lapsing" and "emerging" of substances can lead to a deep understanding of the fundamental characteristics of chemical reactions if a teacher gives students enough time to reflect on questions that, almost inevitably, appear: How can a yellow substance emerge from colourless stuff?—obviously two special substances are necessary. Do the substances disappear while the yellow is emerging?—the yellow was not there before, therefore it is new. But nothing was added or removed. Is it possible the yellow was already there?

Wagenschein's idea of genetic and Socratic teaching and learning is in evidence with this example. The described way of knowledge growing leads to an "enrooting" that is different from knowledge that can be assessed by means of questionnaires. Unfortunately, in Germany students are not allowed to work with lead nitrate, but good chemistry teachers need to find a way to keep up the principles of genetic and Socratic teaching with a similar instructive example.

There are some preconditions for teaching and learning situations aiming at Bildung: concentration on selected topics which have the power to serve as examples to achieve Bildung; reference to historical examples if suitable (because often these developments are similar to students' ways of thinking); sufficient time for the students to try out in experiments what they have conceptualized in order to solve a problem; and phases of metacognitive reflections on the status of one's own learning and knowledge (Gunstone 2001). Knowing that the everyday situation in classrooms and schools very often hinders the realization of such teaching–learning processes, Wagenschein proposed some rules teachers should take into account when striving for the improvement of their teaching:

- *Not always:* First, the simple, elementary (and often boring) topic, then step by step the more difficult topic,
  but often: First, an astonishing, complicated, and problematic case, then the challenge to discover comprehensible and familiar topics.
- First, the phenomenon in nature, then the phenomenon in laboratory.
- First "qualitative", then "quantitative".
- First the phenomenon, then the theory and the models.

Teacher Education that Facilitates Students' Bildung

Questions

What are the consequences of these reflections on students' Bildung for teacher education programs? Which knowledge base is necessary to become a teacher capable of fostering the development of students' Bildung? What other attributes of a teacher besides his/her knowledge are characteristic features of a teacher with high professional expertise? What are the main indicators of different phases of teacher education?

In a profound analysis of the literature on attempts to systematize the various components of a teacher's professional expertise, and as a basis of a research project on mathematical teachers, Kunter et al. (2007) propose a model that describes components seen as being at the core of a mathematical teacher's professional competence. As psychologists the researchers concentrate on variables that can be recorded by questionnaires, and they regard the notion "competence" as being appropriate in this context.

The main aspects of teacher competence Kunter et al. (2007) propose are represented by the following concepts:

- *Knowledge*. For knowledge, the authors adopt a part of Shulman's differentiation between different facets of a teacher's knowledge (Shulman 1986): general pedagogical knowledge, subject-matter content knowledge, and pedagogical content knowledge.
- *Beliefs*. Teachers' beliefs indicate how they think about different conceptions of teaching and learning, about the nature of knowledge, and about their instructional goals.
- *Psychological functioning*. A combination of high engagement and a high capacity to deal with the pressures of school life is crucial for teachers' psychological

functioning. Different motivational variables are defined in order to have instruments to measure these aspects.

In science teacher education, among these aspects it is mainly the knowledge-related components that are subject to efforts to help students develop a basic qualification for their profession. This is so despite the fact that knowledge is not a sufficient (and sometimes not even a necessary) precondition for excellence in teaching. In a later section of this chapter the relationship between knowledge and ability to teach is discussed in more detail. The development of student-teachers' subject matter content knowledge (CK) and pedagogical content knowledge (PCK) are the goals of science teacher education, although aspects of general pedagogical knowledge (PK) are effective within every teaching situation. At least two basic questions need to be carefully considered when designing a study program:

(a) Which topics within science and which themes within science-related pedagogy are necessary parts of a teacher education program?
(b) What are the expectations concerning the influence of teachers' content knowledge (CK) and pedagogical content knowledge (PCK) on their teaching competence?

These two questions cannot be discussed separately. But for analytical reasons, some aspects of each particular question are now considered individually; after this is done the two questions are referred to each other.

Answers: Subject Matter Knowledge

In 1999 the vast majority of the European countries agreed upon a declaration in which they promise to introduce, among other things, a system "of easily readable and comparable degrees", adopting a "system essentially based on two main cycles, undergraduate and graduate" (Bologna 1999).

In the context of subsequent intensive discussions about whether and how to introduce the Bachelor-/Master-System for teacher education studies, two opposing positions were put. The first position advocated a more consecutive model where the dominant idea is that a broad basis of subject matter knowledge—acquired in a first study phase resulting in a bachelor's degree—is a good foundation of various professions. In this case a bachelor's degree is a polyvalent certificate. The second position advocates a model stressing studies oriented towards an integrative design in which subject matter CK studies and PCK studies are referred to each other from the beginning. The goal of this integrative model is to lay the foundation stone for a successful process of professionalization as early as possible.

In Germany, the second model became accepted. For the time being this situation marks the end of an area of many decades in which the following phrase guided the science studies of prospective teachers working at a Gymnasium (high school): "The more excellent a teacher's subject knowledge is the more efficient is his/her teaching." After much questioning of this dictum, the guiding principle is now an optimal interconnection of subject matter and pedagogical (content-related) topics. This means that in teacher education content-related as well as organisational elements of teaching and learning settings are to be integrated, so that they offer alternatives to the traditional modes of teacher training. Self-determined studies, long-term projects, historical references to scientific topics, and a presentation of science that starts from phenomena and holistic approaches then moves to systematic and analytical considerations are some of the elements necessary to prepare teachers for activities with which they foster students' Bildung in schools. The principles of Didaktik applied in teacher education require a study program for prospective teachers that is different from the programs for bachelors or masters students who are to become science researchers. These changes require a longer period of time to develop but more and more changes become visible which can support the claim that teacher education programs are study program "sui generis".

In a study for the German Physics Society (DPG 2006) physicists together with physics educators called for changes of methods and topics in physics teacher education studies. The starting point of the physicists' reflections on appropriate physics studies for student-teachers was the demanding tasks teachers are confronted with in schools.

> The young prospective teachers have to be provided with an optimal instruction and with optimal tools for their performing of the tasks. Practice has shown that teacher training which is—to a considerable amount—just an appendix to subject matter studies in physics, does not meet these requirements. Therefore, student-teachers' studies in physics have to be optimized especially for the demands on teachers. That means student-teachers' studies have to be studies sui generis. (DPG 2006, p. 4, translation: author)

- Methods in Physics courses should be designed in a way that students experience teaching–learning situations which they later as teachers can apply as models of their own lessons in which they teach captivatingly, with enthusiasm, and oriented at students' interests.
- The topics should not be determined by the system of physics, but assigned to themes across different complex areas, e.g. swimming–streaming–flying or earth–weather–environment.

Obviously, these proposals by German physicists do not explicitly define a program to prepare student-teachers for processes of Bildung in physics education in schools. But some elements point to this direction, for example, the focus on students' active participation in lessons and the concentration on topics which are challenging students' engagement (See the quote from Wagenschein given above, at the end of the section Bildung Within Natural Sciences: "…but often: First an astonishing, complicated, and problematic case…"). Physicists at universities have begun accepting that the knowledge standards they expect from future physicists have to be different from the standards they demand of student-teachers. With knowledge about the Lagrange formalism in mechanics, or the Dirac equation in quantum physics, teachers are definitely overqualified. In Germany, prospective teachers have to study (and later on teach) two subjects. During their studies the professionally oriented components (general and subject-related pedagogy, teaching internships) of their qualification cover a third of their whole study program.

There are no empirical investigations about the effectiveness of different teacher education systems, for instance comparing the consecutive Anglo-American system with the integrative system in Germany, but the results of large-scale surveys show that German teachers at all school levels believe they are sufficiently knowledgeable about the science content of their subjects and that they appreciate the early connection of science contents with instructional aspects since it helps them to realize early the pedagogical potential of the science topics to be taught and learnt (Merzyn 2003).

Answers: Pedagogical Content Knowledge

To carry out a lesson according to Wagenschein's example of the inclined plane requires more than subject matter knowledge and pedagogical knowledge. For instance, it is necessary to know something about students' preconceptions concerning motion and acceleration, to know how familiar students are with methods of measurement and how to help students organize group work in physics. There is no doubt that science teachers need content-related pedagogical knowledge. For many decades prior to Shulman's introduction of the notion pedagogical content knowledge, "Fachdidaktik" has been a part of student-teachers' study programs at the Universities of Education in Germany. "Fachdidaktik" combines "Fach"—subject matter—with Didaktik and is closely connected with the conception of Didaktik. One single definition of Fachdidaktik does not exist (as is also the case with PCK) but during a long tradition an understanding developed that became visible in study programs and examination regulations.

The somehow diffuse character of Fachdidaktik needs, however, to be sharpened in investigations where Fachdidaktik (or PCK) is a variable in a research design. In a broad quantitative study a research group in Germany investigated the impact of mathematics teachers' PCK on particular aspects of their mathematics instruction, e.g. on students' cognitive activation. The processes of conceptualization resulted in items forming subtests which covered subfacets of PCK. "Square" is an item of the subfacet "Tasks": "How does the surface area of a square change when the side length is tripled? Show your reasoning." (Kunter 2007, p. 47; see also Krauss et al. 2008)

The researchers' basic assumption is that tasks with multiple solutions are best suited to support students' learning processes. As a consequence, teachers' competence is seen to be reflected in the largest possible number of solutions they are able to depict. In the "Square" item teachers are prompted to show their competence: "Please note down as many different ways of solving this problem (and different reasonings) as possible."

With this example, the problem of a more precise description of PCK becomes evident. How near to a teacher's subject matter knowledge is PCK to be defined? Should the elements of PCK not be closer to a teacher's decision making in the classroom?

The conceptualization of PCK revealed in the above item can be described as overemphasizing the intellectual aspects of teaching because, with this understanding of PCK, acting in classroom is not imaginable without acts of intellectual planning and applying. For the development of students' Bildung, this restricted conception of PCK is lacking aspects which need to be taken into account when reflecting on the relationship between thinking and acting.

Bildung and Technical Rationality

Why is the conception of PCK apparent in the item described above not an appropriate one in order to be a guideline for science teacher education aiming at the development of students' Bildung?

Schön, taking on and developing Michel Polanyi's phrase "tacit knowing" ("we can know more than we can tell", Polanyi 1966, p. 4), has described professionals' "thinking in action". He argued against the idea of a successive progression of thoughts and acts: "Once we put aside the model of Technical Rationality, which leads us to think of intelligent practice as an *application* of knowledge to instrumental decisions, there is nothing strange about the idea that a kind of knowing is inherent in intelligent action" (Schön 1983, p. 50). We often carry out actions without any need "to think about them prior or during their performance" and "we are usually unable to describe the knowing which our action reveals" (Schön 1983, p. 54). Knowing-in-action, therefore, is "the characteristic mode of ordinary practical knowledge" (p. 54). Accordingly, Schön holds that, as a rule, experienced practitioners do not act according to a consecutive model—first the theory, then the practice—but perform in an intuitive-improvisational manner using their knowing-in-action (Schön) or tacit knowing (Polanyi), knowing which is often not accessible either to an observer or to the actor himself/herself.

Under this perspective, pedagogical content knowledge in Shulman's and many other authors' conception focusing on "knowledge" and "understanding" misses some facets out and takes too narrow a view. From experts in general and teachers in particular we expect to have "not mentally stored knowledge, but the ability to perceive, to think, and to act skilfully, to *do* certain things in an expert-like way. We are interested in *knowledge in use* rather than *knowledge as a state*" (Neuweg 2004, p. 2). Similar ideas are expressed by Jones and Cowie in their conclusion to their chapter in this volume (Chap. 4): "The knowledge, skills and practices that teachers describe provide one, and we would suggest, a rather restricted insight into the knowledge an accomplished teacher brings into play in the moment of interaction. Potentially more useful in the long term, but much more demanding in the short term, is the depiction of how and why teachers interact with students and their ideas in particular ways."

In the cases of pre-service teachers and novice teachers, another problem reinforces the separation of stored knowledge on the one hand and orientations of acting on the other. In many studies discrepancies between teachers' intentions to act—based on their knowledge—and their actions in classrooms have been found (Fischler 1994). The interpretation of this dilemma refers to the special demands on teachers' work: "Teachers must learn to weigh difficult dilemmas and to make and implement decisions on the fly; to put their plans into action effectively as well as to alter plans for unforeseen circumstances while they are in the midst of teaching; to respond to children and to represent well the material they are teaching" (Hammerness et al. 2005, p. 370).

In the current mainstream of research projects on teachers' professional development under a cognitive psychological perspective, the ideas of Polanyi, Schön, and of Neuweg (described below) play only a marginal role.

From the perspective of a tacit knowing Neuweg (2004, 2005) portrays the way to help student teachers to make explicit progress in the processes of professionalization. Neuweg specifies four preconditions for the emergence of pedagogical expertise: (1) Experience, (2) Knowledge, (3) Reflection, and (4) Personality.

Neuweg's First Precondition—Experience

In the light of the tacit knowing view, the phrase "knowledge informs action" is not tenable. Intuitive-improvisational acting is not primarily determined by plans but, above all, by a sensitive engagement in a "situation of uncertainty, instability, uniqueness, and value conflict" (Schön 1983, p. 49). Because implicit knowledge (knowing-in-action) cannot be made explicit, a novice is dependent on processes of learning through experience (learning by doing). Modes of apprenticeships presumably are appropriate means to meet these demands. Of course, these modes have to be connected to deep reflections on the relationships between the observed actions and the actor's underlying planning, knowledge, beliefs about teaching and learning, and pedagogical principles. Otherwise student teachers' "apprenticeship of observation" (Lortie 1975, p. 61) in the long period of being students themselves would prevent them from changing their preconceptions about teaching which they have developed through their numerous experiences.

Under this precondition, the following statement is fully justified: "…what we need is not so much theories, articles, books, and other conceptual matters, but, first and foremost, concrete situations to be perceived, experiences to be had, persons to be met, plans to be exerted, and their consequences to be reflected upon" (Kessels and Korthagen 1996, p. 21). Under the perspective of Schön, it is self-evident that he emphasizes the significance of a "reflective practicum" in which a novice has the chance to get to know practitioners with "their conventions, constraints, languages, and appreciative systems, their repertoire of exemplars, systematic knowledge, and patterns of knowing-in-action" (Schön 1987, pp. 36–37). The interactions with practitioners serving as coaches and, sometimes more importantly, with fellow students lead to reflections and learning processes that go "beyond statable rules… by constructing and testing new categories of understanding, strategies of actions, and ways of framing problems" (Schön 1987, p. 39).

Neuweg's Second Precondition—Knowledge

Besides the problematic nature of knowledge that is assumed to guide actions, another category of knowledge is significant for teachers: It is knowledge that prepares their actions in classroom, leads their perceptions in classroom situations, and helps them to justify their classroom decisions. Even though scientific knowledge on its own cannot produce excellent practice, a professional has to be able to show that his/her decisions have been reasonable under a scientific perspective.

Neuweg's Third Precondition—Reflection

In the above mentioned investigation with mathematics teachers one of the results referred to the question of whether or not experienced teachers are more competent in activating students cognitively. No correlation was found. Experience *per se* does not contribute to pedagogical expertise. In order to enable student-teachers to gain experience of high quality it is necessary to offer to them interplay between engagement in practice, reflection on their practice, and again acting and experiencing. In this way a reflective habitus can be developed.

Neuweg's Fourth Precondition—Personality

The personality paradigm does not play a significant role in contemporary research projects on teachers. The variables within the category "psychological functioning" are near to the dimension of personality but not completely in line with it. In teacher training it is important to inform students about the relevance of individual personal characteristics for their professional career and to offer to them possibilities of self-experience.

Consequences

Which consequences should be drawn from the statements, positions, and judgements unfolded above? There does not exist a master plan leading to science teachers' competence to foster their students' Bildung. But on different levels and in various contexts there are elements, facets, and hints about how to approach situations in which student-teachers grasp the idea of Bildung.

Congruence Between Goals and Experience

One of the basic requirements is that student-teachers experience themselves situations they intend to create as teachers in classroom. Instructors and student-teachers have to be aware that they need to not only talk about Didaktik but to permanently generate Didaktik. In a kind of a "pedagogical double-decker" (Wahl 2001, p. 163), the instructor has to demonstrate professional behaviour when talking about it. In Wagenschein's example, it is not sufficient just to inform student-teachers about possible relationships between intervals of lengths and times investigating an accelerated motion and to tell them how to measure these intervals. Student-teachers have to get the chance to explore the experimental problems on their own, to be confronted with ideas they cannot comprehend quickly (as this is the normal case with students in school) and to reflect on the task's potential to contribute to processes towards Bildung.

Subject Matter Studies

The aspect of acting independently in study-phases in which this is a reasonable mode of work applies also to subject matter studies. The statements by the DPG in Germany, discussed above, are not much more than a program at the moment but are more or less a revolution in physicists' minds.

Knowledge

As discussed in previous paragraphs, content knowledge as well as pedagogical content knowledge is not dispensable, because very often it is a necessary precondition for instructional processes that a teacher can justify his/her decisions by means of evidences from the educational sciences. It is generally accepted that both types of knowledge (CK and PCK) are not sufficient for a good teaching practice. But the function of knowledge for teachers' actions has to be considered more cautiously. In most domains, university studies are predominantly shaped by a conception of technical rationality. Experts tell us that this conception has to be generally questioned, and especially in teacher training. Tacit knowing or knowing-in-action requires more careful attention concerning the question how to support its development.

Reflection

Thinking about and working with Didaktik and Bildung permanently necessitates reflections on the goals of science education, on appropriate selections of topics for science instruction, on methods supporting processes of Bildung, and on questions about what the essential features of Bildung are and what relevance Bildung in science still has in the present. Following the ideas of Didaktik a teacher needs to become aware of being constantly challenged to reflect on his/her decision making prior to, during, and after classroom situations. This is an essential precondition for good teaching practice.

References

Bologna (1999). The European higher education area. The Bologna Declaration of 19 June 1999. Joint declaration of the European Ministers of Education. http://en.wikisource.org/wiki/Bologna_Declaration_of_19_June_1999. Accessed 2 Nov 2009.

Buck, P., & v. Mackensen, M. (2006). *Naturphänomene erlebend verstehen [Understanding and experiencing natural phenomena]*. (7th revised edition). Köln: Aulis.

de Vos, W. (2002). Ausschau halten nach einer Versuchsanordnung, bei der nichts die Aufmerksamkeit von der Hauptsache, der Umwandlung von Substanzen, abzieht [Searching an experimental setting in which nothing withdraws the attention from the main issue, that is the conversion of substances]. *chimica didactica, 90*, 209–220.

DPG (2006). *Thesen für ein modernes Lehramtsstudium im Fach Physik [Statements about a modern teacher training in physics]*. Bad Honnef: Deutsche Physikalische Gesellschaft.

Duit, R., Komorek, M., & Wilbers, J. (1997). Studies on educational reconstruction of chaos theory. *Research in Science Education, 27*, 339–357.

Duit, R., Niedderer, H., & Schecker, H. (2007). Teaching physics. In S. K. Abell & N. G. Lederman (Eds.), *Handbook of research on science education* (pp. 599–629). Hillsdale: Erlbaum.

Fensham, P. J. (2001). Science content as problematic—issues for research. In H. Behrendt, H. Dahncke, R. Duit, W. Gräber, M. Komorek, A. Kross, & P. Reiska (Eds.), *Research in science education—past, present, and future* (pp. 27–41). Dordrecht: Kluwer.

Fensham, P. J. (2007). Values in the measurement of students' science achievement in TIMSS and PISA. In D. Corrigan, J. Dillon, & R. Gunstone (Eds.), *The re-emergence of values in science education* (pp. 215–229). Rotterdam: Sense.

Fischler, H. (1994). Concerning the difference between intention and action.—Teachers' conceptions and actions in physics teaching. In I. Carlgren, G. Handal, & S. Vaage (Eds.), *Teachers' minds and actions. Research on teachers thinking and practice* (pp. 165–180). London: Falmer.

Gunstone, R. (2001). The education of physics teachers: Contents plus pedagogy plus reflective practice. In R. Pinto & S. Surinach (Eds.), *Physics teacher education beyond* 2000 (pp. 27–33). Paris: Elsevier.

Hammerness, K., Darling-Hammond, L., Bransford, J., Berliner, D., Cochran-Smith, M., McDonald, M., & Zeichner, K. (2005). How teachers learn and develop. In L. Darling-Hammond & J. Bransford (Eds.), *Preparing teachers for a changing world. What teachers should learn and be able to do* (pp. 358–389). San Fransisco: Jossey-Bass.

Hopmann, S. (2000). Klafki's model of Didaktik analysis and lesson planning in teacher education. In I. Westbury, S. Hopmann, & K. Riquarts (Eds.), *Teaching as a reflective pracice. The German Didaktik tradition* (pp. 197–206). Mahwah: Erlbaum.

Kattmann, U., Duit, R., Gropengießer, H., Komorek, M. (1995). *A model of educational reconstruction*. Paper presented at the annual meeting of the National Association for Research in Science Teaching (NARST), San Francisco.

Kessels, J. P. A. M., & Korthagen, F. A. J. (1996). The relationship between theory and practice. Back to the classics. *Educational Researcher, 25*(3), 17–22.

Klafki, W. (1998). Characteristics of critical-constructive Didaktik. In B. B. Gundem & S. Hopmann (Eds.), *Didaktik and, or curriculum. An international dialogue* (pp. 307–330). New York: Lang.

Klafki, W. (2000a). The significance of classical theories of Bildung for a contemporary concept of Allgemeinbildung. In I. Westbury, S. Hopmann, & K. Riquarts (Eds.), *Teaching as a reflective practice. The German Didaktik tradition* (pp. 85–107). Mahwah: Erlbaum.

Klafki, W. (2000b). Didaktik analysis as the core of preparation of instruction. In I. Westbury, S. Hopmann, & K. Riquarts (Eds.), *Teaching as a reflective practice. The German Didaktik tradition* (pp. 139–159). Mahwah: Erlbaum.

Klette, K. (2008). Didactics meets classroom studies. *Zeitschrift für Erziehungswissenschaft, 10*(Special Issue 9), 101–114.

Krauss, S, Brunner, M., Kunter, M., Baumert, J., Blum, W., Neubrand, M., & Jordan, A. (2008). Pedagogical content knowledge and content knowledge of secondary Mathematics teachers. *Journal of Educational Psychology, 100*, 716–725.
Kunter, M., Klusmann, U., Dubberke, T., Baumert, J., Blum, W., Brunner, et al. (2007). Linking aspects of teacher competence to their instruction. Results from the COACTIV project. In M. Prenzel (Ed.), *Studies on the educational quality of schools* (pp. 39–59). Münster: Waxmann.
Lortie, D. (1975). *Schoolteacher*. Chicago: University of Chicago Press.
Matthews, M. R. (2000). *Time for science education. How teaching the history and philosophy of pendulum motion can contribute to science literacy.* New York: Kluwer.
Merzyn, G. (2003). *Stimmen zur Lehrerausbildung [Voices about teacher education]*. Hohengehren: Schneider.
Neuweg, G. H. (2004). Tacit knowing and implicit learning. In M. Fischer, N. Boreham, & B. Nyhan (Eds.), *European perspectives on learning at work: The acquisition of work process knowledge* (pp. 130–147). Cedefop Reference Series; 56. Luxembourg: Office for Official Publications of the European Communities. http://www.wipaed.jku.at/images/stories/forschung/tacitknowing.pdf. Accessed 2 Nov 2009.
Neuweg, G. H. (2005). Emergenzbedingungen pädagogischer Könnerschaft [Conditions for the emergence of pedagogical expertise]. In H. Heid & G. Harteis (Eds.), *Verwertbarkeit. Ein Qualitätskriterium (erziehungs-)wissenschaftlichen Wissens? [Usability: A criterion for the quality of (pedagogical) scientific knowledge?]* (pp. 205–228). Wiesbaden: VS Verlag für Sozialwissenschaften.
Nordenbo, S. E. (2002). *Bildung* and the thinking of *Bildung. Journal of Philosophy of Education, 36*, 341–352.
OECD (2006). Organisation for economic co-operation and development: Programme for International Student Assessment (PISA): Assessing Scientific, Reading and Mathematical Literacy. A Framework for PISA 2006. http://www.oecdbookshop.org/oecd/display.asp?sf1=identifiers&st1=9789264026407. Accessed 2 Nov 2009.
Polanyi, M. (1966). *The tacit dimension*. Garden City: Doubleday.
Schön, D. A. (1983). *The reflective practitioner. How professionals think in action*. New York: Teachers College Press.
Schön, D. A. (1987). *Educating the reflective practitioner*. San Francisco: Jossey-Bass.
Shirley, D. (2008). The coming of post-standardization in education: What role for the German Didaktik tradition? *Zeitschrift für Erziehungswissenschaft, 10*(Special Issue 9), 35–45.
Shirley, D. (2009). American perspectives on German educational theory and research: A closer look at both the American educational context and the German Didaktik tradition. In K. H. Arnold, S. Blömeke, R. Messner, & J. Schlömerkemper (Eds.), *Allgemeine Didaktik und Lehr-Lernforschung: Kontroversen und Entwicklungsperspektiven einer Wissenschaft von Unterricht* (pp. 195–209). Bad Heilbrunn: Klinkhardt.
Shulman, L. S. (1986). Those who understand: Knowledge growth in teaching. *Educational Researcher, 15*(2), 1–22.
v. Hentig, H. (1996). *Bildung*. München: Hanser.
Waddington, D., Nentwig, P., & Schanze, S. (2007). *Making it comparable. Standards in science education*. Münster: Waxmann.
Wagenschein, M. (2000a). Teaching to understand: On the concept of the exemplary in teaching. In I. Westbury, S. Hopmann, & K. Riquarts (Eds.), *Teaching as a reflective practice. The German Didaktik tradition* (pp. 161–175). Mahwah: Erlbaum.
Wagenschein, M. (2000b). The law of free fall as an 'exemplary theme' for the mathematicizability of certain natural processes. In I. Westbury, S. Hopmann, & K. Riquarts (Eds.), *Teaching as a reflective practice. The German Didaktik tradition* (pp. 285–293). Mahwah: Erlbaum.
Wahl, D. (2001). Nachhaltige Wege vom Wissen zum Handeln. [Sustainable ways from knowing to acting]. *Beiträge zur Lehrerbildung, 19*, 157–174.
Westbury, I. (1998). Didaktik and curriculum studies. In B. B. Gundem & S. Hopmann (Eds.), *Didaktik and, or curriculum. An international dialogue* (pp. 47–78). New York: Lang.

Westbury, I. (2000). Teaching as a reflective practice: What might Didaktik teach curriculum? In I. Westbury, S. Hopmann, & K. Riquarts (Eds.), *Teaching as a reflective practice. The German Didaktik tradition* (pp. 15–39). Mahwah: Erlbaum.

Wimmer, M. (2003). Ruins of *Bildung* in a knowledge society: Commenting on the debate about the future of *Bildung*. *Educational Philosophy and Theory, 35*(2), 167–187.

Chapter 4
Moving Beyond Deconstruction and Reconstruction: Teacher Knowledge-as-Action

Alister Jones and Bronwen Cowie

Introduction

Investigations into the knowledge needed for teaching, particularly those conducted through classroom-based research that also take into account teacher perspectives, have illuminated the complexity of the knowledges teachers bring into play at the moment of teaching. Teachers use an integrated amalgam of understandings about students, the subject and pedagogy that is subject to change, context-specific and linked with personal experience, inside and outside the classroom (Hiebert et al. 2002; Shulman 1992). There is a body of evidence that teacher beliefs and views about students, teaching and the subject of study influence practice with some referring to these as a hidden curriculum. With this research has come the realisation that teaching is a complex practice that cannot be dichotomised into knowledge and action (Boaler 2003). Rather, as Shulman proposed, teacher knowledge is "part of a complex set of interactions involving action, and analysis and affect" (Shulman 2003, cited in Boaler 2003, p. 1–2).

Assessment for learning is acknowledged as a crucial pedagogical approach to enhance learning. It is the process used by teachers and students to recognise and respond to student learning in order to enhance that learning, during the learning (Cowie and Bell 1999). Teachers can plan to undertake assessment for learning but more often it is embedded in and accomplished through interaction. Marshall and Drummond (2006) make a distinction between lessons and teacher actions that embody the "spirit" of assessment for learning and those that conform only to the "letter". They found that the nature and sequence of tasks and especially "high organisation based on ideas" was crucial to the former. The implication is that for teachers to undertake assessment for learning they need a deep understanding of the subject domain, likely student learning pathways and pedagogical practices likely to move student learning forward. Thus, the interplay of knowledge and action is a key issue in assessment for learning.

A. Jones (✉)
University of Waikato, Hamilton, New Zealand
e-mail: a.jones@waikato.ac.nz

D. Corrigan et al. (eds.), *The Professional Knowledge Base of Science Teaching*,
DOI 10.1007/978-90-481-3927-9_4,

Primary teachers generally have a deep knowledge of their students and a broad repertoire of pedagogical practices, particularly practices that involve teacher–student (and student–student) interactions, but they often lack in-depth content knowledge (Appleton 2003) and have a limited repertoire of subject-specific pedagogies. This is almost always a particular issue for their teaching of science where teacher understandings of the nature and purpose of a discipline tends to be limited. Pedagogically appropriate teacher engagement of/with students as part of assessment for learning that is responsive to student ideas and interests requires teachers to have appropriate content knowledge and knowledge of their students and to be able to access and deploy this in the moment. In this context the idea of "knowledge as action" is central.

In this chapter we are deconstructing and reconstructing three constructs associated with teacher knowledge-as-action in science classrooms when exploring assessment for learning practices. These constructs are related to theoretical, pedagogical and improvement aspects. We elaborate on these aspects from three classroom-based research projects. The Learning in Science Project (Assessment) was a two-year project with 12 teachers of Year 7–10 students and their classes (Bell and Cowie 2001). The Learning in Technology Education (Assessment) Project was a three-year study with 12 teachers and Year 1–8 classes (Jones and Moreland 2005). The Classroom Interaction In Science and Technology Education Project was again a three-year project involving 12 teachers and their Year 1–8 classes (Cowie et al. 2008). Data were collected through intensive researcher participation, classroom observation, teacher and student interviews, as well as teacher and researcher collaborative meetings. Representative examples are used to contextualise the deconstruction and reconstruction of teacher knowledge-as-action.

Deconstructing and Reconstructing Theoretical Aspects

Two broad views of learning and knowledge are evident in the science education literature: a constructivist view and a sociocultural view. While the predominant focus has been on student learning there is also evidence that these orientations have shaped how we think about teachers and teacher learning and subsequently the knowledges needed for teaching. A constructivist view of learning postulates that knowledge is a mental representation that is actively built by the learner as part of the process of making sense of their world (Driver 1989). The learner's prior knowledge and experience are considered to both enable and constrain individual sense and meaning making. Learning is seen as an active, rational, individual and somewhat idiosyncratic process for which the learners themselves have the major responsibility. From a cognitive constructivist perspective the recommendation is that teachers serve as conceptual change agents who also foster student metacognitive awareness. To this end, their role is to provide activities to shift student thinking toward that of the target discipline. Activities that generate cognitive conflict and/or development including the use of mental models and analogies are seen as useful in this regard. Just as importantly, teachers need to monitor and respond to the sense students actually make through formative assessment (Bell and Cowie 2001). The implication here is that teachers require extensive knowledge of the content to be taught, of the likely progression of student ideas, of ways for finding out about student ideas and of strategies for moving student ideas forward. Teaching is also a learning process—teachers learn about their students, the subject and the impact of the activities they are using.

What sociocultural views of learning bring to the fore is that any study of learning involves the situated social system as a unit of analysis (Lave and Wenger 1991). Some writers with a social view of learning construe it as an individual process mediated by tools and social interaction; others propose that both learning and what is learned are situated by virtue of being distributed over people, places and things and the changing relations between them (Wells and Claxton 2002). In this latter sociocultural view the practices in which people participate constitute what they learn (Wenger 1998). Knowledge is a matter of competence with respect to those activities valued by the social group of which one is a part. Learning involves the transformation of participation and formation of identity through a process in which the individual and the collective shape each other and experience life and the world as meaningful. In this view, learning is about becoming as well as knowing and identity develops both through individual agency and through social practice.

Schools are a very particular context for the learning of science. The teacher and the setting are integral with the learning that takes place for students. The challenge for teachers is to move student views toward those currently viewed as viable while at the same time supporting students as active and critical meaning makers. Teaching activities need to engage students in practices consonant with the discipline under study. They also need to contribute to positive student identities and identifications with learning and the subject of study in the short and long term. The aim is for students to become "owners…acquirers, users and extenders" of knowledge in a particular domain (Brown et al. 1993, p. 190). The collective learning trajectory is shaped by both teacher and student interests, knowledge and skills and by the resources available in the setting. Teachers developing the class as a "community of scholars" implies that teachers themselves need to be intentional, self-motivated individual and collaborative learners with their students (Brown et al. 1993, p. 190). Put another way, teachers need to manage the interaction of the planned and the emergent curriculum so that teaching and learning interact to "become structuring resources for each other in a way that maximises the negotiation of meaning" (Wenger 1998, p. 14). Therefore, teachers need a breadth and depth of knowledge and repertoire of practices that will enable them to engage their students effectively in the components of science as a discipline.

Subjects as taught in schools are a representation of that subject rather than the subject itself. The nature of a subject or discipline from a sociocultural perspective includes the ways of knowing and knowledge generation. Stetsenko and Arievitch (2002) describe the seminal work of Piotr Gal'perin, one of Vygotsky's students and colleagues, who argued in essence that teachers should organise their work around the coherent principles that characterise a particular domain of knowledge.

These principles are the core conceptual tools, the internalisation of which enables students to think powerfully about a whole range of phenomena. This means that the teachers need to have a sense of the nature of the discipline, its organising concepts and its tools. This includes also cultural notions of language concepts and the mediation of tools and frameworks. Stetsenko and Arievitch (2002) highlight that Gal'perin's theory emphasises that to understand the development of the mind, one needs not only to observe how children participate in practices and make use of cultural tools, but also to construct instructional procedures that specially provide students with tool use, in which the evolving histories and functions of the tools are made explicit.

While research on teacher knowledge has tended to be conducted away from the classroom, a sociocultural perspective focuses attention on the practices in which teachers engage. This view highlights the need to examine teacher knowledge for teaching in situ across multiple contexts. In the next section we address some of the challenges this orientation raises for researchers who seek to understand how teacher knowledges and pedagogical practices might be represented, reported on and improved.

Deconstructing and Reconstructing Pedagogical Aspects

A scan of the literature reveals a multiplicity of different depictions of the knowledges teachers need and use for teaching. In this section we set out to explore some of the complexity of the knowledges that teachers bring into play in the moment in the classroom. In this exploration we clarify content knowledge and pedagogical knowledge, our meaning of pedagogical content knowledge and then discuss PCK as knowledge as action, including affective aspects. Finally, we link PCK to notions of assessment for learning.

Exploring Aspects of Knowledge for the Classroom

Empirical studies that have explored the characteristics of effective teachers (for example, Gipps 1999; Wragg et al. 1998) indicate that effective teachers call on a broad range of knowledge and understandings. Good teacher knowledge of subject content has been found to have a positive effect on decision-making related to changing pedagogical strategies for creating better learning opportunities (Harlen and James 1997). With familiar content teachers are able to focus more on levels of student understanding than "mechanical success or failure" (Gess-Newsome 1999, p. 62). When they move outside their area of content expertise it seems that even teachers with well-developed pedagogical skills experience difficulty in responding appropriately to student thinking. Where teachers' subject knowledge is weak, confidence levels to teach that subject are low, leading to restricted classroom practices (Harlen 1997). Sound content knowledge seems to have a positive effect on planning, assessment, implementation of curriculum and curriculum development. Harlen and James (1997) comment that teachers cannot provide experiences and activities that guide student progress toward understanding of ideas if they themselves do not know what the ideas are. Compartmentalised subject knowledge of the discipline is often not enough though, as this knowledge can be rather fragmentary in nature, particularly in relation to the organisation of knowledge for teaching. Teachers with a strong overview and a structure of inter-related ideas are able to make more connections during teaching and learning episodes.

Pedagogical content knowledge is the distinctive knowledge of teaching. It is a complex melding of pedagogy and subject content and includes aspects related to an understanding of what is to be taught, learned and assessed, an understanding of how learners learn, an understanding of ways to facilitate effective learning and an understanding of how to blend content and pedagogy to organise particular topics for learners (Shulman 1987). Studies by Gess-Newsome (1999), and Magnusson et al. (1999), have reiterated that pedagogical content knowledge includes knowledge of subject matter, students, curriculum and associated pedagogy. That is, pedagogical content knowledge encompasses useful ways of formulating and representing a subject that makes it comprehensible for others. Not only do teachers need to understand content and purpose; they must be able to transform content knowledge so that it becomes pedagogically powerful. Teachers need to develop a clear sense of the conceptual terrain they are exploring and will also need to have a pedagogical sense of the likely understandings the students will bring to a domain. With flexible pedagogical content knowledge, teachers can respond to students productively. If teachers have generally sound pedagogical skills they may rely on them to carry them through difficult aspects of the subjects they teach, but this can limit student learning. When teacher subject matter knowledge is limited this hinders pedagogical decision-making. To choose the most appropriate strategy teachers need to know the understandings students have reached. Transformations of the subject matter as understood by the teacher into actions relevant and applicable to their students constitutes knowledge-as-action.

Knowledge-as-Action

The notion that effective teaching requires knowledge as action has significant implications for research and practice from the transfer of knowledge and skills toward developing understanding of, and in, the practice in question. Knowing about the knowledges identified in the previous section is necessary but not sufficient for their deployment in the classrooms. For example, teachers we have worked with have been clear that in order to respond to the students' learning they need to know about a strategy; understand how it functions and why they might use it; have the skills to use it in action, and to be able to recognise in the moment when it would be useful. One teacher who had flexible grasp of the content she was teaching, summarised the importance of being able to integrate these different knowledges in the moment of action thus:

> The strategies just sit there and wait and when you get to one, you recognise it. But could you do the same thing if it (the strategy) was sitting in a book somewhere and you read it the night before? It's an interesting question. That's what we're talking about here. In that case, I was pulling it out of a repertoire, rather than planning ahead to use it. You've got to have the strategy sufficiently on board so that with the people in front of you can not only think of the strategy, but you can do it.

On the other hand, we noted a number of instances when teachers who were proficient with a range of pedagogical strategies were challenged by student science questions. This was the case for Tayla, a year 7/8 teacher, who described herself as having limited experiences and knowledge in science. During our first two classroom observations Tayla facilitated learning experiences where the students were "doing" activities, rather than necessarily interacting around ideas. Tayla had planned another activity-based lesson for our third observation but had to change it at the last minute. The lesson ended up as a follow-up discussion about the previous day's lesson on physical and chemical changes. Early in the lesson it was evident that while Tayla had some knowledge of physical and chemical changes she was finding the dynamic, unpredictable nature of the class discussion problematic. Tayla confidently explained water freezing as a physical change, using the criterion that the process is reversible. However, when the students suggested other examples such as "a seed growing", "a glass breaking", "an egg boiling" or "chocolate melting in your mouth" she struggled to respond. Although Tayla had learned and read about these concepts herself the students' questions and comments moved beyond the examples and contexts with which she was familiar. This is an example of a primary school teacher who was self-reflective enough to actively improve her own content knowledge prior to teaching, yet her reliance on definition and a restricted set of examples along with her limited experience in applying the concepts restricted her ability to respond to student ideas.

This example highlighted that even with preparation, interacting with students around ideas can be difficult and daunting. Teachers cannot provide experiences and activities that guide student progress toward understanding of ideas if they themselves have limited experience with the relevant science concepts and processes. Effective classroom interactions mean that teachers need rich and flexible knowledge base in order to undertake effective interactions with diverse groups of students in the moment. Black et al. (2001) highlight the interplay between views of learning and knowledge, views of the nature of the subject and teacher selection and articulation of learning goals. In considering the components of a more robust and comprehensive model for pedagogical content knowledge from an assessment for learning lens, we have found it is important to consider the nature of the discipline, the structure of the big ideas (including notions of the progression), the conceptual and procedural knowledge of the subject and technical aspects. With assessment for learning the pivotal aspects is that teachers have a repertoire of practices that might help students progress.

Affective Aspects of Knowledge-as-Action

As with learning, research has indicated that teaching has affective and social dimensions. Teacher confidence and self-efficacy play a role in their practice with teachers adopting a more transmissive approach when they lack confidence in their understanding of a curriculum area. Throughout our studies we have found that teacher confidence is associated with their willingness to engage and probe student understanding. A teacher from the Learning in Science Project (Assessment) (Bell and Cowie 2001), illustrated this aspect when she described how her confidence in her pedagogical knowledge of expansion in metals and her confidence in and knowledge of the skills of her students in discussing ideas contributed to her decision to allow time for class discussion of the effect of heat on solids. Despite this, she noted that her confidence in her own pedagogical content knowledge wavered when the class seemed to be coming to the wrong conclusion. Affect plays a key role in supporting and constraining teacher change. We have found in our research (for example, Bell and Cowie 2001; Jones and Moreland 2005) that, for teachers, changing their practice was as much an emotional as an intellectual challenge.

Linking PCK to Assessment for Learning Interactions

Assessment for Learning is accomplished through effective classroom interactions with knowledge-as-action. Black et al. (2001) indicate that the teacher, the subject and the student, and their various interactions, should be the focus in accounting adequately for what is going on. A focus on either the teacher or students or subject in isolation is inadequate. Our classroom research on assessment in science has found that classroom interactions are dependent on teachers' knowledge-as-action. The teachers' role in providing feedback is crucial to effective learning. Most importantly, assessment must emphasise the skills, knowledge and attitudes thought to be the most important, even if this is technically difficult. Harlen (1997) also asserts the importance of commenting on the substance of the work, rather than its superficial aspects, in order to convey what is important for subsequent learning. What is required is an appropriate setting of challenging goals, a structuring of situations to attain those goals effectively and the provision of feedback relevant to attaining the goals. Students have been shown to benefit from descriptive feedback that identifies the strengths and weaknesses of their work as this enables them to take control of their own learning.

The teachers in our studies indicated that their knowledge and experience mediated their assessment for learning actions. They considered that their content and pedagogical content knowledge influenced their assessment actions and interactions. Teachers need to make decision in the moment about what is relevant and what is not in relation to the learning and their knowledge of what is important for the particular student. One teacher explained:

> Gathering information is not as simple as that…we need a filter…you gather information but then you interpret what you want to do with it, and you have to filter which is relevant, which is not.

The role of the teacher's knowledge of science was noted by another teacher, when she said:

> Yes, well you see that's the whole notion about whether you need to have some science yourself to teach it. I would say you do. I mean, you don't have to have heaps of it, but you have to have some to be able to interact with the students. To be able to recognise and respond, you have to have some scientific knowledge in order to respond to it. Otherwise, you're just talking around on ill-formed ideas and we have a responsibility in teaching the scientific ideas. That's not saying that all students are going to learn it and that's the only goal of the lesson, but the government and parents charge you with responsibility of teaching science.

The teachers considered their professional knowledge allowed them to "take advantage of the one-off situations that sometimes happen". Their experience of teaching a particular concept was considered to be pivotal to their recognising the significance of student actions and comments and in interpreting them as idiosyncratic or widely held alternative conceptions. They claimed this knowledge was critical in their being able to take appropriate actions. For example, when a teacher asked her students to separate the colour pigments in ink, she commented that if the purpose of the activity had been chromatography rather than scientific ways of investigating she would have interpreted the elicited information in a different way. The teachers commented that the importance of their professional knowledge was highlighted for them when they watched student-teachers because they could recognise the implications of certain student comments and actions in terms of how they were thinking and what actions might help them reconsider their views.

Deconstructing and Reconstructing Improvement Aspects

In this section we explore strategies that we have used in our research that enhances teacher knowledges in action to enhance student learning. These strategies include assisting teachers to become aware of the knowledges required to enhance students learning, the use of planning frameworks, the role of collaboration and teacher and researcher meetings, and examining student misconceptions. The value of a whole school approach is also crucial as part of thinking about improvement over time. These aspects acknowledge that teacher change is both an individual and a social process and that teacher knowledge-as-action is shaped by teacher understandings of the social and organisational context of their work.

Individual Teacher Awareness and Knowledge

In our research programme we have always worked with teachers for more than one year (Bell and Cowie 2001; Jones and Moreland 2005; Cowie et al. 2008). As part of this process we have observed their classroom practice, noted their classroom interactions and then discussed what we noticed and the reasons teachers had for particular actions. Typically as part of this process the teachers became more aware of what they did, the knowledges they used, and the flow-on effects on students' actions and learning. For example, two teachers commented:

> If you have a good understanding you can help someone else get a good understanding. You can help them make connections much more than if you don't have PCK.

> The knowledge that I bring to the task, lesson and/or content of what I am doing. It can be information, ideas and skills. It's marrying those up and being able to help students develop their own knowledge and skills. It is about me using my knowledge to help students learn.

The teachers came to appreciate the knowledges they used in teaching and to understand that PCK was about them knowing the ideas and concepts so that they could teach them to their students. They emphasised that coming to know the ideas for teaching was an important step in the teaching process. As another two teachers commented:

> I have to be able to grasp the concepts and learn about what I want to teach before I teach them to my children.

> I have to find out to get a better understanding of the science concepts for teaching. Then I can help students in class understand the concepts.

The teachers also talked about having more confidence and knowledge to respond constructively to student science and technology understandings when they were teaching. They commented that now they deliberately planned to interact with students and question them with the intention of enhancing their learning. For example, another two teachers:

> Previously I would have thought that just getting an answer from them was OK, now I challenge them in their actual understanding.

> My questions got them to think about things, their ideas, so they could change them on the way. They could improve while they were going.

Providing teachers with research publications, for example, working papers from the early LISP projects not only informed the teacher interactions with their students but also could be used to enhance their own concepts in science as well as potential classrooms. Research on student alternative conceptions can also help teachers make sense of student actions.

Planning as Tool to Bridge Individual and Collective Improvement

The use of a planning framework has proved to be an effective means for developing teachers' PCK over the course of our work (Jones and Moreland 2005; Cowie et al. 2008). We used a science-specific planning framework to assist the teacher to articulate and develop the knowledges required to teach primary science effectively. The first layer of the planning framework helped teachers to identify, clarify and phrase science learning outcomes in a manner that would be appropriate for their students. It included a space for articulating the main (big) idea that was the overall purpose for student science learning. Another space was included for teachers to tease out the learning outcomes embedded in the main idea. The first layer also included spaces for unpacking and specifying four categories of more specific learning outcomes: conceptual, procedural, nature of science and technical. Conceptual learning outcomes included knowledge and understanding of relevant scientific concepts and procedures. The procedural learning outcomes involved the strategic application of procedures and processes such as those used in science investigations. The nature of science outcomes related to, for example, what counts as evidence and methods appropriate for the communication of scientific ideas. Technical skill outcomes related to practical techniques and equipment use. The specific design of this framework compelled the teachers to consider all categories resulting in a broadening of the range of possible science learning outcomes and the articulation of specific science ideas.

The teachers believed that clear planning helped them to stay focused on the topic. They indicated that they had changed their interactions to be focused on science.

> When I was teaching I could see the benefit of having a clear outline of what I wanted to achieve. It stopped me from going off the point and the children from doing the same.

> I think I'm more confident to interact with the exploring of things and interacting with the kids about the ideas. And if they say, "Oh, it's wrong, it doesn't work", then that's okay. We'll work on the science aspects.

Role of Collaboration and Teacher and Researcher Meetings

One-to-one, ongoing support in classrooms and the collaborative workshop atmosphere were important. When teachers' foundations were shaken and feelings of uncertainty surfaced, it was crucial that researchers were understanding, supportive and appreciative of their efforts. The support of others was mentioned throughout the project. For example:

> The interchange of ideas with other teachers in the project has been important in terms of conceptual development, knowledge of practice and the development of technological language, Consistently being refocused and supported by the research team has helped implementation and to support risk taking.

The activities shared included specific formative assessment activities; learning activities that created opportunities for the teacher to carry out assessment for learning; and ways to introduce flexibility to the school scheme or curriculum. The sharing of classroom activities, and hence, hearing how other teachers had used an activity, was an important preparation for doing assessment for learning.

Organisational Culture for Improvement

Much of the emphasis in this chapter has been on the individual teacher and what they do in the classroom. However, sociocultural views of learning emphasise the setting in which learning takes place. When teachers are changing their practice they are reliant on their students and the wider school community to both implement the changes and to grow and sustain them over time. In one instance we were able to return to a research school two years after our classroom work to explore what had helped to implement individual classroom changes across the school (Jones and Moreland 2005). Teachers and the principal considered that the culture of the school was an important factor in both sustaining the classroom research project for three years and subsequently implementing changes at the school-wide level. The culture was described as one that allowed teachers to show initiative, take risks, question, examine and reflect. For example:

> The culture in the school allows us to take risks, encourages us to keep thinking and reflecting but I am never satisfied.

The principal was focused on building a trustworthy, supportive school culture focusing on developing curriculum knowledge, self-examination and questioning, risk-taking and reflective attitudes. The staff saw the principal as crucial in that she was focused on being an effective leader and she was focused on enhancing teaching and learning. Although only four teachers were involved in the initial research project, the research findings were incorporated into whole school planning, assessment for learning practices and reporting systems. The culture of sharing information in the school as well as team planning assisted this dissemination process. It was the gains in students and classroom practices that encouraged other teachers to try out some of the ideas from the project.

Conclusion

One challenge for the research community is to find ways to illuminate and develop the unique and particular knowledges teachers need for teaching science. Shulman (1987) initially proposed the notion of teacher knowledges in the mid 1980s when the teaching profession was under attack. The construct PCK sought to encapsulate teacher professional knowledge, that is a knowledge base that is unique to teachers. Subsequently, researchers have sought to explicate what this knowledge looks like. While much of this work has taken place outside the classroom once we adopt a sociocultural view of learning and development it is not really possible to fully explore PCK except in the setting of its use and enactment. The knowledge, skills and practices that teachers describe provide one, and we would suggest, a rather restricted insight into the knowledge an accomplished teacher brings into play in the moment of interaction. Potentially more useful in the long term, but much more demanding in the short term, is the depiction of how and why teachers interact with students and their ideas in particular ways.

Assessment for learning which involves the use of assessment data to adjust teaching and to support students to move their learning forward is central to effective student–teacher interactions. It is not possible for even the most experienced teacher to fully anticipate student responses to any particular task: different students respond to the same task in different ways. Therefore, ultimately it is the understandings and possible actions a teacher is able to generate and enact in the moment that makes a difference to student learning. To us, knowledge-as-action is crucial in this because it encompasses teachers' understanding and enactment of how to help a particular group of students to understand specific subject matter using individual and sequenced activities, representations and social groupings. Teachers developing broad and flexible repertoires of practices that encompass these factors is very challenging.

Teachers are under increasing pressure to demonstrate the impact/outcome of their work in terms of student achievement. In science, this can be more challenging in that there is a related demand to capture and retain student interest and engagement. Primary teachers of science face further challenges given that science may not be an area of expertise or interest for them. Working within a sociocultural orientation it is important to consider teacher personal aspects such as teacher knowledge, awareness and confidence in conjunction with contextual factors including, for instance, planning tools and collaborative meetings that support collection improvement as well as the nature and level of broader organisation support for change.

In conclusion, we have shown in our research that by enhancing teachers' knowledge-as-action gains can be made in their assessment for learning practices and subsequently in the students' learning.

References

Appleton, K. (2003). How do beginning primary school teachers cope with science? Toward an understanding of science teaching in practice. *Research in Science Education, 33*, 1–25.

Bell, B., & Cowie, B. (2001). *Formative assessment and science education.* Dordrecht: Kluwer.

Black, P., Harrison, C., Lee, C., & Wiliam, D. (2001, April). *Theory and practice of formative assessment.* Paper presented at the Annual Meeting of the American Educational Research Association, Seattle.

Boaler, J. (2003). *Studying and capturing the complexity of practice—the case of the 'dance of agency'.* Proceedings of the 2003 Joint Meeting of Psychology of Mathematics Education and Psychology of Mathematics Education North America. Honolulu, Hawaii, 13–18 July.

Brown, A., Ash, D., Rutherford, M., Nakagawa, K., Gordon, A., & Campione, J. (1993). Distributed expertise in the classroom. In G. Salomon (Ed.), *Distributed cognitions; psychological and educational considerations* (pp. 188–228). Cambridge: Cambridge University Press.

Cowie, B., & Bell, B. (1999). A model of formative assessment in science education. *Assessment in Education, 6*(1), 101–116.

Cowie, B., Moreland J., Jones A., & Otrel-Cass, K. (2008). *The Classroom InSiTE Project: Understanding classroom interactions to enhance teaching and learning in science and technology. Teaching and learning research initiative.* Wellington: New Zealand Council for Educational Research.

Driver, R. (1989). Students' conceptions and the learning of science. *International Journal of Science Education, 11*, 481–490.

Gess-Newsome, J. (1999). Secondary teachers' knowledge and beliefs about subject matter and their impact on instruction. In J. Gess-Newsome & N. Lederman (Eds.), *Examining pedagogical content knowledge* (pp. 51–94). Dordrecht: Kluwer.

Gipps, C. (1999). Sociocultural aspects to assessment. *Review of Educational Research, 24*, 353–392.

Harlen, W. (1997). Primary teachers' understanding in science and its impact in the classroom. *Research in Science Education, 27*, 323–337.

Harlen, W., & James, M. (1997). Assessment and learning: Differences and relationships between formative and summative assessment. *Assessment in Education, 4*(3), 365–379.

Hiebert, J., Gallimore, R., & Stigler, J. (2002). A knowledge base for the teaching profession: What would it look like and how can we get one? *Educational Researcher, 31*(5), 3–20.

Jones, A., & Moreland, J. (2005). The importance of pedagogical content knowledge in assessment for learning practices: A case study of a whole-school approach. *The Curriculum Journal, 16*(2), 193–206.

Lave, J., & Wenger, E. (1991). *Situated learning: Legitimate peripheral participation*. Cambridge: Cambridge University Press.

Magnusson, S., Krajcik, J., & Borko, H. (1999). Nature, sources and development of pedagogical content knowledge for science teaching. In J. Gess-Newsome & N. Lederman (Eds.), *Examining pedagogical content knowledge* (pp. 95–132). Dordrecht: Kluwer.

Marshall, B., & Drummond, M-J. (2006). How teachers engage with assessment for learning: Lessons from the classroom. *Research Papers in Education, 21*(2), 133–149.

Shulman, L. (1987). Knowledge and teaching: Foundations of the new reform. *Harvard Educational Review, 57*(1), 1–22.

Shulman, L. (1992). Toward a pedagogy of cases. In J. Shulman (Ed.), *Case methods in teacher education* (pp. 1–29). New York: Teachers College Press.

Stetsenko, A., & Arievitch, I. (2002). Teaching, learning, and development: A post-Vygotskian perspective. In G. Wells & G. Claxton (Eds.), *Learning for life in the 21st century: Sociocultural perspectives on the future of education* (pp. 84–96). Oxford: Blackwell.

Wells, G., & Claxton, G. (Eds.). (2002). *Learning for life in the 21st century: Sociocultural perspectives on the future of education.* Oxford: Blackwell.

Wenger, E. (1998). *Communities of practice: Learning, meaning and identity*. Cambridge: Cambridge University Press.

Wragg, E., Wragg, C., Hayes, G., & Chamberlain, R. (1998). *Improving literacy in the primary school.* London: Routledge.

Chapter 5
Making a Case for Improving Practice: What Can Be Learned About High-Quality Science Teaching from Teacher-Produced Cases?

John Loughran and Amanda Berry

In recent times there has been considerable debate about the nature of teachers' professional knowledge. An important aspect of such debate can be traced back to Dewey (1929) who noted that educational practices themselves must be the ultimate source of the problems to be investigated if we are to build a science of education. Therefore, to better understand that which might be described as teachers' knowledge of practice, a focus is needed on those aspects that teachers themselves "name and frame" as important in informing their practice. Understanding how teachers conceptualise and develop their knowledge has been a contentious issue for some time.

Munby et al. (2001), in their review of teachers' knowledge and how it develops in the *Handbook of Research on Teaching* (Richardson 2001) highlighted the fact that there has long been considerable tension around how teachers' knowledge is conceptualised. The tension has emerged largely as a consequence of differing views about what is knowledge, and what counts as knowledge. Stark lines of difference quickly emerge when a philosophical definition of knowledge is applied in relation to that which teachers know and are able to do.

When Fenstermacher (1994) explored the nature of knowledge in The Knower and the known he described two types of knowledge. The first was formal knowledge which he described as developed using conventional scientific approaches which led to knowledge that could be generalised and applied across contexts. The second was practical knowledge which he described as being derived by teachers as a consequence of their experiences of classroom teaching.

Wideen et al. (1998) reviewed aspects of these two forms of knowledge, and considered not only how knowledge was produced but also who produced the knowledge and how it was used. They drew attention to how, through the concept of utilisation, knowledge was often described from a "producer-user" perspective—producers develop knowledge and users implement it in their practice. From a "producer-user" perspective, formal knowledge is developed by researchers, external to the classroom who produce propositions intended to direct what and how teachers

J. Loughran (✉)
Monash University (Clayton campus), Clayton, VIC, Australia
e-mail: john.loughran@monash.edu

D. Corrigan et al. (eds.), *The Professional Knowledge Base of Science Teaching*,
DOI 10.1007/978-90-481-3927-9_5, © Springer Science+Business Media B.V. 2011

should teach (with the assumption that knowledge produced by outside experts will be used by teachers). In contrast, using an interpretive perspective, knowledge is considered to be problematic, actively developed by the knower, idiosyncratic and contextually bound. Hence the notion of teachers' professional knowledge in and of itself creates an expectation that the knower and the known are inextricably linked and that such linking is embedded in their experiences of practice.

By drawing on the work of Carter (1993), Wideen et al. suggested that adopting an exclusive view of knowledge from either perspective limited the possibilities for teacher development and change by ignoring the importance of how teachers use knowledge. Following on from this, exclusivity also impacts understandings of the value of knowledge, as it tends to create a competition borne of perceived importance. Therefore, paying careful attention to how knowledge is created and used offers interesting possibilities for developing approaches to teacher learning that might influence the manner in which teachers might inquire into their own practice and share the learning of those inquiries with others (as well as how others interpret and use that knowledge in their own practice).

Cochran-Smith and Lytle (1999) offered one way of thinking about linking knowing and doing through their description of teachers' professional knowledge in terms of: knowledge for practice; knowledge in practice and, knowledge of practice. Through each of these three forms of teachers' knowledge they illustrated the importance of conceptualising teaching as being based on inquiry (a reminder of Dewey's point about educational practice as the ultimate source of inquiry) and, in so doing offer richer understandings of how:

> knowledge [is] generated in inquiry communities, how inquiry relates to practice and what teachers learn from inquiry. ...it involves making problematic the current arrangements of schooling; the ways knowledge is constructed, evaluated and used; and teachers' individual and collective roles in bringing about change. (pp. 288–289)

If teaching is understood as a form of inquiry, then teachers' professional knowledge must be actively developed, assessed and adjusted in response to teachers' experiences of practice. Making that clear for oneself and others requires an ability to articulate such knowledge. Therefore, if teachers' professional knowledge of practice matters, it must be the teachers themselves who make that knowledge explicit; but that is not as simple as it might sound. An important factor underpinning the difficulties associated with making explicit teachers' professional knowledge is its location. Knowledge developed through inquiry is located both within the individual teacher as well collectively, within communities of inquiry. The dual location of this knowledge makes it difficult to collect and represent since it "both transcends and shapes the knowledge of individual participants" (Wilson and Berne 1999, p. 186). At the same time, it has been long been recognised that teachers' knowledge of practice is largely tacit (Clark and Peterson 1986; Richardson 1997). Teachers are not used to articulating their knowledge of practice and typically, "know more than they can say". Thus, the shift from knowledge of practice as tacit to explicit creates a challenge for teachers; even though the response offers possibilities for new insights into the expertise inherent in practice.

This chapter explores one such response through the outcomes of the Science Teaching and Learning (STaL) project that have been explicated through the use of a cases methodology (Shulman 1992). The study draws on data from cases written by science teachers (primary and secondary), who participated in a one-year Science Teaching and Learning program run by Monash University in collaboration with the Catholic Education Office Melbourne. The program was designed to stimulate new understandings of science teaching and improve students' science learning. The program concludes with a writing day in which each participant constructs a case intended to illustrate, through a classroom situation, an aspect of their professional learning as a science teacher. Through the analysis of these cases (n = 32), insights into the ways in which participants explored their professional knowledge and practice is offered as one way of explicating the professional knowledge of these participants.

Science Teaching and Learning Project

The Science Teaching and Learning project (STaL) was initiated in 2005. The Catholic Education Committee of Victoria (CECV) focused on the key learning area of Science (among others) for improvement in terms of overall schooling with particular interest in student performance (CECV 2005). As a consequence of this focus, a Science Reference Group was established to report on specific issues and concerns related to school science teaching and learning. The Reference Group noted a number of issues central to student performance in science, three in particular included:

1. Teachers' confidence in teaching science (particularly at the primary level) and the role and status of the science coordinator and its impact on professional learning, mentoring and extension opportunities for science staff.
2. Science curricula may not incorporate up-to-date content and pedagogy and science teachers may not always be aware of the range of resources available and the science activities possible.
3. Teaching practices and curriculum are not always engaging, nor do they necessarily make links to relevant real-life situations (for students) or cater to different learning styles.

As a consequence of considering these issues, the Catholic Education Office Melbourne (CEOM) engaged members of the Science Education Research group at Monash University to develop a science professional development program as one way of responding to their concerns. However, as discussions about the program evolved, the notion of professional development was increasingly questioned. The difficulty with the notion of professional development that arose was related to issues associated with traditional forms of professional development as some form of "add on" or "top up" to teachers' practice.

Distinguishing between Professional Development (PD) and Professional Learning (PL) became an important touchstone in ensuing discussions about the conceptualisation of the STaL program, both in terms of teaching expectations and anticipated learning outcomes. This distinction was salient to discussions because, as Mockler (2005) noted, PD tends to be delivered in a "spray-on" manner, in that typically, teachers are expected to attend a PD day in order to return to school to implement the PD ideas in their own practice. In contrast, a Professional Learning (PL) approach is based on understandings of practice as something to be developed and refined as a consequence of teachers' needs, issues and concerns derived from their individual situations. Therefore, supporting teachers' professional learning stands out as important not only in terms of fundamentally enhancing science teaching but also for developing teachers' professional knowledge of practice (Berry et al. 2007). In terms of the STaL project then PL was seen as doing that by:

> Emphasiz[ing] practices that are: sustained over time; aligned with the specifics of school and classroom contexts; underpinned by research and practice-based evidence; and, supported by professional learning communities and collaboration. (Loughran et al. 2009, p. 1)

In adopting a PL approach for the STaL project it was decided that there was a need to place participants in learning contexts through which their understandings of science could be probed and challenged in ways that might reflect the very practices anticipated as teaching and learning outcomes in their own classrooms on their return to school. In addition, the program was organised so as to provide a space for participants to draw insights from their experiences of the program and to document selected insights in the form of a case (using a cases methodology as described by Shulman 1992). The program was therefore developed as a residential experience (2×2 days, +1 day) distributed over the course of a school teaching year with appropriate school-based activities and ongoing face to face and other forms of support between sessions.

> The STaL program [is designed to]...value teachers firstly as professionals, and secondly, as owners and constructers of their own knowledge and wisdom of teaching and learning. The organization of the STaL program takes these aspects into account through providing a comfortable working environment (the program is conducted as a residential experience in a Melbourne hotel) during the day, rather than as an added demand after school. Being away from the usual distractions of school and having an overnight stay gives teachers the opportunity to engage in sustained professional dialogue with colleagues from other schools and sectors that they normally do not have time for in their busy day to day work. ...the STaL program provides opportunities for teacher professional growth through highlighting differences between the ways in which teachers plan, talk about and enact their pedagogy...[and] aims to provide experiences that lead teachers to reflect on, reframe and reconsider these aspects of practice. ...[STaL is designed to highlight] a great deal about the complex nature of teachers' professional learning and the considerable wisdom that can be generated when teachers are given the opportunity to develop their own practice. (Keast and Berry 2009, pp. 5–6)

The final day of the program is a case writing day at which participants share insights from their learning about their science teaching and learning through the development of cases as a form of portrayal of their knowledge of practice.

Case Writing

Case writing creates opportunities for teachers to reflect on, and begin to articulate aspects of practice specific to their own needs and contexts. The case writing approach used in STaL revolves around reading and later writing cases. Participants are introduced to cases (written by previous STaL participants) in a variety of ways during the first four days of the program. Cases are used as a program resource to offer another way of bringing science teaching and learning issues to the fore using real classroom examples. As part of the process of learning about cases, participants engage in case discussions. This has a two-fold purpose: they are introduced to case writing as a genre and, through discussion of the case issues, participants may be stimulated to see into practice in new ways.

The final day of the program (case writing) is based on the view that in creating an opportunity for teachers to reflect on their practice through a case writing experience, that further insights about practice may be generated and alternative perspectives and approaches to science teaching and learning might be explored. During the day, participants develop drafts of their case and share these with colleagues in order to gain constructive feedback that can help to refine the case and increase its readability for other teachers. Hence, case writing is integral to the design of the STaL project as both a process and product. Cases serve as a vehicle for challenging participants' science teaching and learning practices and for encouraging the development of new possibilities for, and insights into, their practice. In so doing, STaL has an explicit aim of finding ways for participants to seriously focus on their science teaching so that they can see themselves as producers of sophisticated knowledge of teaching and learning, not just users.

The case writing approach adopted in the STaL project is similar to that described by Shulman (1992) and Lundeberg (1999) whereby participants are encouraged to build their case based around the following key prompts:

- Title/Topic: what is this a case of? The title should invite the reader into the topic and make clear the issue under investigation.
- Set the scene: succinctly outline the context.
- Explain and offer insights into the author's role or actions.
- Create a rich narrative about what happened by including: dialogue; feelings; reactions; prejudices and examples of evidence.
- Closure is important: A good case often ends with the reader having something to do/think about/pursue/apply to their own situation.
- Drafting and redrafting to refine ideas is crucial.

One of the STaL participants, Colquhoun (2006), described her experiences of case writing in the following way:

> From sharing our experiences with the group, we were empowered to recognise and affirm the collective knowledge that we, as practising teaching professionals, could also contribute by sharing our knowledge drawn from our classroom encounters with others. This led to engaging in teacher research and the formulation of cases that documented our classroom experiences. I have to say, that doing cases does appear to be very daunting at first but

> the essential thing to note with formulating a case is that it is really only formalising and extending that which we already do as professionals on a day to day basis. We are constantly refining our craft, not every change is a monumental teaching moment but anything that causes you to think, "Yes, that was a good lesson, getting the students to re-create a scenario in groups allowed them to see the idea more clearly", is a case waiting to be written. Often the staff room debriefs with a colleague on an informal level provide excellent food for thought and avenues for writing something that may help other teachers. …we wrote about personal powerful experiences which gave us insights into the teaching and learning of our students with the intention that, as others read them, they might be encouraged to do something similar. (p. 12)

It is this type of participant response that reinforces for us the value of case writing as an important aspect of supporting teachers' professional learning.

Case Analysis

The STaL project has run for four years (2005–2008) and at the time of writing, another cohort has begun the program (2009). The cases from each year cohort are compiled and published (year 1, 22 cases; year 2, 30 cases; year 3, 28 cases; year 4, 32 cases). These publications offer a wealth of insight into participants' understandings of science teaching and learning. The cases have been analysed and worked with in different ways (see, for example, Berry et al. 2009). One important outcome of this work that has emerged for us, as Professional Learning leaders working with teachers in the STaL program, is that we have learnt more about how to support teachers in their case writing endeavours. As a consequence, the nature of the cases developed by cohorts over the years highlight different aspects of participants' learning and foreground issues that occupy their thinking at the time of writing. An emphasis on particular aspects of teaching and learning emerged most strongly through analysis of the most recent set of cases (Berry and Keast 2009).

Table 5.1 Cases by major theme

Category & brief description	Number of cases (N=32)
1. Documenting professional practice: articulating knowledge of learning of students and teachers and providing evidence of change	7
2. Valuing student decision making: approaches to supporting and encouraging students to become more intellectually active, independent learners	7
3. Making real-world links: providing opportunities to connect with students' understandings of the world outside of school	4
4. The purpose of practical work: approaches to practical investigations that engage students in more purposeful thinking about what they are doing and why	2
5. Implementing new strategies: learning about practice that emerges when teachers explore different approaches to teaching and learning	10
6. Working from student ideas: taking a view that learning science must involve the ways of thinking and knowing that each student brings to class	2

The 2009 case book (published from the cases of the 2008 program) was constructed following an analysis of the central issue/dilemma/concern considered to be at the heart of the case. Analysis was conducted by two academics from the STaL team who read through all of the cases independently, before conducting a thematic analysis based on their academic interpretation of the dominant themes of the text. The two academics then met to compare the results of their analysis and to develop a common final set of categories that accounted for their perceived similarities and differences. Through this process, six major categories emerged (see Table 5.1).

Although it could well be argued that all of the categories relate in some way to documenting professional practice (category 1), of the agreed upon final six categories, five are derived of specific aspects of participants' concerns for science teaching and learning (categories 2–6). Throughout all of the cases, participants demonstrate aspects of their learning about professional practice, and the value of so doing, as Corrigan (2009) notes:

> The process of articulating thinking to others helps to clarify and distil the ideas and insights gained. Being able to name and share such learning is a critical part of teaching…[and such a] documenting process [shows] the richness of learning that goes on in each classroom [and how that] can become visible and understood [so] that further learning paths can be developed. (p. 17)

Moving beyond documenting practice per se (i.e. category 1), the overwhelming majority of the cases demonstrate a concentration on a particular aspect of participants' learning about science teaching and learning in ways not quite so evident in the previous three volumes of cases. We explore these aspects of learning in detail, in the following sections, using extracts from the cases.

Valuing Student Decision Making

It seems reasonable to assert that good teaching should support the development of active, responsible learners. Too often though, science teaching is characterised as the delivery of facts and information. Students need to be offered real opportunities to accept responsibility for their own learning but saying that and doing it are not the same thing. The natural tendency for many teachers is to make decisions for their students; but not because they purposely plan to do so. Rather, it could be seen as a necessary response to the need to keep things moving in such a way as to manage the many competing demands of the classroom quickly and efficiently.

Osler and Flack (2008) highlighted how quickly their students became more active and responsible learners when decision making was handed over to them, i.e. when the responsibility for decision making shifted from the teacher to the student. Osler and Flack's descriptions (and examples of students' changed learning behaviours as a consequence) are compelling. In a similar way, Wood's STaL case of making the shift in "who makes the decisions" in class illustrates how student learning quickly develops when the responsibility for decision making changes (Wood 2009).

Wood's case (see Fig. 5.1) demonstrates well what can happen when a teacher accepts the challenge of making changes in approaches to science teaching that

CREATING THINKING STUDENTS

We are beginning a new unit in the Year 7 class, 'states of matter'. It's the last lesson on Friday afternoon and I decide it is worth the risk of introducing a new learning approach, a definition activity.

> "Okay girls, write your definition of solid, liquid and gas. Then share your definitions in small groups. After you come up with a group definition, then we'll have a class discussion about your ideas."

Before the blink of an eye, seventy minutes have flown past! Every student in the room has been busily occupied and every student in the room has spoken out loud either to their small group or the whole class. The ideas that they are coming up with help me know that they are really on track.

> "I've heard of little things called particles."
>
> "This is like … I saw at home."

I force them to stop their discussion – it's difficult, they just want to keep talking! Then I ask them to write some class definitions before they go home. As they walk out of the room I reflect on the quality of the science vocabulary they have came up with; atoms, particles, condense, evaporate, and all of it generated by themselves and explained by themselves.

The girls are enthused! I hear reports from other staff members that students are asking to do the 'fun stuff' like Mr Wood's class. I feel pleased but also confused. Aren't we all doing basically the same thing? The girls are excited to come into class - a huge bonus -and best of all I seem to be answering fewer questions as the girls are confidently asking and answering each other's questions. The quality of their questions is also impressive,

> "I was thinking about what we did last term last term, and why does...?"

This student was making a link between this topic and something we studied months ago!

Creating better thinkers

My school is in the process of introducing the middle years program (MYP), the precursor to the International Baccalaureate. MYP is focused on creating global thinkers and places high value on how students learn and less value on covering vast amounts of content. It was fortunate timing that I joined the STaL programme and began thinking more deeply about my teaching and learning approach, as the STaL philosophies marry so well with MYP.

I have been able to develop a process methodology with much more confidence so that when students ask questions such as, "Why do leaves go brown?" I scratch my head and ask in return,

Fig. 5.1 Wood's case

"Why were they green in the first place? Go and find out."

At progress evenings, I notice that parents are starting to talk about the topics we have been studying. It seems that not only are the students engaged in class, but they are also discussing what we are doing with their families. Fantastic!

In our classes there is an increased focus on students being able to use knowledge in unfamiliar situations. They need to be able to think on their feet and apply their learning in unfamiliar situations, and to be confident enough to articulate a question and know how, and where, to seek out the answers. After many years of teaching, I can see quite a change in the learning and I'm proud of the depth of understanding that my students are developing.

Assessing progress

After making these changes to the teaching and learning in my classes, I felt that I needed some way of checking how the knowledge and skills of my group compared with other Year 7 groups. I noticed that on recall tests the results from students in my class are significantly lower than the other class. Yet in terms of applying their thinking experiences in the real world, my students are achieving well. When it comes to posing their own scientific questions, designing their own experiments and researching information my class excels. I am relieved to see that my Year 7's final reports stand up well.

A learning journey for all

I have no doubt that the value of the new approaches developed through my participation in the STaL programme has brought renewal to my teaching and a sense of joy and ownership to my students in science. I am glad that I have been able to change the focus of my classes so that thinking really matters! (Wood 2009, pp. 48–49)

Fig. 5.1 (Continued)

overtly demonstrate real expectations of students as responsible learners. As his case shows, he also found renewed professional satisfaction as a consequence of developing a more engaging science teaching and learning environment.

Making Real-World Links

It has been well documented that students' need to know is an important driver of interest and engagement in science. One way of developing that need to know is by making real-world links. Four of the cases written for the STaL project focused on this particular aspect of student engagement and each raised interesting perspectives on how such links enhanced student learning of science. For example, one of the cases explored the value to students in contextualising scientific discoveries by encouraging an understanding of such things as the personal, professional, political and economic reasons "of the time" that drove the development of scientific knowledge. This approach highlighted for the teacher new perspectives on engaging students in the processes associated with building scientific knowledge and how such things as war, money and politics can influence discovery. Another interesting real-world link was documented through attempting to confront students' concerns about physics. In this case, the teacher, Emma Rhodes, asked her students to bring broken/unused electronic devices from home and to pull them apart in class in order to make the familiar unfamiliar as a way of catalysing links between physics and the world around them.

> …The teacher next door peered through the glass as she shut the door connecting [our] two rooms. We were making a lot of noise. …Then, just as I was going to ask them to pack up…
> "No way, check this one!" screamed Nina.
> Nina's group had broken into the tape deck. All of the girls came running over to look.
> "Look at all the transistors!"
> "No they are resistors. See all the coloured stripy things."
> "Can you pull out the speakers? Cool! Check out the switch."
> "Hey," called Shirleen from another desk. "I think I found some LED's, they're green and on there is a pink one."
> I was smiling so much my face was starting to hurt. (Rhodes 2009, p. 59)

The final two cases that focused on making real-world links did so through implementing a translation activity introduced to participants during the STaL program (using a toy car to depict the movement portrayed in a graph, see Mitchell and Mitchell 1997). Although the students being taught were dramatically different in age (4-year olds and 14-year olds), by linking their experiences to moving vehicles, each group of students was able to work through the scenarios in meaningful ways and build a language of motion that supported their learning. In each, the students were able to relate their everyday experiences to those portrayed in the graphs of movement described through distance versus time graphs.

The Purpose of Practical Work

Considering the concerns so often documented in the literature about science practical work (recently highlighted again by Goodrum et al. 2001) it is interesting to note that only two cases were based on the purpose of practical work. It may well be that this is an indication of the difficulties faced by teachers in trying to move students beyond a superficial view of, and approach to, practical work in science. This challenge was well portrayed in Beale's case:

> "What are we doing in science today?"
> "Are we doing work?"
> "I don't want to do any work today. Can we do prac?"
> Every lesson as I greet my Year 8 students at the classroom door I am bombarded with requests for no work and lots of experiments. And yet when they engage in practical work I notice that conversation is rarely on the experimental process, the observations being made

> or even what these results might mean, and mostly on the activities conducted with friends last weekend.
> "Why don't my students see experiments as work?"
> "Why when assessing prac reports do student record poor results and little analysis?" (Beale 2009, p. 72)

Yet it may be that a teacher's approach to practical work may be more an issue than the practical work itself. When that possibility emerges, opportunities for change stand out:

> In science practical classes, my usual approach is to explain the scientific theory we are studying then get students to carry out an experiment to prove the theory. So, they begin the experimental work knowing already the results that should be expected. Although they are developing their practical skills in these classes, I began to doubt whether they were developing any skills in analysing and reaching conclusions about results. …I decided to make a change…we were studying light…I gave students a title for the lesson and explained the steps to carry out the experiment, but I didn't give them any indication of the expected outcome. The students then had to carry out the experiment, record their results, then try to explain the theory to me, and each other, based on their experiment evidence. (Seago 2009, p. 69)

The two cases that focused on the purpose of practical work both raised similar issues with regard to engaging students in learning science. They both illustrated how easy it is for students to become routinised into an approach to practical work that diminishes thinking and engagement and how easy it is for teachers to inadvertently reinforce that approach. The challenge that confronts teachers is to move beyond practical work as simply doing activities and to reconceptualise practical work in ways that might genuinely invite inquiry so that students take seriously these experiences as important to their learning, not just as tasks to be completed.

Implementing New Strategies

Perhaps it is not so surprising that 10 of the 32 cases were based on implementing new strategies. As Appleton and Kindt (1999) noted, teachers need "activities that work". Therefore, any new ideas or approaches to classroom teaching that appear to work very quickly become a source of interest to teachers who are expert at adapting and adjusting ideas to suit their classrooms. However, differentiating between activities that work and engaging students in learning are not necessarily the same thing. As the cases in the purpose of practical work demonstrated, students may well be occupied and/or having fun but not engaged in their learning.

An important aspect of case writing is well captured through those cases categorised as implementing new strategies. Each illustrates how the process of case writing can help authors move beyond a simplistic description of the activity itself and examine the underlying principles of (science) teaching and learning. The process of uncovering the bigger pedagogical issues is important because a teacher's natural response when hearing about a new strategy is to attempt to grasp quickly the essence of the steps necessary for the strategy to work. Hence, there is an inevitable need for teachers to have a language for sharing based on succinct descriptions of what is done. Less common is a desire to know why a strategy works, i.e. the pedagogical underpinnings of effective teaching and learning; typically that task is left to the individual and is sometimes viewed as over theorising. Yet it is through questioning the "why" behind an activity that real insights into (science) teaching and learning can occur.

The ten cases that pay attention to this theme highlight what happens when teachers experiment with their practice through: risk taking; trusting students to be independent learners and, relinquishing control. In reflecting on their experiences of implementing new strategies, these cases illustrate how these teachers explore more deeply their knowledge of practice. Each teacher moved beyond the "what" of teaching and began to develop deeper understandings of the "why"; not only of their teaching but also how that teaching influenced their students' science learning.

In essence, these cases illustrated how an initial focus on oneself as a teacher created new views of students as learners. For example:

> I find myself looking into the fishbowl [teaching procedure] and realising that I am controlling the proceedings…Suddenly, it hit me. This is not a discussion. They're [students] talking to me, not each other. …Can students take on board what it is they are supposed to be learning when they are working so hard to find and supply what the teacher wants or expects? (Kozera 2009, pp. 78–79)

> Now, I had unravelled new challenges and needed more time to resolve them, but at what cost? At the same time, this doubt was fostering a desire in me to take more risks and trial a few more activities…seeing the students work so well made me think, this has to be worth pursuing. (Carboon 2009, pp. 80–81)

> Taking the risk to work from the "too hard basket" has been an amazing learning experience for me and my students. I have developed insights into these students' understandings in far more depth that I had expected. (Chiodi 2009, p. 95)

> I have learnt a lot about myself as a teacher. I thought I was a teacher who was all about an enquiry approach to teaching. I discovered later that in fact, I was controlling, teaching the students what I felt they needed to know. I was having a hard time "letting go" and giving students the opportunity to guide their own learning as their own questions. (Maloney 2009, p. 101)

Working from Student Ideas

The final theme that emerged from analysis of the set of cases was that of teachers working with the science ideas and experiences that their students bring to the classroom. Two cases explicitly dealt with this theme; each highlighting a different aspect of how teachers undertook to pay more careful attention to their students' prior knowledge, and the consequences of doing so, for their planning and teaching. However, taking a view that the development of science knowledge is a personal process of active mental construction adds new and challenging demands to teaching—for instance, how do you explore and follow the ideas of a class group of students? How do you challenge students' ideas while at the same time maintaining levels of trust that enable students to feel comfortable to articulate their thinking?

Each case in this section explored a different aspect of this question, with interesting consequences. Stead came to recognise that although he believed he was working from student ideas, in fact, his teaching approach reinforced a different view—that his ideas mattered most. As he began to change his practice in order to incorporate learners' ideas he found himself wresting with new questions about his teacher role.

> Starting the topic of Solids, Liquids and Gases with my two Year 7 classes led me to see that they carried many such misconceptions:
> "All solids are hard."
> "Liquids are all like water."
> "All gases are invisible."
> Previously when moving through this topic I would have addressed the misconceptions that I thought students had. Rarely did I take the time to find out what these misconceptions actually were or, more importantly, why they believed them. I was too focused on what I wanted them to do, without nearly enough emphasis on involving the students in the learning process. In hindsight, I realised that this approach had led to some students having very superficial understandings of the content. Students could recall what I had told them but did not understand how or why. I needed to think about how I could help them to develop a deeper understanding of the topic. Just telling them the answers didn't work. So, what if I stepped back, didn't tell them what I thought they needed to know and let them work out the answers for themselves? To do this, I needed to change my role in the classroom, to become more of a learning guide, rather than the fount of all knowledge. This presented quite a challenge for me and I wondered whether I was up to it. (Stead 2009, p. 109)

Working from students' prior views is a complex task, not only because of the complexities of working with multiple ideas, but also because the teacher must be sensitive to the demands of the learning environment that requires students to find and use language to express their thoughts. Long experienced a dilemma as she worked with students from a non English speaking background and who were experiencing learning difficulties. She realised that the language demands associated with this approach were too great and caused anxiety for these students.

> While the class were working on their task, I became acutely aware of a small group of students who seemed distressed...They showed their discomfort with their eyes, their body language, their hesitation, and their attempts to write.
> While I thought I was providing an opportunity for all students to clarify and extend their ideas, this group of students found the experience confusing and difficult—and ultimately, embarrassing.
> **My dilemma**
> I have reflected on this experience several times since because I feel so guilty about hurting these vulnerable students. Could I have found a different way to do this task? Was there a strategy that relied less on language? Was it too early in the year to try deep thinking tasks with Year 7s? Should I have tried it with a different class first? (Long 2009, p. 113)

Long's case illustrates well how the feelings of both the teacher and the learner contribute to the quality of the learning that takes place. Teaching and learning is about a relationship that goes beyond the content and the procedures and must take into account the personal dimensions of all participants.

Learning About Facilitating Case Writing

Through our experiences of working with teachers to develop their cases, our learning about the process of facilitating case writing has also been enhanced. We have come to learn that many teachers are fearful about the idea of writing a case, partly because they typically believe that their case should be of a ground breaking success story, and they may not have such "outstanding successes" to report, and partly because the case writing genre is quite different from their usual writing experiences (reports, curriculum documents, etc.), and therefore presents a daunting task. They also have concerns about their skill to write about their practice in this way because they are not typically expected to do so in their normal way of functioning as a teacher.

To address these concerns, we have introduced short, structured journal writing opportunities into the workshops (we provide a journal to each participant as part of the program), so that teachers can write notes about their thinking about their experiences during the workshops as well as implications for classroom practice. This journal approach is designed to encourage participants to document their feelings and ideas around particular instances of teaching and learning that they experience and to act as an *aide de memoir* when it comes to reflecting on their experiences. Teachers discuss their journal entries in the workshops, then revisit and add to their journal notes back at school. The journal also provides a basis for discussion when the STaL team member visits teachers in their schools between the formal program days. This in-school meeting also serves as a starting point for beginning to pinpoint the "small stories" of success and failure that might serve as the basis for a case—in an attempt to challenge the perception that cases need to be about big success stories. Developing a sense of confidence in themselves both as teachers who have something to say about their pedagogy and as writers who can communicate those ideas in engaging manner, unfolds over the year in different ways for each participant. The case writing day itself is often experienced as a sense of relief, as the teachers come to view their case writing as a positive and empowering as they share drafts, provide feedback to each other and notice similarities in the issues with which they are grappling.

Conclusion

The set of cases examined in this chapter illustrate attempts by science teachers to change their practice, and their growing awareness of the implications for themselves and their students of the nature of change as a complex and multidimensional process. Many of the cases report higher levels of student engagement in their learning through teachers shifting their orientations towards teaching science in ways that begin to bridge the gap between school science and the science in the worlds that students inhabit. Altogether, the cases illustrate these teachers' efforts to better understand the nature of science teaching and learning and how the relationship between teaching and learning might be addressed through better understanding more about the details of what happens in their classrooms.

An important outcome of this work is that when a professional learning approach is adopted to support teachers as producers of knowledge the outcomes can genuinely be seen as impacting students' science learning in positive and meaningful ways as a direct consequence of the same happening to teachers' understanding of a pedagogy of science.

Acknowledgements This project was conducted through the support of an Australian Research Council (ARC) Linkage Grant (Loughran et al. 2009).

References

Appleton, K., & Kindt, I. (1999). Why teach primary science? Influences on beginning teachers' practice. *International Journal of Science Education, 21*, 155–168.

Beale, A. (2009). Interpretive discussion. In A. Berry & S. Keast (Eds.), *Looking into practice: Cases of science teachers' professional growth* (pp. 72–74). Melbourne: Monash University and the Catholic Education Office Melbourne.

Berry, A., Clemans, A., & Kostogriz, A. (Eds.). (2007). *Dimensions of professional learning: Identities, professionalism and practice*. Dordrecht: Sense.

Berry, A., & Keast, S. (Eds.). (2009). *Looking into practice: Cases of science teachers' professional growth*. Melbourne: Monash University and the Catholic Education Office Melbourne.

Berry, A., Loughran, J. J., Smith, K., & Lindsay, S. (2009). Capturing and enhancing science teachers' professional knowledge. *Research in Science Education, 39*, 575–594.

Carboon, D. (2009). Timely risky business. In A. Berry & S. Keast (Eds.), *Looking into practice: Cases of science teachers' professional growth* (Vol. 1, pp. 80–82). Melbourne: Monash University and the Catholic Education Office Melbourne.

Carter, K. (1993). The place of story in the study of teaching and teacher education. *Educational Researcher, 22*, 5–12, 18.

CECV (2005). Improving science teaching and learning. Seminar 6, June, 2005, http://www.cecv.melb.catholic.edu.au/reseminar/seminars5-8/seminar6.pdf. Accessed 12 Dec 2010.

Chiodi, K. (2009). Too hard basket. In A. Berry & S. Keast (Eds.), *Looking into practice: Cases of science teachers' professional growth* (Vol. 1, pp. 93–95). Melbourne: Monash University and the Catholic Education Office Melbourne.

Clark, C., & Peterson, P. (1986). Teachers' thought processes. In M. C. Wittrock (Ed.), *Handbook of research on teaching* (3rd ed., pp. 255–296). New York: MacMillan.

Cochran-Smith, M., & Lytle, S. (1999). Relationships of knowledge and practice: Teacher learning communities. In A. Iran-Nejad & P. D. Pearson (Eds.), *Review of research in education* (Vol. 24, pp. 249–305). Washington: American Educational Research Association.

Colquhoun, Y. (2006). Cases: A teacher's perspective. In J. Loughran & A. Berry (Eds.), *Looking into practice: Cases of science teaching and learning* (2nd ed., Vol. 1, pp. 11–14). Melbourne: Monash University and the Catholic Education Office Melbourne.

Corrigan, D. (2009). Documenting professional practice. In A. Berry & S. Keast (Eds.), *Looking into practice: Cases of science teachers' professional growth* (Vol. 1, pp. 16–17). Melbourne: Monash University and the Catholic Education Office Melbourne.

Dewey, J. (1929). *The sources for a science of education.* New York: Liveright.

Fenstermacher, G. D. (1994). The knower and the known: The nature of knowledge in research on teaching. In L. Darling-Hammond (Ed.), *Review of research in education* (Vol. 20, pp. 3–56). Washington: American Educational Research Association.

Goodrum, D., Hackling, M., & Rennie, L. (2001). *Research report: The status and quality of teaching and learning of science in Australian schools.* Canberra: Department of Training and Youth Affairs.

Keast, S., & Berry, A. (2009). Looking into practice, sharing professional wisdom: Introduction to this volume. In A. Berry & S. Keast (Eds.), *Looking into practice: Cases of science teachers' professional growth* (Vol. 1, pp. 5–7). Melbourne: Monash University and the Catholic Education Office Melbourne.

Kozera, C. (2009). I am your audience. In A. Berry & S. Keast (Eds.), *Looking into practice: Cases of science teachers' professional growth* (Vol. 1, pp. 78–79). Melbourne: Monash University and the Catholic Education Office Melbourne.

Long, S. (2009). Pre-existing concepts, pre-existing trauma. In A. Berry & S. Keast (Eds.), *Looking into practice: Cases of science teachers' professional growth* (Vol. 1, pp. 112–113). Melbourne: Monash University and the Catholic Education Office Melbourne.

Loughran, J. J., Berry, A., Clemans, A., Keast, S., Miranda, B., Parr, G., et al. (2009). *Exploring the nature of teachers' professional learning.* Paper presented at the bi-ennial conference of the International Study Association on Teachers and Teaching (ISATT), Rovaniemi-Lapland, July 1–4, 2009.

Lundeberg, M. (1999). Discovering teaching and learning through cases. In M. A. Lundeberg, B. B. Levin, & H. Harrington (Eds.), *Who learns what from cases and how?: The research base for teaching and learning with cases.* Mahwah: Lawrence Erlbaum.

Maloney, S. (2009). Who's running the show? In A. Berry & S. Keast (Eds.), *Looking into practice: Cases of science teachers' professional growth* (Vol. 1, pp. 99–101). Melbourne: Monash University and the Catholic Education Office Melbourne.

Mitchell, I., & Mitchell, J. (Eds.). (1997). *Stories of reflective teaching: A book of PEEL cases.* Melbourne: PEEL.

Mockler, N. (2005). Trans/forming teachers: New professional learning and transformative teacher professionalism. *Journal of In-Service Education, 31*, 733–746.

Munby, H., Russell, T., & Martin, A. K. (2001). Teachers' knowledge and how it develops. In V. Richardson (Ed.), *Handbook of research on teaching* (4th ed., pp. 877–904). Washington: American Educational Research Association.

Osler, J., & Flack, J. (2008). *Whose learning is it?: Developing children as active and responsible learners.* Rotterdam: Sense.

Rhodes, E. (2009). The P-word. In A. Berry & S. Keast (Eds.), *Looking into practice: Cases of science teachers' professional growth* (Vol. 1, pp. 58–60). Melbourne: Monash University and the Catholic Education Office Melbourne.

Richardson, V. (1997). Constructivist teaching and teacher education: Theory and practice. In V. Richardson (Ed.), *Constructivist teacher education: Building a world of new understandings* (pp. 3–14). London: Falmer.

Richardson, V. (Ed.). (2001). *Handbook of research on teaching* (4th ed.). Washington: American Educational Research Association.

Seago, C. (2009). Practically speaking, less is more. In A. Berry & S. Keast (Eds.), *Looking into practice: Cases of science teachers' professional growth* (Vol. 1, pp. 69–71). Melbourne: Monash University and the Catholic Education Office Melbourne.

Shulman, J. H. (1992). *Case methods in teacher education.* New York: Teachers College Press.

Stead, D. (2009). Do they understand what I am not telling them? In A. Berry & S. Keast (Eds.), *Looking into practice: Cases of science teachers' professional growth* (Vol. 1, pp. 109–111). Melbourne: Monash University and the Catholic Education Office Melbourne.

Wideen, M., Mayer-Smith, J., & Moon, B. (1998). A critical analysis of the research on learning to teach: Making the case for an ecological perspective on inquiry. *Review of Educational Research, 68*, 130–178.

Wilson, S., Berne, J. (1999). Teacher learning and the acquisition of professional knowledge: An examination of research on contemporary professional development. In: A. Iran-Nejad, P. D. Pearson (Eds.), *Review of Research in Education* (Vol. 24, pp. 173–209). Washington: Sage Publications.

Wood, M. (2009). Creating thinking students. In A. Berry & S. Keast (Eds.), *Looking intopractice: Cases of science teachers' professional growth* (Vol. 1, pp. 48–49). Melbourne: Monash University and the Catholic Education Office Melbourne.

Chapter 6
An Approach to Elaborating Aspects of a Knowledge Base for Expert Science Teaching

Deborah Corrigan and Richard Gunstone

Introduction

Our recent involvement with a multi-year collaborative project to develop research-based resources to support science teachers in their classroom teaching has caused us to re-examine our own understandings of aspects of both pedagogy and of selected ideas in science. In this chapter we document and discuss some of the changes resulting from this re-examination, and ways in which this self-analysis might contribute to our conceptualising of expert science teaching. As science teachers and educators with many years of teaching experience at both school and university levels, it became clearer to us through the re-examination that we can sometimes take for granted ideas about science and the teaching of science that we should continue to question. While we (and others) may use the same language to communicate these ideas and articulate our expertise, our interpretations of that language remain quite individual and our understandings of ideas can be less well formed and less consistent with the ideas of others than we imagined.

As part of the process of development of the research-based resource for teachers we had clear records of our discussions and drafts that emerged through the development process. We have used these records of our individual and joint understandings being challenged to construct this chapter—a chapter that in essence is an exploration of what may be necessary in order to generate a shared "expert" knowledge base for aspects of the teaching of science. Please note that we use the term "expert" here with absolutely no intent of hubris, but rather in recognition of the fact that both of us are teachers of science who are recognised as being skilled and as being forms of authority in at least substantial aspects of the teaching and learning of science.

This chapter, then, is a self-analysis. It is also an attempt to articulate what some would call "PCK in action" (Loughran et al. 2001), but we describe an example

---

D. Corrigan (✉)
Centre for Science, Mathematics & Technology Education, Monash University (Clayton campus), Clayton, VIC, Australia
e-mail: debbie.corrigan@monash.edu

D. Corrigan et al. (eds.), *The Professional Knowledge Base of Science Teaching*,
DOI 10.1007/978-90-481-3927-9_6,

of a process of learning where expert knowledge was an essential beginning. This process of professional learning resulted in not only a product that could be shared with others, but also a deeper understanding for each individual "expert" as they experienced the process.

The project involved the progressive development of a web-based resource for the Victorian Department of Education and Early Childhood Development (DEECD) in Australia by members of the science education group in the Centre for Science, Mathematics and Technology Education (CSTME) at Monash University. In simple terms, DEECD wanted this resource to give support to teachers to effectively implement the current P-10 Science curriculum for Victorian schools (known as the Victorian Essential Learning Standards [VELS] in Science). The resource is labelled the "Science Continuum P-10". In developing this resource for teachers, we have realised how important a collaborative approach to curriculum development and pedagogy is if we are to have clear intentions about what students need to learn and be able to do as they study science.

We begin this chapter by outlining relevant aspects of broad curriculum policy and consequences of this for the VELS in Science because these are relevant to the somewhat different structure for the resource Science Continuum P-10. Then we briefly discuss the research contexts from which the Continuum evolved, research contexts with which members of the development team had considerable and long-term direct involvement, and outline the structure and features of the Science Continuum. A more complete account of these issues of origins and development, including illustrative examples from the science curriculum, is in Isaacs et al. (2008).

We then turn to our purpose for the chapter. We use two of the entries in the final Science Continuum P-10 resource to illustrate how the process of generating those entries led us to examine our own understanding of selected focus ideas in science, and then consider how this process provides an elaboration of aspects of a knowledge base for expert science teaching.

The Policy Context Brought to the Development of the Science Continuum P-10

The *Blueprint for Government Schools* adopted by the Government of Victoria in 2003 addressed three priority areas for reform:

- recognising and responding to diverse student needs;
- building the skills of the education workforce to enhance the teaching–learning relationship and
- continuously improving schools.

A focus on developing a "Student Learning Strategy" was an important aspect of recognising and responding to diverse students' needs. This learning strategy provides a broad framework of essential learnings to be articulated through the development of a new curriculum and reporting procedures. Defined assessment measures of student achievement and principles of teaching and learning to be applied throughout the school system are also included. The Science Continuum P-10 provides an evidence-based resource for teachers to support the teaching and learning of the new science curriculum. It was with this background, and in the context of other science education policy influences, that the specifications for the Science Continuum P-10 were developed through negotiation between DEECD and the science education team from CSTME.

The Science Continuum P-10 incorporates "Concept Development Maps", to illustrate progression in student ideas in multi-faceted ways rather than in linear ways. It focusses on student alternative conceptions and links to the Science Standards in VELS. The "Concept Development Maps" are the "Conceptual Strand Maps" from the American Association for the Advancement of Science; these are designed to "show how students' understanding of the ideas and skills that lead to literacy in science, mathematics, and technology might develop from kindergarten through 12th grade" (AAAS n.d.; see also AAAS 2001, 2007). As implied above, The Science Continuum P-10 also emphasises a learning focus which includes responding to students' needs, assessment that encompasses "as", "of" and "for" learning and therefore links teaching and assessment practices in ongoing ways, and other perspectives not discussed here. In line with DEECD's emphasis on the development of digital resources, the Science Continuum, and all supporting documents, are only published online.

The Research Context Brought to the Development of the Science Continuum P-10

The CSMTE science education team has for many years pursued three coherent strands of research of particular relevance to the Continuum and which have consequently influenced its development. These interrelated strands are: research specifically on alternative conceptions; research on broader issues of learning and teaching (including such ideas as conceptual change, metacognition, alternative representations of learning, e.g. relational diagrams); research on teacher learning and teacher change (represented through work such as the Project to Enhance Effective Learning [PEEL]). More recently, our work on teacher learning and teacher change has expanded to embrace the exploration of pedagogical content knowledge (PCK) and the values of science and science education in relation to teacher learning. All three strands are closely interwoven in the work of the group. So, for example, while our early research work on alternative conceptions may have been a focus, at the same time attention was already being given to ways this research impacted on how we understood and then researched the other two strands.

The CSTME's early work on alternative conceptions in science, conceptual change and metacognition is discussed in Gunstone et al. (1988). Very early on in our work in all three strands, we also recognised the need for strong links between our research and practice, and worked at building these (e.g. Gunstone and Northfield 1988). Our work on the Science Continuum thus began with a rich body of relevant wisdom in all of the conventional broad domains of science (biology, chemistry, earth science, physics, nature of science), and much of it involving first-hand experience that synthesised research and practice.

There were several aspects to this wisdom. We briefly note six here. The first was that it was possible to teach successfully for conceptual change, and that such teaching often resulted in better performance on conventional assessment as well as changes in student conceptual understanding, student engagement, class dynamics and classroom learning environments (Gunstone 2000). We had developed a range of teaching procedures designed to promote aspects of the thinking needed for conceptual change (eliciting, clarifying, challenging, reflecting, etc.) (e.g. Mitchell 2000; White and Gunstone 1992).

A second aspect of this wisdom relates to broader issues of learning and teaching—we developed considerable understanding about the influence of content *per se* on significant aspects of teaching and learning. For example, we have been aware for a long time that in many content areas it is not possible to develop an acceptable scientific view solely from experimental challenges to existing beliefs because central components of the concepts involved are not observable (e.g. the particle model, energy).

The third wisdom we note here, because it was crucial to our development of the Science Continuum, was the need to rethink key "ideas" in a topic in terms of science *learning*. Text-books and curriculum documents describe science ideas such as the particle model in ways that flow from the *logical structure* of the discipline, e.g. "tiny particles that are arranged differently in different states of matter". However, these descriptions are often inconsistent with our knowledge of *learning difficulties* for the idea. For the particle model, two significant examples are that the particles have quite different properties to very small bits of the (macroscopic) matter and that there is nothing in between particles—i.e. matter is not continuous. At its heart, this issue is a specific and fundamental example of the distinction drawn by Ausubel over four decades ago between *psychological meaning* and *logical meaning* (e.g. Ausubel 1968).

A fourth aspect of this wisdom emphasised the need to build qualitative explanations of ideas before introducing mathematical representations and manipulations (e.g. Gunstone and Mitchell 1997; Gunstone et al. 2005). This awareness had a strong effect on our initial selection of "Focus Ideas" for the Continuum; and explains why the Continuum occasionally contains explicit statements of an epistemological nature such as "science is a human and creative activity" (Science Continuum, level 5, The Work of Science). Other aspects of the thinking associated with quality learning were informed by research, for example, "poor learning tendencies" (Baird 1986) and the work of PEEL (Mitchell and Baird 1985). PEEL has a strong agenda of promoting metacognition; these metacognitive perspectives relate closely to dimensions of VELS (with "Managing Personal Learning" being a significant component of the "Personal Learning" domain in the VELS cross-curriculum perspectives).

A fifth aspect of wisdom focussed on the promotion of engagement of students in science learning, building high levels of understanding and expertise in teachers about student science learning and engagement, implications of this for teaching (Corrigan et al. 2008; Cooper 2008) and the development of new ways for science teachers to explore their teaching and their students' learning (Loughran and Berry 2006; Berry and Keast 2009).

The sixth aspect of wisdom involved research focussing on teacher learning and teacher change that has extended our exploration of the importance of content in other quite different contexts. These directions include exploring the relationship between content and teaching, most particularly through a consideration of pedagogical content knowledge (PCK) in science teachers (Loughran et al. 2006) and in pre-service chemistry teachers (Corrigan 2009). This research has given us insights into ways in which science content can be (and is) deconstructed and reconstructed by teachers. This in turn has assisted us to better understand how PCK can be promoted, developed and articulated by and for science teachers in ways that promote better teaching and professional learning among these teachers at different stages of their careers.

The breadth and depth of the research background of the CSTME science education team has made us sensitive to a range of pedagogical issues central to student learning and engagement which have implications for teaching science. Importantly, it has also enhanced our understandings of what areas/ideas of science content *can* be and should *not* be considered at particular school year levels. The development of the science continuum reflects a great deal of this wisdom. Of central significance for this chapter is that our collaborative development of the Continuum has gradually led us to the capacity to articulate aspects of this wisdom in more cohesive ways. It is this articulation that we seek to present in this chapter.

The Structure of the Science Continuum

The Science Continuum resource is structured around "Focus Ideas". Each Focus Idea consists of four components: (1) a brief account of relevant research about learning—the ideas and beliefs that students bring to the study of the Focus Idea ("Student everyday experiences"); (2) an acceptable (age-appropriate) account of the science central to the Focus Idea ("Scientific view"); (3) our view of the appropriate and specific purposes the teaching of the Focus Idea should aim to achieve ("Critical teaching ideas") and, (4) some pedagogical approaches for developing students' understanding of these critical teaching ideas ("Teaching activities"). The Science Continuum is at an open access website: http://www.education.vic.gov.au/studentlearning/teachingresources/science/scicontinuum/default.htm.

Our development of these four aspects of the Science Continuum has been significant in leading us to examine our own understandings of selected Focus Ideas in science, and then to consider how this process provides an elaboration of aspects of a knowledge base for expert science teaching. However the processes of first considering the most appropriate "Focus Ideas" for development and then creating the "Critical Teaching Ideas" for each of these Focus Ideas have been particularly important in leading us to re-examine our own understanding of science concepts. We have had to combine ideas and beliefs from research, practice and personal experiences in coming to an agreed position about the significant aspects associated with each Focus Idea. The process provides some possible insight into the elaboration of aspects of a knowledge base for expert science teaching.

Focus Ideas

There are five broad areas of science under which we have developed Science Continuum P-10 resources: Force & Motion, Living Things, Matter & Energy, Earth & Space and Science Skills. While these labels suggest the five areas would look like many science topics in the curriculum, the Focus Ideas within each are a small number of concepts and/or forms of relationship and/or a significant proposition intended to cumulate to an engaged understanding. It is not intended that the Focus Ideas collectively cover all aspects of the VELS Science standards. Rather Focus Ideas are conceptualised as examples of how different science ideas can be considered in ways that account for what students bring to the learning of the idea (alternative conceptions), why students may think like this (impact of everyday experiences), what might be scientifically acceptable thinking about this idea, how to focus one's teaching in ways that recognise what students bring and what teachers can do to challenge their students to shift their thinking to a more acceptable view.

In this chapter we illustrate our development of our own understandings by considering in detail the development of two Focus Ideas. The first is "The Work of Science" (for students aged 12–13 years; closely related other Focus Ideas are "Introducing scientific language" [age 5–7], "Doing Science authentically" [age 8–9], "Scientific models" [age 10–11], "Science and decision making" [age 14–15]). The second is "Electrostatics" (age 10–11; closely related other Focus Ideas are "Pushes and pulls" [age 5–7], "What is a force?" [age 8–9], "Forces without contact" [age 10–11], "Electric circuits" [age 12–13], "Making sense of voltage" [age 14–15]).

Student Everyday Experiences

Each Focus Idea begins with an account of what research has to tell us about relevant student alternative conceptions. The accounts are written with the intended audience (teachers at the grades level[s] of the Focus Idea) in mind. A small number of references to the research are given for those teachers with motivation and time to read them.

Scientific View

As with the Focus Ideas, this section is written with the intended audience of teachers at the year level for which the Focus Idea is constructed. The Scientific view is also written in ways that recognise important matters of both student and teacher understanding of the particular Focus Idea. Hence, we sometimes describe this component as "age-appropriate" statements about the acceptable scientific view.

Critical Teaching Ideas

These are indications of the ways we see the significant points on which teaching should focus, a "psychological" (i.e. seriously considers issues of students constructing meaning) rather than "logical" (presenting the scientific view as a *fait accompli*) analysis of the Focus Idea. The Critical Teaching Ideas also allow teachers to more clearly see ideas/teaching foci that should be revisited across Focus Ideas and levels if students are to understand an acceptable Scientific view. Usually, there is elaborating commentary following the Critical Teaching Ideas.

Teaching Activities

In each case these are examples of teaching activities that (a) reflect the Critical Teaching Ideas, (b) have, in simple terms, a focus on helping students move from "everyday experience" towards an acceptable scientific view and (c) are examples of one of 11 different "Pedagogical Purposes" that underpin every teaching activity in the Science Continuum. These pedagogical purposes represent what we believe to be some of the important characteristics of science, such as proposing possible explanations, testing competing ideas, refining explanations, looking for consistency across a range of situations and recognising that simple scientific explanations or classifications may not account for all phenomena (e.g. the concepts of dead/alive).

The Impact of the Development Process on Our Own Understandings

The collaborative approaches we have used in developing each Focus Idea in the Continuum have demanded that we examine our own individual understandings of both science and pedagogy. The form of the Continuum has also meant that in this developmental experience we have had to combine ideas and beliefs from research, practice and personal experiences in coming to an agreed position about each of the four sections that make up each Focus Idea. Hence we believe the process provides insight into the elaboration of aspects of a knowledge base for expert science teaching, and so now consider our own knowledge development for two quite different Focus Ideas.

The process of development of each Focus Idea first involved a smaller group of CSMTE science education academics producing drafts, then the further development to a final product via successive considerations by a larger CSMTE group, with a potential final additional modification in response to a critical reading by a staff member of DEECD (the client).

Below we give accounts of the development of two Focus Ideas. These accounts draw on hand-written records of discussions and successive electronic drafts, including track changes and inserted comments. We have deliberately chosen two different types of Focus Ideas. The first is centred around science skills while the second is a more familiar "content" area; more importantly for our purposes in this chapter, these two Focus Ideas represent quite different approaches to producing a similar product and hence gave rise to quite different developmental experiences and considerations for us.

Focus Idea: "The Work of Science" [for Grades 7/8] and the Development of Our Understanding

The development of the Focus Idea "The Work of Science" began with an initial brainstorming session involving the two authors of this chapter and a visiting Canadian science education academic (Professor Glen Aikenhead) who has written extensively in the broad area. The three of us brainstormed the questions "What was important for students to learn?" and "What do students already know [about 'The Work of Science']?" Our intent was to focus on relevant "big ideas" rather than on individual items of propositional knowledge.

Of course, in such an approach all of us had to make explicit our own thinking around this idea, in terms of all of our own subject matter knowledge, our pedagogical knowledge (particularly in terms of what we knew about students' ideas—both from research and our own experience), our pedagogical content knowledge (built up over time from teaching this idea to a range of people), and our views of how such ideas can be represented in the classroom by others through specific teaching approaches.

In participating in this initial brainstorm we found that we were making our private knowledge relevant to this Focus Idea quite public to two others also expert in the field. This process required each of us to feel both sufficient confidence in our ideas so as to advance them and sufficient trust in our two colleagues so as to take the risks inherent in doing this. There was strong recognition that the sharing of knowledge contributes to our individual understanding as well as to the group. In essence we were unpacking our own understanding of these ideas in order that our personal understanding would evolve, and so then would the "public" understanding represented by the three of us as a group.

What follows is an account of the experience of developing this Focus Idea for ultimate publication on an open website. The analysis of this process is framed by five of the knowledge domains identified by Shulman (1986, 1987):

1. Subject matter knowledge
2. Pedagogical knowledge
3. Pedagogical content knowledge
4. Knowledge of Context
5. Knowledge of Curriculum

The other two knowledge domains identified by Shulman—Knowledge of Students and Knowledge of Educational Purposes and Ends—are not used here. The Continuum is a resource for teachers, it is therefore conceptualised and constructed for teachers alone. While there is consideration of what students already know, of the everyday experiences they have likely already had, of our past experiences in teaching these ideas to specific children, etc., any knowledge of specific students in specific classrooms is not part of the resource. Hence "Knowledge of Students" is not specifically considered in our account here. Similarly, while knowledge of educational purposes and ends (which includes assessment of achievement) is considered in broad terms, this domain is not a focus of the resource. So, while aspects of these domains are present in the subsequent discussion, they are not considered in any depth.

After the initial brainstorming session the first few versions of this Focus Idea were generated by the first author of this chapter and discussed and further developed in conjunction with the second author. Later versions evolved from discussions involving both of us and a larger group of CSTME science educators. The discussion below draws on six versions of the Focus Idea. In order to help frame this discussion we have used the knowledge domain ideas of Shulman, and so include his descriptions to provide content for our thinking.

Subject Matter Knowledge (SMK)

> To think properly about content knowledge requires going beyond knowledge of the facts or concepts…it requires understanding of the structures of the subject matter…[which] include both the substantive and the syntactic structures…substantive structures are the variety of ways in which the basic concepts and principles of the discipline are organized to incorporate its facts. The syntactic structure of a discipline is the set of ways in which truth or falsehood, validity or invalidity, are established. (Shulman 1986, p. 9)

During the brainstorming session a number of areas of subject matter knowledge (SMK) were identified as important aspects of the Focus Idea "The Work of Science". These aspects included an understanding of the concept of evidence and its relationship to data, models, modelling and intuition and the nature of scientific investigation. Here only one of these, the concept of evidence and data, is explored to illustrate how our SMK developed.

An Understanding of the Concept of Evidence: Notes taken from this initial brainstorm (now called Version 1) do no more than just record the issue of "the concept of evidence", as though the concept was universally understood by all three involved. As it later transpired, such a shared understanding needs to be developed rather than assumed. Subsequent writing about evidence included the following two extracts from later versions:

> Scientists' work is focused on trying to improve their understandings of and explanations about the natural world. They construct their explanations based on evidence, by using data, imagination and logic, and the feedback on emerging explanations that comes from peer and public review…Scientists share some fundamental assumptions about evidence, data, and logic, but they differ in their talents, imagination, intuition and courage…Evidence is data that has been subjected to some form of validation so that is it possible to assign "weight" when coming to an overall judgement. For example, are some data more valid, reliable, etc. than others. In considering the "weight" of data it is important to consider the quality of the experiment, the conditions under which it was conducted, its reproducibility and the practicality of implementing outcomes of the evidence. (Version 2)

> The concept of "evidence" is complex and multi-faceted. Evidence and data are not equivalent terms. Data are directly recorded, often through observation. Not all data are equally valuable. For example, "outliers" are data that are regarded as less valuable (and less valid). Data that has occurred only once, generated from investigations that might be flawed or not able to be reproduced, are regarded as less reliable. Data (and sets of data from multiple sources) that are considered reliable and valid are regarded as evidence…When the evidence supports outcomes that are, for example, impractical or expensive to implement, a higher level is demanded of the quality (validity and reliability) of that evidence. (Version 5)

What became clear from our discussions around the concept of evidence is that a distinction needed to be made between evidence and data. And before this issue was considered, it was necessary to generate a need for evidence—"what is its purpose?" Hence, the ideas of generating knowledge and creating explanations for our natural world as necessary precursors to considering the idea of evidence were made explicit.

One of the crucial differences between evidence and data was the value judgements made about data in order to accept these as contributing to a body of evidence (for example, are data reliable and valid?). Differentiating between evidence and data highlights the important differences between the substantive structures of knowledge (e.g. establishing that there *is* a difference between evidence and data) and the syntactic structures of knowledge (which place value judgements on particular types of data which can be aggregated to form evidence). In Version 1, such a distinction was implied but not well articulated, via brief notes about Milliken's experiments and his log book, leading to discussion of how data are generated, as seen from the excerpts below.

> Milliken's log book—details that he ignored outliers and indicated that some data "are beautiful". The notion of what data are appropriate and what are not is a decision made by the scientists. (Version 1)

> The notion of pooling data—20 pieces of data are more like 20×20 pieces of data due to the effect of the experimenters. (Version 1)

What is apparent from these excerpts is the important role played by the context in which this content is applied. For example, in Version 1, a great deal of discussion centred around the role of personal experience on a scientist's work and the important role played by culture in determining what observations and subsequent inferences are seen as acceptable:

> Inference = observations + culture based past experiences and knowledge and sometimes creativity. Usually therefore a person cannot deviate from observation, and cannot deviate from past experience and knowledge unless you exercise unusual creativity. Thus inferences are guided by whatever a scientist already knows or believes. (Aikenhead, using the work of Einhoft—Version 1)

> Past experience and knowledge = ideas & metaphors & conventions found within a person's culture: e.g. culture of science, family culture. Therefore inferences in science are culture based. Each culture has a traditional way of describing and explaining nature (way of knowing nature). "Science" taught in conventional schools express a Eurocentric way of knowing nature. There are other credible ways: common-sense, neo-indigenous (e.g. Islamic science, Japanese science, etc.), indigenous, Eurocentric. (Aikenhead—Version 1)

> The notion of data and evidence is also dependent on culture, e.g. [the] public view of mean, average, median, etc. A concept of evidence is different in science and history and from a social perspective. When do you have enough evidence? (Version 1)

This discussion focussed on the importance of the cultural context in which science is situated. We saw this position as a significant reframing of our knowledge around evidence and the nature of the work of scientists. Hence, in Version 2, these perspectives were represented quite strongly in the scientific view, and incorporated some of the perspectives presented by Bybee (2006):

> Scientists' work is focused on trying to improve their understandings of and explanations about the natural world. They construct their explanations based on evidence, by using data, imagination and logic, and the feedback on emerging explanations that comes from peer and public review. Science is a human and creative activity. Scientists share some fundamental assumptions about evidence, data, and logic, but they differ in their talents, imagination, intuition and courage. Given the same results, two (groups of) scientists may form different explanations. (Bybee 2006, p. 3—Version 2)

This version persisted to the final form, with only very minor alterations in the order of presentation. Our strongly held view in writing this Focus Idea was to not only present the concepts of evidence and data, but also to indicate that it was critically important that our conceptual understandings of evidence and data recognised the cultural context in which these were situated. This view, again, highlights both the substantive structures (i.e. ways the concepts and principles of science are organised) and syntactic structures (i.e. the ways truth and falsehood, validity and invalidity are established) of science, but also recognises that in school science syntactic structures are almost always ignored.

In the final version of the Focus Idea, "The Work of Science", there were additional areas of SMK explored such as the nature of scientific investigation. These areas are not considered here. The areas are clear in the final form of this Focus Idea; this can be accessed at the website indicated above.

Pedagogical Knowledge (PK)

> …general pedagogical knowledge, with special reference to those broad principles and strategies of classroom management and organization that appear to transcend subject matter. (Shulman 1987, p. 8)

Shulman gave relatively little attention to pedagogical knowledge in his 1987 model, probably because of his focus on reinstating content as a critical facet in teacher knowledge. Morine-Dershimer and Kent (1999) have expanded on Shulman's original description of the domain of pedagogical knowledge to acknowledge the importance of classroom organisation and management, instructional models and strategies and classroom communication and discourse. They have also included another facet that they call "personal pedagogical knowledge", which they describe as incorporating personal beliefs and perceptions and personal practical experience. What is important in this facet is the role of reflection in promoting the interplay between general pedagogical knowledge (derived from research and scholarly literature) and personal pedagogical knowledge. Figure 6.1 presents Morine-Dershimer and Kent's diagrammatic representations of their ideas about pedagogical knowledge and pedagogical content knowledge (PCK). In this figure the authors of the present chapter have indicated, at least in simple terms, how they see these two Morine-Dershimer and Kent models linking together by inserting an additional (and much thicker) arrow. While this additional arrow is intended to both emphasise the interrelated but complex nature of the two originally separate models and capture Morine-Dershimer and Kent's notion that context-specific pedagogical knowledge is a precursor to pedagogical knowledge, this representation does over-simplify some connection between these two models. For example, it could be argued that some Facets of Pedagogical Knowledge (in particular "Classroom Management", "Instructional Models and Strategies" and "Classroom Discourse") are individually linked with some Categories contributing to Pedagogical Content Knowledge (in particular "Educational Ends etc.", "Assessment/Evaluation" and "Knowledge of Learners") as important components of pedagogical practice.

However, importantly, Morine-Dershimer and Kent (1999) have suggested that it is the development of "context specific pedagogical knowledge (CSPK) that helps to guide teachers' decisions and actions" (1992, p. 23). It (CSPK), they argue, provides the context for science teachers' considerations of educational ends, goals, purposes and values, as well as assessment procedures and evaluation of outcomes in conjunction with the knowledge of learners. More importantly here, these other factors are also argued to be important precursors to the development of pedagogical knowledge. In this sense, Morine-Dershimer and Kent's two models support the present authors' use of a single arrow to connect the two models (see Fig. 6.1). While this chapter adopts Morine-Dershimer and Kent's models as a way of fram-

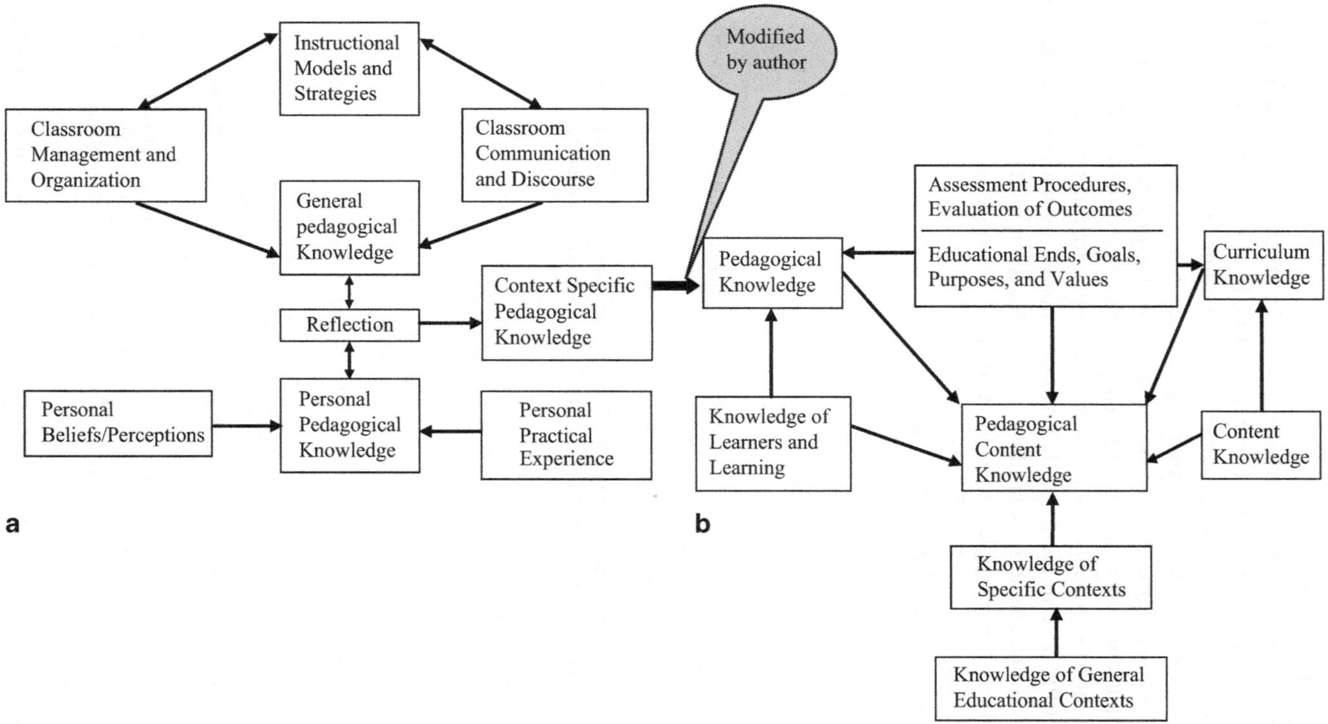

Fig. 6.1 Modified version of two models proposed by Morine-Dershimer and Kent (1999). **a** Facets of pedagogical knowledge. (Morine-Dershimer and Kent 1999, p. 23). **b** Categories contributing to Pedagogical Content Knowledge. (Morine-Dershimer and Kent 1999, p. 22)

ing issues, a more substantial discussion of these two models in comparison with other conceptualisations of PCK such as Magnusson et al. (1999) can be found elsewhere (Corrigan 2009).

In developing the Science Continuum, we have taken elements of both general and personal pedagogical knowledge into account. In our approaches, it was appropriate to draw on research about students' prior beliefs and everyday experiences and to take account of our own experiences of teaching the Focus Ideas (our own professional knowledge). Our use of the term PK (professional knowledge) here recognises that our considerations of general pedagogical knowledge necessarily occur in a context of science. While some may argue that this represents PCK, we use the term PK here to highlight that we are considering many of the things that generally confront teachers, regardless of the context in which they are teaching. As many other researchers have found in the past, we observe that the distinction between PK and PCK can be arbitrary.

In the Focus Idea, The Work of Science, the following general pedagogical knowledge was indentified in Version 1:

> High school science has nothing to do with real world science. (Aikenhead–Chin)
>
> Science is a list of facts and this is absolute, rather than creates answers. (Driver)
>
> Science is all about practical work? (Armstrong)
>
> Students think scientists are male, balding, have glasses and a beard and wear a white lab coat. (Fleer et al. p. 6; Skamp p. 54—Version 1).

These thoughts were further elaborated in Version 2:

> Many students believe that science is all about dramatic discovery, rather than discovery that involves repetitive work and the painstaking collection of evidence. (Skamp 2008, p. 55)
>
> When conducting science investigations students rarely think about the "thinking behind the doing" and the need for evidence that will be believable and acceptable to others. (Gott and Duggan 1995, http://www.dur.ac.uk/richard.gott/Evidence/cofed.htm)
>
> Students often view practical work in science as the means for verifying a fact or supporting a theory already presented in class. (Berry et al. 1999)
>
> Students frequently view school science as belonging only to the classroom. (Tasker 1991, p. 22)
>
> For most students, science content is not directly usable in science-related everyday situations (Chin et al. 2004). Consequently, most students see science as having little or no personal or social relevance. (Aikenhead 2006, p. 29) and they often see science as a "foreign culture".
>
> Added to this perception of lack of relevance is the persistent image that students have of scientists. Students often think scientists are male, balding, wear glasses, have a white laboratory coat, are manic looking and have scientific instruments in their pockets. (Schibeci 1987, p. 1; Skamp 2008, p. 54)

At this stage, the personal pedagogical knowledge of the authors also played an important role in modifying Version 2 into a synopsis of student everyday experiences that captured both general and personal pedagogical knowledge—

Students' experience of science at school can lead to a narrow,restricted and non-authentic view of science and the way it is done. Developing notions of who does science, where it is done and what it means to work in scientific ways is the important focus here.

Many students believe that science is all about dramatic discovery, rather than generating knowledge and creating explanations through repetitive work and the painstaking collection of evidence.

When conducting science investigations, students rarely think about the 'thinking behind the doing' and the need for evidence that will be believable and acceptable to others.

Students do not distinguish between data and evidence. Students often believe a single datum is all the evidence required to support a conclusion. Many school science experiences support this idea. For example conducting one experiment is often presented as proving Boyle's Law.

Students often view practical work in school science as the means for verifying a fact or supporting a theory already presented in class. Students frequently view school science as belonging only to the classroom, and science content as not directly usable in everyday situations. Adding to this perception of lack of relevance is the persistent image that students have of scientists as male, balding, wearing glasses and a white laboratory coat with scientific instruments in their pockets, and looking manic.

Research: Skamp (2008), Gott & Duggan (1995), Berry, Mulhall, Gunstone & Loughran (1999), Tasker (1991), Chin et al (2004), Aikenhead (2006) (Version 6).

Fig. 6.2 Final version of the Focus Idea: "The Work of Science" [for grades 7/8], as published on the Continuum website

what Morine-Dershimer and Kent (1999) term "context-specific pedagogical knowledge". In this particular case, the context is "teachers wanting to teach this idea in their own classroom". While the task is situated within the realm of teaching science, it is seen as PK because it is advice to teachers about the discourse and instructional strategies that they may employ in teaching this Focus Idea.

The final version (Version 6) of the Focus Idea that was published on the Continuum website shows further modification. The extract in Fig. 6.2 uses shading to highlight changes made as a result of us adding personal pedagogical knowledge (and PCK):

The interplay between our general pedagogical knowledge (our understanding of the research literature) and our own combined personal pedagogical knowledge via numerous discussions has added a robust dimension to our pedagogical knowledge for this Focus Idea. Through this collaborative process we were, it seems, developing the precursor to a form of collective pedagogical content knowledge, which is represented below. (Note: we do not intend any implication in this comment that PK is somehow a pre-requisite to PCK, only that for us in this particular task this is the mode of evolution of our ideas.)

Pedagogical Content Knowledge (PCK)

> ...the particular form of content knowledge that embodies the aspects of content most germane to its teachability...the ways of representing and formulating the subject that make it comprehensible to others...also includes an understanding of what makes the learning of specific topics easy or difficult: the conceptions and preconceptions that students of different ages and backgrounds bring with them to the learning of those most frequently taught topics and lessons. (Shulman 1986, p. 9)

It could be argued that the previous section on PK incorporated "the conceptions and preconceptions that students of different ages and background bring with them to the learning of those most frequently taught topics and lessons" (Shulman 1986, p. 9), and so might be seen as PCK. However we do not; these ideas are not included in this section on PCK as the context and purpose of this whole analysis is to provide advice for other science educators (researchers, teachers, etc.). Hence there are no "students" as such in this context. Rather, the authors are examining their own PCK for the public consumption of others. So in the context of developing the Science Continuum, PCK, in this instance, focusses on our (the authors') PCK and how it has changed within this context.

The representation of our pedagogical content knowledge lies in the "Critical Teaching Ideas" (CTIs), the elaborating comments to accompany these CTIs, and to some extent in the teaching approaches and attached pedagogical purposes articulated in the Focus Idea. Again this has been an iterative approach, with numerous versions of the CTIs, the refinement of any explanatory notes that have been included, and indeed the refinement and addition of teaching activities that draw out these CTIs. In order to articulate our PCK in this iterative process, our own understanding of the content and the pedagogical knowledge we have brought to these considerations have also been reframed. Importantly, the CTIs could not be developed until we had discussed and clarified our SMK and PK.

Our first attempt at the CTIs was:

> Common practices in science include modelling (physical modelling, hypothesizing, predicting, analyzing, sampling, control experiments).
> Scientific use of words such as model, modelling and intuition differ from the common day use of these words.
> There is a need for identification of scientists in our community. How do they use evidence? How do they use basic sampling procedures? (Perhaps this can be done by interview.)—(Version 1)

At this point, the CTIs were relatively naïve and did not reflect well some of the thinking that had gone it to our collective understandings of the SMK and PK. Again the rethinking involved in subsequent iterations led to an evolution of the CTIs to the final version:

> The purposes of scientific investigations are to generate knowledge and create explanations.
> Science investigations are conducted in multiple contexts and by a range of people.
> Scientific investigations are conducted in multiple ways that rely on the collection of a range of types of evidence. The ways an investigation can be validly conducted depend on the specific problem being investigated.

> The concept of what is evidence needs to be developed and should consider its credibility, acceptance, bias, status, appropriateness and reasonableness.
> Verification and reproducibility of investigations is another important process of science—one experiment is not enough. (Version 6)

Explanatory comments to accompany the CTIs included ideas about common practices in science and how these practices are integrated with scientific knowledge and contexts, what it means to investigate scientifically and different approaches that can be taken, and some explanation of the characteristics of "evidence". Links to others ideas about observations were also highlighted.

The teaching activities were designed to challenge existing student ideas, give practice using and building scientific models, help students to work out some scientific explanations for themselves and encourage students to identify phenomena not explained by the scientific model that is the focus of their learning.

In initially considering our SMK and PK, we came to examine and articulate our own PCK. Our shared common understanding and redefined personal SMK, PK and PCK enabled development of the final version of the Focus Idea. The iterative and reflective process has been critical to this development and articulation of a shared understanding in all Focus Ideas.

Curricular Knowledge

> …the full range of programs designed for the teaching of particular subjects and topics at a given level, the variety of instructional materials available…and the set of characteristics that serve as both indications and contraindications for the use of particular…materials in particular circumstances…In addition…I would expect a professional teacher to be familiar with the curriculum materials under study by his or her students in other subjects…[and] familiarity with the topics and issues that have been and will be taught in the same subject area during preceding and later years in school, and the materials that embody them. (Shulman 1986, p. 10)

At one level it may seem that the teaching activities outlined for each Focus Idea in the Continuum are "stand-alone" forms of curricular knowledge as defined by Shulman above. However, it is a significant dimension of the structure of the Continuum that these activities go beyond usual outlines of teaching approaches as there is explicit linking of these activities with a specific pedagogical purpose. This linking is designed to help elicit what the group of science educators who created the Continuum see as the PCK that needs to be appreciated by teachers if their students are to develop a rich understanding of the content knowledge. The interweaving of all of the ideas indicates the complex nature of teaching, however, by focussing on how these different knowledge domains interplay, significant gains can be made in developing the teachers' (or in our case science educators') own understanding of science.

The knowledge domain of context has not been considered specifically above. Nevertheless, it was an important component of our thinking throughout the development process. Given the demands of this context—in particular the ways that this form of writing for teachers demanded of us that we re-examine our own science understanding—then our SMK understanding in particular has been challenged and enhanced, an impact of substantial significance.

Our second example of a Focus Idea development is very different to the example of The Work of Science, in terms of both the nature of the content focus in the Idea and our approaches to the development. One very illuminating difference is in the different ways our SMK both shaped the development and was in turn shaped by the development as we began to question our initial certainty of understanding. Because of space limitations we give much less detail.

The Focus Idea "Electrostatics" [Grades 5/6] —and the Development of Our Understanding

This Focus Idea which was developed via a quite different sequence, yet again led to significant rethinking of our SMK and other categories of our knowledge. Our initial planning was to include the Focus Idea "Electrostatics" at grades 3/4. The first work was a written piece from the second author of this chapter rather than a brainstorm or other collaborative beginning point. This written beginning point then developed through several iterations via work with a different CSMTE academic, then the final versions evolved from discussions in the same group of academics as worked on latter versions of "The Work of Science".

The initial document contained four short sections. It began with the relevant extract from the intended curriculum (VELS Science Standards) headed "What is stated in Level 3 [grades 3/4] that is relevant", then noted content that therefore could not be part of the Electrostatics Focus Idea (including "atoms contain charged particles"). Then "a very first pass at a logical analysis of what could be pedagogically appropriate at this level" (a first draft of CTIs, not yet in "level-appropriate" language), then, separately, "student everyday experiences" and "alternative conceptions". Given the significance that we came to see for developing CTIs via a psychological rather than a logical approach (Ausubel 1968), such a beginning point reflects how this was early in our thinking.

At the end of this document was a description of a decision we needed to make:

> So—I think we have two possibilities we must make a decision between–
>
> 1. Do electrostatics at a higher level (maybe even level 5)
>   OR
> 2. Do electrostatics at level 3, and introduce the "undefined" notion of charge (i.e. just accept that "all atoms are composed of charges" does come later, and we work within that limitation)
>
> I favour 2 I think. (Version 1)

In terms of Shulman's knowledge domains, such a decision involves curricular knowledge as it is focussing on "the set of characteristics that serve as both indications and contraindications for the use of particular materials in particular circumstances" (1986, p. 10).

The logic of this first draft was to begin with VELS, then give some CTIs, then student experiences/alternative conceptions, and then a fundamental choice that was needed about the direction of the Focus Idea. There was very little concern with the content per se, and indeed no indication in the document of SMK issues. Indeed, the author of the initial draft was, at the time of writing this draft, quite confident of having a comprehensive and deep understanding.

The second draft had moved to the more familiar format. It began with "Student everyday experiences" (alternative conceptions), now including some references to research, then a draft of "Scientific view", and then the same CTIs as in Version 1, now reworked in more appropriate language. Also now in the CTIs was "an Important Note for Teachers" which asserted that it was not appropriate at this broad age to try to explain attraction arising from induced charges. This was the first recognition of content complexity (even difficulty) at this level.

The next substantively changed draft was headed "Complete Draft 1", perhaps a naïve suggestion on the part of the second author of this chapter given that the "Teaching Activities" section was incomplete and matters of content were just beginning to emerge as significant. Changes in the next substantively different draft included a query to the other academic regarding cell v battery ("should we discriminate?"), some additions in "Scientific view" about non-contact forces, a comment about links needed to the glossary for force, and a second "Important Note" after the CTIs (the need to avoid days with a lot of moisture in the air when teaching these ideas). More interesting for this present chapter is that a note to the other academic at the beginning of this draft includes "I am still not certain we can do this at this level—I have had to continually fudge and fiddle to avoid attraction via induced change—level is still to be discussed I think!!", and a Comment at the beginning of "Teaching Activities" that reads "The need to avoid anything that involves induced charge is a MAJOR problem for teaching activities at this level [all the standard stuff like picking up paper scraps with a rubbed comb is out]—what I have in this section needs more additional thought and suggestion than I had hoped—sorry!!" (And the next small iteration of this draft was no longer headed "Complete Draft 1".)

What is emerging from the process at this point is that as well as issues associated with pedagogical considerations about what strategies to use and the language that is appropriate in providing advice for the teaching of this focus idea (PK), there are also the beginnings of the academics rethinking their own SMK. In attempting to articulate their PCK (through the CTIs) and their PK (through the Teaching Activities), they have realised their SMK is at least a little less certain (and perhaps more confused) than they had believed. This realisation is not only relevant to consideration of the appropriate student age/level for this Focus Idea, it is much more significant for the present context as it clearly illustrates the underpinning significance of both Shulman's categories of SMK and curricular knowledge.

In the development of this Focus Idea there now followed discussions between the two academics leading to two more (and small) iterations, before another substantive shift. The iterations and the major shift are now summarised together. All the substantive changes being summarised result from a decision that the Focus Idea would have to include issues relating to induction. This decision shows that the academics were beginning to grapple with some of the SMK issues that had arisen for them in this iterative process.

In the changes "Student experiences" had had some minor editing, and the addition of a paragraph beginning "For many students, the most common experience of the effect of electrical charges is seeing lightning and hearing thunder" and describing common student (and adult) experiences of and beliefs about thunder and lightning. Such a shift indicates consideration of PK issues associated with the classroom discourse that might occur in teaching this idea. "The Scientific view", representing the SMK of the academics, now included statements about the unknown nature of charge, that the terms "positive" and "negative" should not be conceptually linked with arithmetic operations, and an attempt to explain lightning and thunder that was recognised as being much too long. CTIs now included one relating to action-at-a-distance (two charged objects exerting forces on each other without touching), extended commentary about issues relating to teaching this topic at this level, and some common examples of observations that are explained by induction. In this sense, the PCK of the academics had been articulated more explicitly than in previous drafts, due largely to the reconsideration of SMK. "Teaching Activities" now included some POEs (Predict-Observe-Explain) relating to induction phenomena, and a number of websites for exploring the origins of lightning and thunder. This is a pedagogical response to the authors' changed understanding in SMK.

There then followed a number of drafts in which there were reductions in length (many arising from dramatic reductions of the amount written about lightening in both "Student everyday experiences" and "The scientific view") which resulted from progressively greater clarity in SMK for the academics, considerable copy editing and a number of additional "Teaching Activities", again as pedagogical responses to this changed/clarified SMK. The issue of induction was eventually resolved in the *final* version of the Focus Idea, by shifting it to the next level (grades 5/6) where the key knowledge for thinking about induced charge (that "atoms contain charged particles") was part of the intended VELS curriculum, again highlighting the important role of curricular knowledge as a knowledge domain.

Conclusion

In developing the Science Continuum P-10, the CSMTE academics have all learnt much more from the experience, and about different things, than we expected. The kinds of questions we asked in our development of the Continuum resource and the order in which we asked them reflect an approach to curriculum design that was different to that often used by curriculum developers and teachers in planning. However, the significantly longer time that it took us, a group with very high levels of expertise, to develop each Continuum Focus Idea indicates that we underestimated some of the complexities of designing teaching that is highly responsive to student learning. In part (and only in part), this was another example of the difficulties of recognising and articulating tacit knowledge in that there were sometimes significant aspects of our own classroom experience that we were drawing on that we could not articulate well. More significantly, it is clear we needed to think much more carefully about the issues of content, teaching and learning and their interactions than we had initially expected.

One reason why the CSMTE group participated in this project was that it intended to produce a different kind of resource that connected research and practice in ways potentially facilitated by an electronic environment. The resource has features that are different (at least in emphasis and in some cases in conceptualisation) from earlier resources. The most important differences lie in the purposes and nature of each the four sections of each Focus Idea and the ways the deep links between these sections lead to an integrated whole that is each final Focus Idea. The resource not only exposes teachers to common student conceptions and supports them with relevant acceptable science, it also suggests significant rethinking of key ideas and links teaching advice to this rethink. While this resource may share some similarities with others, such as the CLIS materials (Driver and Oldham 1986), it also differs significantly in three main ways: the Focus Ideas are each an integrated whole that intends to validly represent the rich interconnections within each ideas; the "Critical Teaching Ideas" and the reshaping of the teaching focus these led to are substantially different to the more conventional "aims" that are usual; the explicit Pedagogical Purposes that underpin the Teaching Activities describe the type of thinking required for conceptual learning and the teaching needed to achieve such thinking. In the complexities of considering these different features and their interactions, we were also consistently confronted with the considerable extent to which school science ignores the syntactic structures of science. At times this also provoked us to reassess the extent to which our own science understanding embraced syntactic structures. From the perspective of the Victorian Department of Education and Early Childhood Development (who contracted the CSMTE group to create and develop the Science Continuum), a key difference from past teacher resources is that the Continuum resource gives teachers advice which is contestable—or at the very least invites a range of reactions; previous Department resources allowed little room for contestation.

The last section of the previous paragraph noted matters about the Continuum in terms of *product.* Of greater importance for the purposes of this chapter are matters of *process*. The analysis of our thinking presented in this chapter has highlighted some important processes that, at least for us, have not previously been well articulated. Our choice of these particular two specific Focus Ideas to illustrate something of these processes was of course substantially influenced by the rather differing matters highlighted by each.

At one level, each Focus Idea had a different beginning point. For "The Work of Science" we began by brainstorming the questions "What was important for students to learn?" and "What do students already know?" For "Electrostatics" our beginning was a written statement focussed on the relevant intended curriculum. However, these two different approaches both represented the same broad process—the need to clarify personal ideas in order to advance them as "public knowledge" to a (small) group of peers. There was certainly a stronger need for confidence (even "certainty") in committing to paper the public knowledge that was the beginning for "Electrostatics" (the written document) and a lesser (but still real) demand for confidence for the interactions with two other experts that we each undertook in the initial brainstorm for "The Work of Science". The assertion of greater certainty for the "Electrostatics" Focus Idea is clearly supported by the written document beginning with "what does the curriculum say we have to teach?" That is, the uncertainty of "what" and "how" to teach that were central aspects of why we began with brainstorming for "The work of Science" were not initially of any concern for "Electrostatics"; indeed, the author of the initial drafts of "Electrostatics" (the second author of this chapter) initially saw the "what" and "how" to teach as quite unproblematic.

The issue of significance here with "Electrostatics" is that the process changed our SMK during the development of this Focus Idea, and that this SMK change came through the progressive and iterative interactions of content/learning/teaching for a particular and appropriate age-level. One way of describing this process might be that our SMK was affected by a change in our age-specific PCK. More importantly, however, the change illustrates well that the links between these two constructs (SMK and PCK) are two-way and interactive and iterative, not unidirectional (and certainly not as is sometimes implied unidirectionally causal). We believe it is clear that this complexity of relationships is broadly the same for all the other relationships in the various schematic representations of teacher knowledge that have been argued in the literature. That is, however one wishes to conceptualise teacher knowledge, the components that are argued to exist will be interrelated in ways that are two-way and interactive and iterative; these relationships are dynamic.

Similarly, in the development of "The Work of Science" Focus Idea, there was constant reframing of both our individual ("private"), and collective ("public") knowledge in the domains of SMK, PK and PCK; these were dynamically changing through the iterative process we engaged in as described above. For example, in our collaborative considerations of our collective PCK, the contestations about SMK and PK provoked each of us to reframe our individual SMK and/or PK. The significance of our collaborations and discussions occurring in the environment of professional trust and challenge as they did cannot be overstated. The complete sense of trust in which we worked was central to this dynamic reframing. Indeed, such trust is very commonly a fundamental requirement for any learning that involves deep personal challenge (something that most learning of significance involves).

In many models created to represent the range of knowledge domains considered in this chapter (e.g. Magnusson et al. 1999; Morine Dershimer and Kent 1999) the relationships between the domains are often represented as hierarchical or unidirectional. We contend that we need to rethink the relationships as dynamic and that until we do we will struggle to elaborate these knowledge domains in ways that allow examination of their influence on expert science teaching for promoting student understanding. The processes outlined here highlight the importance of the reframing of these domains in an iterative process that demands articulation of the domains so that they become explicit rather than implicit. The iterative nature of the process also promoted the challenge of ideas in dynamic ways so that the interplay between these domains becomes central to the reframing process.

This reflective analysis of our professional understandings does not consider the issue of whether such a process translates for use by other science education experts (including teachers), an issue of obvious importance and one that merits further investigation. An important aspect of such further investigation may be to explore how science teachers have engaged with ideas presented in the Science Continuum and whether or not they appreciate the importance of making ideas explicit in such a public way. It is also through exploring the use of the Science Continuum and engaging science teachers in such an iterative process that insights into the other knowledge domains which were not a focus in this discussion may also be explored.

References

AAAS, Project 2061. (2001). *Atlas of scientific literacy*. Washington: American Association for the Advancement of Science.

AAAS, Project 2061. (2007). *Atlas of scientific literacy* (Vol. 2). Washington: American Association for the Advancement of Science.

AAAS. (n.d.). Atlas of Science Literacy (Vols. 1–2: Mapping K–12 science learning). http://www.project2061.org/publications/atlas/default.htm. Accessed 8 April 2009.

Ausubel, D. P. (1968). *Educational psychology: A cognitive view*. New York: Holt, Rinehart & Winston.

Baird, J. R. (1986). Improved learning through enhanced metacognition: A classroom study. *European Journal of Science Education, 8*, 263–282.

Berry, A., & Keast, S. (2009). *Looking into practice—Cases of science teachers' professional growth*. Melbourne: Monash Print Services.

Bybee, R. (2006). Scientific inquiry and science teaching. In L. B. Flick & N. G. Lederman (Eds.), *Scientific inquiry and nature of science: Implications for teaching, learning and teacher education* (pp. 1–14). Springer: Dordrecht.

Cooper, R. (2008, September). *Big ideas in science*. Paper presented at the XIIIth Symposium of the International Organization of Science and Technology Education, Kusadasi Turkey, 382–387.

Corrigan, D. J. (2009). Chemistry teacher education to promote understanding of learning through effective reflective practice. *Chemistry Education Research and Practice, 10*, 14–23.

Corrigan, D., Mitchell, I., & Lancaster, G. (2008, September). *Sustaining a disperse network of science teacher researchers*. Paper presented at the XIIIth Symposium of the International Organization of Science and Technology Education, Kusadasi Turkey, 171–178.

Driver, R., & Oldham, V. (1986). A constructivist approach to curriculum development. *Studies in Science Education, 13*, 105–122.

Gunstone, R. F. (2000). Constructivism in the classroom. In D. C. Philips (Ed.), *Constructivism in education: Opinions and second opinions on controversial issues* (99th Yearbook of the National Society for the Study of Education Part 1) (pp. 254–280). Chicago: University of Chicago Press.

Gunstone, R. F., & Mitchell, I. J. (1997). Metacognition and conceptual change. In J. J. Mintzes, J. H. Wandersee, & J. D. Novak (Eds.), *Teaching science for understanding* (pp. 133–163). San Diego: Academic Press.

Gunstone, R. F., & Northfield, J. R. (1988, April). *Inservice education: Some constructivist perspectives and examples*. Paper given at the meeting of the American Educational Research Association, New Orleans. (20 pp.) (ED 292 787)

Gunstone, R. F., White, R. T., & Fensham, P. J. (1988). Developments in style and purpose of research on the learning of science. *Journal of Research in Science Teaching, 25*, 513–529.

Gunstone, R., McKittrick, B., & Mulhall, P. (2005). Textbooks and their authors: Another perspective on the difficulties of teaching and learning electricity. In K. Boersma, M. Goedhart, O. de Jong, & H. Eijkelhof (Eds.), *Research and the quality of science education* (pp. 435–445). Dordrecht: Springer.

Isaacs, M., Corrigan, D., & Mitchell, I. (2008, July). *Science curriculum resources: the interface between policy, research and practice.* Paper given at the annual conference of the Australasian Science Education Research Association conference, Brisbane.

Loughran, J., & Berry, A. (2006). *Looking into practice: Cases of science teaching and learning* (Vol. 1). Melbourne: Monash Print Services.

Loughran, J., Milroy, P., Berry, A., Gunstone, R., & Mulhall, P. (2001). Science cases in action: Documenting science teachers' pedagogical content knowledge. *Research in Science Education, 31*, 289–307.

Loughran, J., Berry, A., & Mulhall, P. (2006). *Understanding and developing science teachers' pedagogical content knowledge*. Rotterdam: Sense.

Magnusson, S., Krajcik, J., & Borko, H. (1999). Nature, sources and development of pedagogical content knowledge for science teaching. In J. Gess-Newsome & N. G. Lederman (Eds.). *Examining pedagogical content knowledge: The construct and its implications for science education* (pp. 95–132). Dordrecht: Kluwer.

Mitchell, I. J. (2000). Returning to the discussion. *PEEL SEEDS, 49*, 10–11.

Mitchell, I. J., & Baird, J. R. (1985). A school-based, multi-faculty, action-research project to encourage metacognitive behaviour. *Research in Science Education, 15*, 37–43.

Morine-Dershimer, D., & Kent, T. (1999). The complex nature and sources of teachers' pedagogical knowledge. In J. Gess-Newsome & N. G. Lederman (Eds.), *Examining pedagogical content knowledge: The construct and its implications for science education* (pp. 21–50). Dordrecht: Kluwer.

Shulman, L. S. (1986). Those who understand: Knowledge growth in teaching. *Educational Researcher, 15*(2), 4–14.

Shulman, L. S. (1987). Knowledge and teaching: Foundations of the new reform. *Harvard Educational Review, 57*, 1–22.

White, R. T., & Gunstone, R. F. (1992). *Probing understanding*. London: Falmer.

Chapter 7
Towards a Cultural View on Quality Science Teaching

Glen Aikenhead

Introduction

To judge the quality of science teaching in grades 6–12 requires, at the very least, evidence that encompasses:

- What do teachers teach and not teach (the intended, enacted, hidden, and null curricula)?
- What self-identities are formed or strengthened in students (a foundation for meaningful learning)?
- What can students do with what they have learned (knowledge/procedures/insights-in-action)?
- What identities, belief systems, and knowledge-in-action have teachers developed and embraced?
- What do teachers do inside and outside the classroom (pedagogy and classroom culture)?
- What relationships and responsibilities are established between teachers and students (classroom culture)?

This chapter primarily explores the first three points above, from a cultural perspective, and gives special attention to the intended, enacted, hidden, and null curricula, because collectively these constitute a contentious issue within science education today. The term "null curriculum" refers to what is not taught, but could be (Hildebrand 2007).

Evidence of quality science teaching requires concomitant *indicators* of quality science teaching. These indicators will guide the collection and interpretation of data that could reflect the extent to which quality science teaching may be occurring. This chapter develops 15 such indicators. They are formulated on the basis of *educational soundness*.

But educational soundness often conflicts with *political reality* in science education (Aikenhead 2006). For example, Barnes and Barnes (2005, p. 62) concluded,

G. Aikenhead (✉)
University of Saskatchewan, Saskatoon, Canada

D. Corrigan et al. (eds.), *The Professional Knowledge Base of Science Teaching*,
DOI 10.1007/978-90-481-3927-9_7,

"inquiry-focused science instruction, deemed foundational to science teaching…is problematic to implement, even for master teachers". These researchers found that the educational soundness of guided inquiry conflicted with the political reality of a belief system about school science held by many teachers.

In general, this belief system seems to revere the memorization of facts, abstractions, and algorithms that, according to Alberts (2000), "would seem to make preparation for life nearly indistinguishable with the preparation for a quiz show, or the game of trivial pursuit" (p. 3). This outcome tends to occur, for instance, as a result of political pressure to follow a traditional science curriculum and to prepare students for high-stakes assessments and university courses. It would be politically expedient, therefore, to favour memorization over inquiry-focused instruction. Malcolm (2007) clarified a major part of the problem:

> It is an extraordinary political achievement for scientists, science educators and teachers that in spite of reforms over the past 40 years such as the child as scientist, learner-centred education, problem-based learning, STS [science-technology-society], constructivism, outcomes-based education, critical competences and assessment standards, science in schools remains positivist and authoritarian, with details of content and sequence so widely accepted—and so fixed in time—that international tests such as TIMSS (and its predecessors over 30 years) are possible. (p. 71)

Malcolm did not blame policy documents and government education departments, but instead he blamed "the *culture of science education* that we as science educators, scientists and teachers have produced" (p. 72, emphasis added). Therefore, the task of identifying key indicators of quality science teaching requires a *cultural perspective* on science teaching.

In the real world of school science culture, political expediency invariably outranks educational soundness (Aikenhead 2006). Yet, no matter how influential this political expediency may be, it is not an indicator of quality science teaching. Quite the contrary.

Along similar lines of reasoning, we can establish other indicators of "what quality science teaching *is not*". Consequently, the task of developing indicators of quality science teaching includes the identification of indicators that represent major failures in current practice, indicators that need to be avoided. Accordingly, this chapter begins by identifying five general consequences of traditional science teaching that define indicators of "what quality science teaching *is not*". The contribution of university science and engineering departments to these major failures is discussed.

Failures in current practice lead *prima facie* to a critical question: "Why would students want to engage in school science in the first place?" Guided by this fundamental question, the chapter next explores a culture of school science (grades 6–12) that differs from traditional school science because it is designed to avoid "what quality science teaching *is not*". Such a culture of school science harmonizes with the everyday world of science and technology imbedded in local, national, and global communities; in which people are employed in science-related occupations (e.g. technicians, medical personnel, entrepreneurs, industrial managers, scientists, and engineers), and in which savvy citizens deal with science-related events or issues.

In my exploration of this alternative to the culture of traditional school science, I focus primarily on curriculum content in order to derive nine indicators of quality science teaching for the vast majority of students.

What Quality Science Teaching Is Not

Most students do not identify with the ideology of a scientist-oriented, academically focused, school science program in which students are expected to think like a scientist and to believe what scientists are purported to believe (Eisenhart et al. 1996). This traditional school science has failed in terms of educational soundness in a number of ways, all documented by considerable research (Aikenhead 2006). Traditional school science has succeeded politically, however, as evidenced by its current high status in education (Rennie Chap. 2). Five major failures define indicators of "what quality science teaching is not", in addition to the indicator of political expediency.

Declining Student Interest and Enrolment

While students generally continue to value science in their world outside of school, the chronic decline in interest and enrolment in school science and tertiary education is alarming (e.g. Fensham 2007; Schreiner and Sjøberg 2007). Osborne and Dillon (2008) recently lamented:

> The irony of the current situation is that somehow we have managed to transform a school subject which engages nearly all young people in primary schools, and which many would argue is the crowning intellectual achievement of European society, into one which the majority find alienating by the time they leave school. (p. 27)

Most students (including some science-proficient students) claim: (a) transmissive pedagogy is boring, which makes school science unchallenging; (b) decontextualized science content is irrelevant to their lives, now and in the future; (c) school science dismisses the legitimacy of students' life-worlds and career goals, which makes science classes both irrelevant and impersonal; (d) school science lacks creative thinking and individual expression, which makes it sterile; and (e) the topics studied are intellectually too difficult, which makes science classes frustrating (Aikenhead 2006). This viewpoint was more or less verified by prominent citizens listed in *Who's Who in Australia*, who rejected the vapid knowledge accumulation they associated with their own school science experiences (Symington and Tytler 2004). We cannot blame the foolishness of youth for their negative views of traditional science teaching.

Reasons behind students' decisions on whether or not to take more science courses are complex and idiosyncratic. Based on a 3-year longitudinal in-depth study of a cross-section of secondary students, Cleaves (2005) concluded, "There is an interplay of self-perception with respect to science, occupational image of working scientists, relationship with significant adults and perceptions of school science" (p. 471). An OECD (2006) policy study reported, "Student choices are mostly determined by their image of S&T professions, the content of S&T curricula and the quality of teaching" (p. 2). Lyons (2006) gathered data on science-proficient year 10 students; particularly data concerning their interactions with family, peers, the mass media, and school science. On the one hand, Lyons noted, "…the students' decisions involved the complex negotiation of a number of cultural characteristics with their school science and family worlds" (p. 285). But on the other hand, he concluded, "…the most cogent single force acting against the choice of physical science courses was the culture of school science itself. …[I]nstead of considering why clever students are no longer taking science courses, it may be more pertinent to ask, '*Why should they?*'" (p. 308, emphasis added). In addition, all of these studies found that many students choose to enrol in science courses for reasons of political expediency alone; to acquire credentials needed for higher education. Enrolment inspired by political expediency, however, does not indicate quality science teaching.

Although science educators and teachers do not have control over all factors that influence each student's decision, a decrease in student interest and enrolment in science is certainly a clear indicator of "what quality science teaching is not".

Because the culture of school science has become a focal point in this chapter for understanding what is amiss in science education, I need to consider the key role played by university science and engineering departments in formulating this school culture. Fensham (1993) described three ways by which this occurs. First, as professors enculturate prospective science teachers into scientific disciplines, they teach them *about* science and they convey values held by their department. "All newcomers must display the attitude of subservience incumbent to their position and, with this, demonstrate not only that they subscribe to the game but also possess practical knowledge of the implicit rules of the game in which they intend to play" (Larochelle 2007, p. 713). In short, professors influence prospective teachers' formation of a science-identity. Second, science and engineering professors usually insist that the high school science curriculum be a simplified version of their introductory undergraduate courses, thereby establishing for schools both the goal of preprofessional training and the nature of student assessment. And third, science and engineering departments require certain science credentials for university entrance, thereby establishing a gate-keeping role for schools.

One critical statistic needs to be highlighted: The drop-out rate for students enrolled in university undergraduate science and engineering programs in the United States is about *twice* that of high school science programs (Frederick 1991). Thus, the rate-determining step (to borrow a chemistry metaphor) towards gaining a science or engineering degree occurs at the university undergraduate level. Therefore, any problem associated with insufficient science personnel for business and industry to compete globally rests at the feet of university science and engineering departments (Tobias 1990). An insufficient number of science workers is not a problem for high schools to solve at the present time.

It is instructive to read the research into "Why is it that science departments are unable to attract and retain those who are the most qualified to excel in science education and related professions?" (Adamuti-Trache and Andres 2008, p. 1578). The answer deals with the practices, attitudes, or culture of university departments that harbour an *entitlement* to accept and to discourage students; rather than the responsibility to attract and retain students. Some exceptions exist, of course. But generally speaking, universities need to enact quality science or engineering programs that do not dissuade or marginalize qualified students from finishing their program. According to a number of research studies, university success depends, in part, on students being able to "negotiate a culture characterized by white, masculine values and behavioral norms, hidden within an ideology of meritocracy" (Carlone and Johnson 2007, p. 1187). As a result, Johnson (2007) concludes, "Without the need to impute ill will, prejudice, or discrimination, Black, Latina, and American Indian women are being disadvantaged" (p. 818). These and other research findings at the university level (see also Brandt 2008; Daniell 2006; Malone and Barabino 2009; for example) describe a non-encouraging culture in many university science and engineering departments where future science teachers hone their self-identities and belief systems, as well as forge allegiances to the culture of academic science.

As a consequence, the culture of university science and engineering departments—a culture that tends to discourage certain groups of students—becomes reproduced as the culture of school science. The net effect is a declining interest and enrolment in school science, and an ethos of discrimination and alienation of students who historically have been marginalized in school science.

Discrimination and Alienation of Students

School science continues to privilege White middle-class males (Bianchini et al. 2000). People who belong to certain cultures, subcultures, or socio-economic groups are significantly underrepresented in university science and engineering programs and related careers. Problems in equity and social justice are experienced by visible minorities, by women, and by economically depressed groups (Brotman and Moore 2008). Discriminatory goals in school science emerged unexpectedly from research that revealed systemic exclusion of adolescents (no matter their gender or ethnicity) who existed outside the cultural power structures that sustain schooling and traditional school science (Tobin et al. 1999).

Much more subtle is the alienation of White male students of Euro-American ancestry whose worldviews differ, to varying degrees, from the worldview conveyed by traditional school science. (This type of alienation equally applies to the marginalized students listed just above.) Science, these students tend to say, is like a foreign culture. This happens in spite of supportive influences on student learning. The research on this issue has been synthesized as follows by Aikenhead (2006, supporting citations are omitted):

> Discordant worldviews create an incompatibility between, on the one hand, students' self-identities (e.g. who they are, where they have been, where they are going, and who they want to become) and, on the other hand:
>
> - students' views of Western science, school science, or their science teacher, and
> - students' views of the kind of person they think they must become in order to engage in science.
>
> Students who do not feel comfortable taking on a school science identity (i.e., being able to talk, think, and believe like a scientist) represent the vast majority of any student population. (pp. 107–108)

Alienation from school science does not motivate students to learn much from their science teachers.

A parallel conclusion was reached in Scott et al.'s (2007) review of research into learning science content. They considered epistemological differences between scientific and everyday ways of thinking (e.g. generalizable models versus context-specific ideas) and ontological differences (e.g. energy as a mathematical tool versus energy as a concrete entity). They concluded:

> Learning science involves coming to terms with the conceptual tools and associated epistemology and ontology of the scientific social language. If the differences between scientific and everyday ways of reasoning are great, then the topic in question appears difficult to learn (and to teach). (p. 49)

Alternatives to alienation do exist. For instance, Medina-Jerez's (2008) research showed that what matters is "the acknowledgement of cultural differences in the classroom that provides the needed attention to each student in coping with his/her strengths and weaknesses as they feel integrated into the cross-cultural scenario of the classroom" (p. 209).

Science teaching that ignores cultural alienation and does not acknowledge cultural differences (including epistemological, ontological, and axiological differences) is an indicator of "what quality science teaching is not".

Failure to Learn Content Meaningfully

One indicator of "what quality science teaching is not" certainly must include students' failure to learn science content meaningfully. What does meaningful learning mean? In the context of exploring alternatives to school science as simply transmission and remembering, Berry et al. (2007) characterize meaningful learning as "an effective grasp of the intended content knowledge in personally meaningful ways" (p. 151); in other words, the integration of science content into students' everyday thinking. Similarly, Scott et al. (2007) recognize that students must start to read, talk, and think "with the scientific social language(s) if they are to engage with [scientific conceptual tools and semiotic resources] meaningfully" (p. 50). Meaningful learning is *more than* remembering scientific content; it is using that content effectively when a pertinent situation arises.

When I reviewed decades of research into students' learning academic science, I was led to one conclusion: (Aikenhead 2006; citations omitted; emphasis in the original):

> *Most students tend not to learn science content meaningfully*. …Many research programs in science education have attempted in different ways to solve this lack of meaningful learning. However, even for students preparing for science-related careers (e.g., nursing), very few of them integrate science curriculum content into their own thinking when employed in science-rich workplaces, and this ability tends to be unrelated to their success at passing science courses. A corpus of research suggests that learning canonical science content meaningfully is simply not achievable for the vast majority of students in the context of traditional school science. (pp. 27–28)

This conclusion was recently supported by a 10-year longitudinal study in which only 20% of the participants achieved meaningful learning of the molecule concept (Löfgren and Helldén 2009).

Occurrences of meaningfully learning, however, can readily be found in out-of-school contexts where people have a personal reason for using it in their everyday world (i.e. knowledge-in-action) and are given sufficient time to master it (Aikenhead 2006; Rennie Chap. 2).

Of course, meaningful learning in traditional school science tends to be achieved by science-proficient students whose worldviews generally harmonize with a worldview endemic to academic science (Aikenhead 2006) and "who would want to understand the world scientifically" (Brotman and Moore 2008, p. 992). Scientists and engineers, science educators, and science teachers were this kind of student when they went to school. For instance, they enjoyed learning decontextualized content purported to be value-free. This small subgroup of students will respond well to scientist-oriented innovations in classrooms (e.g. teaching aimed at constructivist-types of learning academic science content). Their pre-test/post-test gain scores are usually large enough to cause the entire classroom mean to increase in statistically significant ways; thus conveying a sense of success for the whole class. Therefore, statistically significant higher mean scores do not necessarily indicate that a majority of students have achieved meaningful learning. As a consequence, such evidence fails as a valid indicator of quality science teaching.

For the vast majority of students, however, little or no meaningful learning usually occurs. Von Aufschnaiter et al. (2008) offer one avenue of explanation:

> Thus, from the current status of research, it can be concluded that major issues of the processes by which conceptual development takes place are still theoretically and empirically vague. Generally, we cannot yet explain in detail why teaching strategies that attempt to promote conceptual change are often unsuccessful. …It would be necessary to trace how students create meaning out of the learning experiences they are offered and how they deploy their own knowledge and understanding in tasks and problems. (p. 104)
>
> …
>
> We should stress that it is not so much that we believe that no learning of new knowledge has occurred here, but rather that learning is a slow and gradual process and a product of extended interaction and reflection. (p. 127)

Out-of-school contexts often provide students and adults with extended interaction and reflection (Rennie 2007; Chap. 16).

Another avenue that explains the paucity of meaningful learning in traditional school science deals with students' failure to form or strengthen a science identity (Carlone 2004). "We need to consider how learning science can change students' identities by changing their ability to participate in the world" (Brickhouse 2001, p. 288). In other words, to learn science meaningfully is to engage in identity work. Some science educators recommend that teachers foster positive school science identities by getting students to talk and think like scientists (Brown et al. 2005). If students begin to talk and think like scientists, then others will identify them as competent science students. However, clashes between students' self-identities and a school science identity can cause many students to feel alienated, and consequently they resist forming an academic science identity; that is, they resist meaningful learning. Rather than becoming scientifically literate, they become scientifically indifferent.

Forgeries of Meaningful Learning

As mentioned above, political expediency encourages many students to pass science courses to acquire credentials for post-secondary opportunities. "Empirical evidence demonstrates how students and many teachers react to being placed in the political position of having to play school games to make it appear as if significant science learning has occurred even though it has not" (Aikenhead 2006, p. 28). For example, Loughran and Derry (1997) investigated students' reactions to a science teacher's concerted effort to teach for "deep understanding" (meaningful learning).

> The need to develop a deep understanding of the subject may not have been viewed by [the students] as being particularly important as progression through the schooling system could be achieved without it. In this case such a view appears to have been very well reinforced by Year 9. This is not to suggest that these students were poor learners, but rather that they had learnt how to learn sufficiently well to succeed in school without expending excessive time or effort. (p. 935)

Their teacher lamented, "No matter how well I think I teach a topic, the students only seem to learn what they need to pass the test, then, after the test, they forget it all anyway" (p. 925).

This age-old problem was systematically studied by Larson (1995) when conducting research into students' unintended learning. Students in a high school chemistry class told her the rules they followed to pass Mr. London's chemistry class without really understanding much chemistry. Larson called these rules "Fatima's rules", her pseudonym for the most articulate informant in the class. Two simple rules are rote memorization and going through the motions of learning without being intellectually engaged. The nemesis of meaningful learning is rote memorization in which one's competence at superficial communication replaces meaningful learning. Fatima's rules can include such coping or passive-resistance mechanisms as accommodation, ingratiation, evasiveness, and manipulation. When students are focused on grades and achieving credit, student motivation plummets (Nieswandt and Shanahan 2008). Therefore, playing Fatima's rules contributes to declining student interest and enrolments, to student alienation, and to students' failure to learn content meaningfully.

In an elaborate quantitative study of 271 teachers and 6855 students, Wood et al. (2009) similarly concluded: "We believe that the [student] culture of 'dealing' [i.e. dealing with school science by jumping though a series of hoops] is an underlying cause of the deterioration in the achievement of US students and the failures of so many reform efforts to bring about substantial and lasting change" (p. 437). Playing Fatima's rules characterizes the common student cultural norm of "just deal with it".

Also from a cultural perspective, Fatima's rules provide specific rituals and practices within a science classroom. These are usually staged by teachers, not students. Tobin and McRobbie (1997) documented a teacher's complicity in playing Fatima's rules: "There was a close fit between the goals of Mr. Jacobs and those of the students and satisfaction with the emphasis on memorisation of facts and procedures to obtain the correct answers needed for success on tests and examinations" (p. 366).

Costa (1997) synthesized the work of Larson (1995) and Tobin and McRobbie (1997) with her own classroom research with Mr. Ellis, and concluded:

> Mr. Ellis' students, like those of Mr. London and Mr. Jacobs, are not working on chemistry; they are working to get through chemistry. The subject does not matter. As a result, students negotiate treaties regarding the kind of work they will do in class. Their work is not so much productive as it is political. They do not need to be productive—as in learning chemistry. They only need to be political—as in being credited for working in chemistry. (p. 1020)

The three teachers (Ellis, London, and Jacobs) did not instil meaningful learning. The resultant credentials earned by their students are a political symbol—a forgery of meaningful learning. Playing Fatima's rules constitutes a clear indicator of "what quality science teaching is not".

Dishonest and Mythical Images Conveyed

The last major educational failure of traditional school science deals with "the many myths and falsehoods perpetuated by traditional science curricula and by the popular media and often promoted by stories drawn from the history of science" (Hodson 2006, p. 302; citations omitted). Such false images about science and scientists are also conveyed and reinforced by teachers, textbooks, and the way students are assessed, particularly in high-stakes testing (Abd-El-Khalick et al. 2008; Aikenhead 2006). Milne and Taylor (1998, pp. 31–43) identified such myths as: "observations of reality correspond exactly to an external reality," and "language is transparent and has no influence on the interpretation of data generated from observations of the natural world".

Many students, similar to their teachers and, in turn, similar to university science and engineering departments, appear to embrace naïve realism and a positivistic ideology of technical rationality (Habermas 1972). "Mainstream Western Modern Science and its product, school science, portray science as the discovery of universal truths based on evidence gained through objective, reproducible experiments stripped of emotion, cultural contexts, and values" (Chinn 2007, p. 1251). Hence, students rarely appreciate the nature of the scientific enterprise, such as: its diversity; its intellectual presuppositions; and its social context of paradigms, economics, national security, and corporate profits (Hodson 2009).

As a consequence, some science-proficient students (including those from under-represented groups) lose interest in taking further science classes because they are discouraged by the positivistic and mythical images that for them represent the science profession. They never become Ph.D. graduates. On the other hand, some students become interested in a science career because they are attracted to such false images. Some become science teachers.

There are implications for the calibre of savvy citizens living in a scientific technological society. When students become adult citizens; some take on key positions in government and industry; and if they make decisions predicated on myths about the scientific enterprise, then the consequences will be mediocrity or worse.

The act of conveying dishonest and mythical ideas about science and scientists is an indicator of "what quality science teaching is not".

Summary

Here is a list of "what quality science teaching is *not*". It is the type of teaching that:

1. subscribes to political expediency over educational soundness,
2. decreases student interest and enrolment in school science,
3. causes student alienation due to neglect of cultural and identity differences between students and school science,
4. leads to a large majority of students not learning science content meaningfully,
5. encourages students to play Fatima's rules, and
6. conveys dishonest and mythical images of science and scientists.

I now turn to the task of developing some indicators that purposefully avoid such failures.

Avoiding the Failures of Traditional School Science

When prominent citizens (listed in *Who's Who in Australia*) were interviewed by Symington and Tytler (2004), all favoured a "widespread understanding of the important part that science is playing in the lives of individuals and in society" (p. 1409). They talked about "an education 'for science in life'", rather than an education for scientific "knowledge building" (p. 1415).

Reiss (2007) sensibly urged science educators to accept a plurality of aims for school science:

> There are two main reasons for favouring, or at any rate accepting, a number of even incommensurate aims for science. One is that, pragmatically, attempting to insist on just one aim is unlikely to succeed. The second is the possibility that different aims may suit different audiences. (p. 25)

Of the many fundamental ideas that speak to this plurality of aims and that encourage quality science teaching, I shall explore from a cultural and value-based perspective the issue of what teachers teach and do not teach. I do this because this is a central issue in the educational failures of traditional school science. My aim is to reconceptualize *school science content* so it will engender quality science teaching rather than inhibit it.

Rethinking School Science Content

Science content is taught in schools and in university programs; it is also learned and used outside those domains. But all of these differing domains have one feature in common: They are Euro-American ways of knowing nature. The content was shaped by its Eurocentric origins and Euro-American evolution (Aikenhead 2006). Thus, this knowledge system can be identified by the phrase *Eurocentric sciences* (plural) to capture its ethnicity and heterogeneity. Eurocentric sciences are first and foremost anchored in culture (Sillitoe 2007). Students who do not feel comfortable with this scientific culture, as mentioned above, tend to become alienated and resist learning science unless their teacher helps them deal with the differences between their cultural self-identities and the culture of the science classroom; ultimately to make students feel at ease in the foreign culture of Eurocentric school science.

Students living outside mainstream Euro-American cultures are certainly welcome to join the cultures of Eurocentric sciences as long as they successfully complete the enculturation process by graduating from qualified university science or engineering departments, in which students learn to play by the rules of the game (Larochelle 2007). Scientists collectively work within a subculture (a paradigm) that frames their thinking and practice. Naturally, different subcultures deal with different science content.

The science content found in schools and undergraduate university programs, for instance, differs in cultural ways from the science content observed in science-related occupations and everyday events and issues (Aikenhead 2006; Munby et al. 2007). The science content encountered in the culture of school science is invariably abstract academic content that serves such purposes as knowledge accumulation and gate-keeping for university departments. On the other hand, science content in the cultures of the non-academic world is invariably content-in-action that serves the purposes of "science for life," the phrase favoured by Australia's *Who's Who*. For example, Munby et al.'s (2007, p. 130) research uncovered a school science metaphor of "knowledge as entity", to be contrasted with a science-rich workplace metaphor of "knowledge as activity". The culture of academic school science differs from the culture of out-of-school science in terms of the type of scientific knowledge found in each domain: abstract academic descriptions and explanations, versus knowing-in-action.

Although science content resides in both cultural contexts (academic school science and out-of-school science), its *relevance* to students' lives differs substantially. Obviously, a teacher is on educationally sound ground when teaching science content identified with out-of-school contexts of science-related occupations or everyday events and issues, because this content is both valid science and relevant to the vast majority of students. Yet political expediency can dictate teaching non-relevant science content. Will university departments continue to exercise pervasive power over what is relevant for school science, or will other stakeholders prevail? The answer will determine what content is taught in schools.

School science content can be conceptualized in terms of two related principles: *relevance* and *who decides* what is relevant for students, today or in their future. The two principles are depicted in Table 7.1. The left-hand column of the table describes who decides what is relevant; and the right hand column represents the

Table 7.1 Who decides on relevance, and the resulting types of science content. (Modified from Aikenhead 2006, p. 32)

Who decides what is relevant?	Type of science content
Academic scientists, education officials, and science teachers, who invariably confirm the traditional curriculum's academic science content	Wish-they-knew science
People mainly in science-related occupations and savvy citizens. Research has identified a wealth of general and specific educational outcomes not normally found in traditional school science but found in science-related occupations and everyday events and issues	Functional science
Science-related experts who interact with the general public on real-life events, and who know the problems the public encounters when dealing with these events	Have-cause-to-know science
The general public who has faced real-life problems or decisions related to science. What science content did they need to know to resolve their problem or make their decision?	Need-to-know science
People who produce the media and internet sites, and who draw upon sensational and controversial aspects of science and technology to achieve motivational value for readers and viewers	Enticed-to-know science
Students themselves express an opinion on what would be of interest to study. What are they curious about?	Personal-curiosity science
Interpreters of culture, who can determine what aspects of science, and what aspects of local knowledge, comprise features of a local, national, and global culture. This category can include a combination of categories above. It is exemplified by, but not restricted to, socioscientific issues	Science-as-culture

type of school science content that would be taught as a consequence. This scheme recognizes seven groups of people who currently decide, or who could reasonably decide, what will be included in or excluded from school science content. The categories are not discrete but overlap and interact in various ways.

To work towards achieving science-in-action rather than abstract academic science, curriculum developers and teachers will draw from several of these categories. The resulting curricula (the intended, enacted, hidden, null, and learned curricula) will most likely be comprised of different combinations of categories (illustrated below). This is one path towards delineating quality science teaching.

Academic Science: A traditional school science curriculum emerges from the first category in Table 7.1. The content is called *wish-they-knew science* because professors and teachers will say of students entering their class, "I wish they knew this, I wish they knew that."

Research provides a reality check on how relevant wish-they-knew science turns out to be for achievement at university (Aikenhead 2006). Little correlation exists between high school science success and first year university success in the sciences ($R=0.133$ in Sadler and Tai's (2001) study); and other studies have found no correlation (e.g. Yager and Krajcik 1989). In the extreme case where university students have not studied the physics or chemistry prerequisites in high school, motivated students achieve as well as their counterparts who had the high school credentials. Moreover, wish-they-knew science is rarely directly transferable to out-of-school contexts (Aikenhead 2006). Therefore, educational arguments that support teaching only wish-they-knew science in schools for pre-professional training or for general scientific literacy are, at best, extremely weak. However, wish-they-knew science continues to be relevant to a very small minority of science-proficient students whose worldviews generally harmonize with a worldview endemic to academic science. Defining school science content primarily on the basis of this small minority of students is a political decision, not one based on educational soundness.

Relevant Science for Most Students: The other six categories in Table 7.1 reflect in various ways the work world of employers and employees as well as the everyday world of citizens. In both worlds, science content pertains to phenomena and events not normally of interest to most university science and engineering professors (scholarly academics). These other six categories tend to represent science-in-action and citizen science, by and large, which most students find highly relevant to varying degrees depending on the student and the topic. Systematic research has produced a wealth of general and specific results related to each category in Table 7.1 (Aikenhead 2006). Space limits me to a short introduction to each of these other six categories.

Functional science is the science content that has functional value to science-rich employment and to science-related everyday events. For example, industry personnel placed "understanding science ideas" at the lowest priority for judging a recruit to their industry. Why?

The content actually used by people in the workplace is so context specific that it has to be learned on the job. High school and university academic science content is rarely drawn upon. Hence, an important quality valued by both employers and employees in science-related employment is the *capacity to learn* science content on the job; that is, knowing *how to learn* science and the ability to put that knowledge into action. Learning *how to learn* school science content develops a capacity for life-long learning. In spite of this educationally sound goal, Canadian researchers Chin et al. (2007) found that "There are few explicit references to the world of work in the majority of science curriculum documents" (p. 123). One example of a curriculum based on functional science is a work-study project for high school science students who learned science content found in science-rich workplaces (Chin et al. 2004).

Students' preparation for science-related occupations, or for being savvy citizens in a scientific technological world, must certainly include learning science concepts. But the choice of concepts can be a functionally relevant choice, rather than a scholarly academic choice that leads to "wish-they-knew science". The science content that underpins local and global contexts of interest to students works well for teaching students *how to learn and use* science as needed (Aikenhead 2006; Bennett et al. 2006; Calabrese Barton and Tan 2009; Prins et al. 2008). Contexts chosen by teachers and curriculum developers, however, are not usually effective, unless there is a shared view of purpose among teachers and students, unless there are changes in the nature of the dialogue between them, and unless students compare and contrast their personal understanding with scientific views (Rodrigues 2006). Functional science also includes concepts and procedural knowledge for understanding and acting upon scientific evidence—"concepts of evidence" (Duggan and Gott 2002)—in the world of work and in everyday situations.

Have-cause-to-know science represents science content identified by science and technology experts who consistently interact with the general public on real-life matters pertaining to science, and who know the problems citizens encounter when interacting with experts (Law 2002). I assume these experts are better situated than academic university scientists and engineers to decide what is worth learning for life in today's changing scientific and technological world. One of Law's curriculum development projects in China shows how to generate have-cause-to-know science content. Another example is Naughton et al.'s (2008) work on how curriculum policy can be formulated by drawing upon several municipal agencies conversant in air pollution chemistry and experienced in dealing with the public.

Need-to-know science arises from what people needed to know when faced with a science-related event or issue in their lives. In 2001 Ryder analysed 31 case studies to find out what science content was relevant to such people. He concluded, "Much of the science knowledge relevant to individuals in the case studies was *knowledge about science*, i.e. knowledge about the development and use of scientific knowledge rather than scientific knowledge itself" (p. 35, emphasis in original). This type of content about the scientific enterprise is known as the nature and social aspects of science. Knowledge *about* science includes the savvy to decide which science expert to trust (Kolstø 2001). Lee (2008) recently added to the encyclopaedia of need-to-know science by investigating what people needed to know about SARS (severe acute respiratory syndrome) to cope with the 2003 epidemic.

Enticed-to-know science, by its very nature, is engaging because its content tends to be controversial and sensational. This science content is found in the mass media and on the internet where writers/producers entice readers/viewers into becoming engaged. Several curriculum projects and instruction strategies have been based on this type of content (Aikenhead 2006; Jarman and McClune 2007).

Personal-curiosity science is content chosen by students. Various data collection methods yield interesting content. The ROSE (Relevance in Science Education) project, a major international questionnaire study, is informing many countries about what topics have motivational value for students (Schreiner and Sjøberg 2007). In another study, an internet site was instrumental in accumulating students' questions over a period of several years, and allowed for longitudinal analyses (Baram-Tsabari et al. 2006). Chin and Osborne (2008) reviewed studies that investigated students' questions of curiosity posed during science classes, which in turn were usually incorporated into the enacted curriculum.

And lastly, *science-as-culture* deals more broadly with the enculturation of students into everyday society; not the enculturation into a scientific discipline of wish-they-knew science. "The meaning making we call science happens in a way that is distributed over the society spatially and temporally" (Weinstein 1998, p. 492). "Science emerges at the intersection and through contestation of multiple groups within and outside of the enterprise [of science]" (Weinstein 2008, p. 395). The *Who's Who in Australia* interviewees called for "science in life". Fleshed out, science in life likely corresponds to science-as-culture.

The category science-as-culture flexibly embraces other categories in Table 7.1. Teaching science-as-culture will draw upon community resources and local culture to stimulate a combination of, for instance, functional, have-cause-to-know, and need-to-know science. It will contextualize science content found in, for instance: (a) the corporate economic world, (b) social movements around diseases, (c) science-related moral/ethical issues, (d) environmental issues, and (e) the political world of regulations and policy making. The field of socioscientific issues (SSI) in school science exemplifies science-as-culture because SSI often transcends several categories of relevant science as well as transdisciplinary knowledge beyond Eurocentric science (Chap. 2; Chap. 17; Sadler 2009).

Combining Categories of Relevant School Science Content: Innovative projects illustrate how different categories in Table 7.1 have been combined, for instance:

- An anatomy-physiology curriculum (Fowler et al. 2009) that mainly combined science-as-culture with wish-they-knew science.
- Teaching materials produced for science-related social practices (Bulte et al. 2006) that mainly combined functional and personal-curiosity science.
- *Active Physics* mainly combined wish-they-knew, functional, and personal-curiosity science (Carlone 2004).
- Kortland's (2001) research program into students' learning how to make decisions in the context of a waste management module, which mainly combined

functional, have-cause-to-know, personal-curiosity, and wish-they-knew science.
- An AS-level textbook in the United Kingdom, *AS Science for Public Understanding* (Hunt and Millar 2000) that mainly combined enticed-to-know, have-cause-to-know, and wish-they-knew science.
- The Science, Technology, Environment in Modern Society project in Israel (Dori and Tal 2000) that mainly combined functional, have-cause-to-know, and wish-they-knew science.
- A grade 10 textbook, *Logical Reasoning in Science & Technology* (Aikenhead 1991) that mainly combined functional, have-cause-to-know, wish-they-knew, and personal-curiosity science.

Many of these projects include wish-they-knew science because the culture of school science in the project's educational jurisdiction required it.

However, evidence from several recent studies demonstrates the negative impact of wish-they-knew science on students' perceptions of innovations such as those listed above: "Students saw the same activities [innovations] as a simple extension of what ordinarily transpires in science classrooms" (Sadler 2009, p. 36); a conclusion verified by a very extensive study of students' and teachers' perceptions of what transpires in their science classrooms (Wood et al. 2009) and verified in a review of research into students' identity in science learning (Shanahan 2009). Perhaps an innovation needs to change the culture of school science before it is perceived by students as a significant innovation.

Indicators of Quality Science Teaching

Indicators of quality science teaching that arise from rethinking school science content will certainly include the following four:

1. Acknowledgement of the degrees of cultural differences between students' cultural self-identities and the culture of their science classroom, and recognition that each student needs help when negotiating this cross-cultural classroom environment.
2. An enacted curriculum predominantly comprised of relevant science content outside the category of wish-they-knew science, but not ignoring that category.
3. An emphasis on the outcome: Teaching students *how to learn and use* science as the need arises in specific contexts.
4. Student assessment formulated in terms of monitoring students' learning how to learn and how to use science and technology as needed.

These four fundamental indicators suggest ways to transform the culture of traditional school science: (a) by taking seriously the fact that many students experience school science as a cross-culture event; (b) by dispelling values, myths, and routines associated with academic science content; and (c) by insisting that educational soundness and relevancy be the main criteria for selecting school science content, and not the criterion of political expediency.

The most efficient and effective way to initiate and achieve such educational change arises from the actions of an entire school system rather than from classroom teachers alone (Elmore 1996). The most powerful political unit of change is the school system. Thus, other indicators that buttress the four listed above include:

5. a school system's support for teachers to develop a cultural perspective on teaching science;
6. a school system's professional development program that encourages and requires teachers to engage in life-long learning into, for instance, science-as-culture and functional, have-cause-to-know, and enticed-to-know science related to their academic science background;
7. a school system's assessment policy for monitoring students' capacity to learn science-in-action and citizen science as the need arises, thereby avoiding the memorization of facts, abstractions, and algorithms;
8. a school system's ability to engage a community's human resources in planning and teaching aspects of the enacted curriculum; and
9. teachers' satisfaction with their school system's support in creating changes to the culture of school science that facilitates quality science teaching.

Concluding Remarks

As mentioned above, my project is to reconceptualize *school science content* so it will engender quality science teaching rather than inhibit it. Accordingly, I have advanced evidence-based arguments concerning what types of content teachers should and should not teach, while at the same time being mindful of students' self-identities. I was guided in part by some major failures of conventional practice and by the two related principles: relevance of school science content and who decides what is relevant. The intricate, context-laden, interplay between relevance and students' self-identities explicitly answers the question: Why would students want to engage with school science in the first place?

This chapter's cultural view on quality science teaching has afforded a partial perspective on school science that represents a plurality of aims proposed by Reiss (2007). This perspective revealed a host of alternatives to traditional school science that promise to displace the hegemony of *academic* science and its wish-they-knew science content. These alternatives were warranted by their educational soundness and supported by empirical evidence.

Along the way I offered nine indicators of what quality science teaching is, and six indicators of what it is not. These indicators relate to evidence that could reflect the extent to which quality science teaching may be occurring in a classroom, in a school, and in a school system.

Perhaps some readers associate the idea of evidence-based indicators with a recent slogan in education, *evidence-based practice*, borrowed from evidence-based medicine, in which administrators and doctors make decisions about the care of individual patients based on systematically derived, universal, quantitative principles. In medical science, a person is objectified and reduced to a physical entity by ignoring his or her emotional, intellectual, ethical, and self-identity aspects; a stance supported by a plethora of medical journals. Interestingly, the variable "interaction between doctor and patient" is virtually absent in evidence-based medicine.

I reject this evidence-based-practice movement in education and its attempt to hijack the meaning of "evidence" to be purely theoretical and quantitative. Rather than relinquish the term "evidence" to the slogan people imprisoned in a quantitative paradigm, I stand by the conventional meaning of "evidence" in qualitative or quantitative social science research: *trustworthy* evidence (Mishler 1990). Consequently, my suggested indicators of quality science teaching relate to evidence-based practice in the social science sense of trustworthy evidence.

With the demise of a "positivist and authoritarian school science" (Malcolm 2007, p. 71), a different culture of school science will take hold; one that affirms an inclusive and relevant experience for students, while refraining from political expediencies that cause traditional school science to fail educationally.

References

Abd-El-Khalick, F., Waters, M., & Le, A-P. (2008). Representations of nature of science in high school chemistry textbooks over the past four decades. *Journal of Research in Science Teaching, 45,* 835–855.

Adamuti-Trache, M., & Andres, L. (2008). Embarking on and persisting in scientific fields of study: Cultural capital, gender, and curriculum along the science pipeline. *International Journal of Science Education, 30,* 1557–1584.

Aikenhead, G. S. (1991). *Logical reasoning in science & technology.* Toronto: Wiley of Canada.

Aikenhead, G. S. (2006). *Science education for everyday life: Evidence-based practice.* New York: Teachers College Press.

Alberts, B. (2000). Some thoughts on science as inquiry. In J. Minstrell & E. H. van Zee (Eds.), *Inquiring into inquiry learning and teaching in science* (pp. 3–13). Washington: American Association for the Advancement of Science.

Baram-Tsabari, A., Sethi, R. J., Bry, L., & Yarden, A. (2006). Using questions sent to an ask-a-scientist site to identity children's interests in science. *Science Education, 90,* 1050–1072.

Barnes, M. B., & Barnes, L. W. (2005). Using inquiry processes to investigate knowledge, skills, and perceptions of diverse learners: An approach to working with prospective and current science teachers. In A. J. Rodrigues & R. S. Kitchen (Eds.), *Preparing mathematics and science teachers for diverse classrooms: Promising strategies for transformative pedagogy* (pp. 61–86). Mahwah: Lawrence Erlbaum Associates.

Bennett, J., Lubben, F., & Hogarth, S. (2006). Bringing science to life: A synthesis of the research evidence on the effects of context-based and STS approaches to science teaching. *Science Education, 91,* 347–370.

Berry, A., Loughran, J., & Mulhall, P. (2007). Values associated with representing science teachers' pedagogical content knowledge. In D. Corrigan, J. Dillon, & R. Gunstone (Eds.), *The re-emergence of values in science education* (pp. 149–163). Rotterdam: Sense Publishers.

Bianchini, J. A., Cavazos, L. M., & Helm, J. V. (2000). From professional lives to inclusive practice: Science teachers and scientists' views of gender and ethnicity in science education. *Journal of Research in Science Teaching, 37,* 511–547.
Brandt, C. B. (2008). Scientific discourse in the academy: A case study of an American Indian undergraduate. *Science Education, 92,* 825–847.
Brickhouse, N. W. (2001). Embodying science: A feminist perspective on learning. *Journal of Research in Science Teaching,* 38, 282–295.
Brotman, J. S., & Moore, F. M. (2008). Girls and science: A review of four themes in the science education literature. *Journal of Research in Science Teaching, 45,* 971–1002.
Brown, B. A., Reveles, J. M., & Kelly, G. (2005). Scientific literacy and discursive identity: A theoretical framework for understanding science education. *Science Education, 89,* 779–802.
Bulte, A. M. W., Westbroek, H. B., De Jong, O., & Pilot, A. (2006). A research approach to designing chemistry education using authentic practices as contexts. *International Journal of Science Education, 28,* 1063–1086.
Calabrese Barton, A, & Tan, E. (2009). Funds of knowledge and discourses and hybrid space. *Journal of Research in Science Teaching, 46,* 50–73.
Carlone, H. B. (2004). The cultural production of science in reform-based physics: Girls' access, participation and resistance. *Journal of Research in Science Teaching, 41,* 392–414.
Carlone, H. B., & Johnson, A. (2007). Understanding the science experiences of successful women of color: Science identity as an analytic lens. *Journal of Research in Science Teaching, 44,* 1187–1218.
Chin, C., & Osborne, J. (2008). Students' questions: A potential resource for teaching and learning science. *Studies in Science Education, 44,* 1–39.
Chin, P., Munby, H., Hutchinson, N. L., Taylor, J., & Clark, F. (2004). Where's the science? Understanding the form and function of workplace science. In E. Scanlon, P. Murphy, J. Thomas, & E. Whitelegg (Eds.), *Reconsidering science learning* (pp. 118–134). New York: RoutledgeFalmer.
Chin, P., Zanibbi, M., Dalgarno, N., Poth, C., & Ayala, G. (2007). Teaching science for the workplace? An analysis of Canadian science curriculum documents. *Canadian Journal of Science, Mathematics and Technology Education, 7,* 107–132.
Chinn, P. W. U. (2007). Decolonizing methodologies and Indigenous knowledge: The role of culture, place and personal experience in professional development. *Journal of Research in Science Teaching, 44,* 1247–1268.
Cleaves, A. (2005). The formation of science choices in secondary school. *International Journal of Science Education, 27,* 471–486.
Costa, V. (1997). How teacher and students study 'all that matters' in high school chemistry. *International Journal of Science Education, 19,* 1005–1023.
Daniell, E. (2006). *Every other Thursday: Stories and strategies from successful woman scientists.* New Haven: Yale University Press.
Dori, Y. J., & Tal, R. T. (2000). Formal and informal collaborative projects: Engaging in industry with environmental awareness. *Science Education, 84,* 95–113.
Duggan, S., & Gott, R. (2002). What sort of science education do we really need? *International Journal of Science Education, 24,* 661–679.
Eisenhart, M., Finkel, E., & Marion, S. (1996). Creating the conditions for scientific literacy: A re-examination. *American Educational Research Journal, 33,* 261–295.
Elmore, R. F. (1996). Getting to scale with good educational practice. *Harvard Education Review, 66,* 1–26.
Fensham, P. J. (1993). Academic influence on school science curricula. *Journal of Curriculum Studies, 25,* 53–64.
Fensham, P. J. (2007). Values in the measurement of students' science achievement in TIMSS and PISA. In D. Corrigan, J. Dillon, & R. Gunstone (Eds.), *The re-emergence of values in science education* (pp. 215–229). Rotterdam: Sense Publishers.

Fowler, S. R., Zeidler, D. L., & Sadler, T. D. (2009). Moral sensitivity in the context of socioscientific issues in high school science students. *International Journal of Science Education, 31,* 279–296.
Frederick, W. A. (1991). Science and technology education: An engineer's perspective. In S. K. Majumdar, L. M. Rosenfeld, P. A. Rubba, E. W. Miller, & R. F. Schmalz (Eds.), *Science education in the United States: Issues, crises and priorities* (pp. 386–393). Easton: The Pennsylvania Academy of Science.
Habermas, J. (1972). *Knowledge and human interests.* London: Heinemman.
Hildebrand, G. M. (2007). Diversity, values and the science curriculum. In D. Corrigan, J. Dillon, & R. Gunstone (Eds.), *The re-emergence of values in science education* (pp. 45–60). Rotterdam: Sense Publishers.
Hodson, D. (2006). What we should prioritize learning *about* science. *Canadian Journal of Science, Mathematics and Technology Education, 6,* 293–311.
Hodson, D. (2009). *Towards scientific literacy: A teachers' guide to the history, philosophy and sociology of science.* Rotterdam: Sense Publishers.
Hunt, A., & Millar, R. (2000). *AS science for public understanding.* Oxford: Heinemann.
Jarman, R., & McClune, B. (2007). *Developing scientific literacy: Using news media in the classroom.* Maidenhead: Open University Press.
Johnson, A. C. (2007). Unintended consequences: How science professors discourage women of color. *Science Education, 91,* 805–821.
Kolstø, S. D. (2001). 'To trust or not to trust, …'—pupils' ways of judging information encountered in a socio-scientific issue. *International Journal of Science Education, 23,* 877–901.
Kortland, J. (2001). *A problem posing approach to teaching decision making about the waste issue.* Utrecht, The Netherlands: University of Utrecht Cdβ Press.
Larochelle, M. (2007). Disciplinary power and the school form. *Cultural Studies of Science Education, 2,* 711–720.
Larson, J. O. (1995, April). *Fatima's rules and other elements of an unintended chemistry curriculum.* Paper presented to the American Educational Research Association Annual Meeting, San Francisco.
Law, N. (2002). Scientific literacy: Charting the terrains of a multifaceted enterprise. *Canadian Journal of Science, Mathematics and Technology Education, 2,* 151–176.
Lee, Y-C. (2008). Exploring the roles and nature of science: A case study of severe acute respiratory syndrome. *International Journal of Science Education, 30,* 515–541.
Löfgren, L., & Helldén, G. (2009). A longitudinal study showing how students use a molecule concept when explaining everyday situations. *International Journal of Science Education, 31,* 1631–1655.
Loughran, J., & Derry, N. (1997). Researching teaching for understanding: The students' perspective. *International Journal of Science Education, 19,* 925–938.
Lyons, T. (2006). The puzzle of falling enrolments in physics and chemistry courses: Putting some pieces together. *Research in Science Education, 36,* 285–311.
Malcolm, C. (2007). The value of science in African cultures. In D. Corrigan, J. Dillon, & R. Gunstone (Eds.), *The re-emergence of values in science education* (pp. 61–76). Rotterdam: Sense Publishers.
Malone, K. R., & Barabino, G. (2009). Narrations of race in STEM research settings: Identity formation and its discontents. *Science Education, 93,* 485–510.
Medina-Jerez, W. (2008). Between local culture and school science: The case of provincial and urban students from eastern Colombia. *Research in Science Education, 38,* 189–212.
Milne, C. E., & Taylor, P. C. (1998). Between myth and a hard place. In W. W. Cobern (Ed.), *Sociocultural perspectives on science education* (pp. 25–48). Boston: Kluwer.
Mishler, E. G. (1990). Validation in inquiry-guided research: The role of exemplars in narrative studies. *Harvard Educational Review, 60,* 415–442.
Munby, H., Taylor, J., Chin, P., & Hutchinson, N. L. (2007). Co-op students' access to shared knowledge in science-rich workplaces. *Science Education, 91,* 115–132.

Naughton, W., Schreck, J., & Heikkinen, H. (2008). Seeking evidence for "curricular relevancy" within undergraduate liberal arts chemistry textbooks. *Journal of Research in Science Teaching, 45,* 174–196.

Nieswandt, M., & Shanahan, M-C. (2008). "I just want the credit!"—Perceived instrumentality as the main characteristic of boys' motivation in a grade 11 science course. *Research in Science Education, 38,* 3–29.

OECD (2006). *Evolution of student interest in science and technology studies policy report.* Paris: OECD.

Osborne, J., & Dillon, J. (2008). *Science education in Europe: Critical reflections.* London: The Nuffield Foundation.

Prins, G. T., Bulte, A. M. W., van Driel, J. H., & Pilot, A. (2008). Selection of authentic modelling practices as contexts for chemistry education. *International Journal of Science Education, 30,* 1867–1890.

Reiss, M. (2007). What should be the aim(s) of school science education? In D. Corrigan, J. Dillon, & R. Gunstone (Eds.), *The re-emergence of values in science education* (pp. 13–28). Rotterdam: Sense Publishers.

Rennie, L. J. (2007). Learning science outside of school. In S. K. Abell & N. G. Lederman (Eds.), *Handbook of research on science education* (pp. 125–167). Mahwah: Lawrence Erlbaum.

Rodrigues, S. (2006). Pupil-appropriate contexts in science lessons: The relationship between themes, purpose and dialogue. *Research in Science & Technological Education, 24,* 173–182.

Ryder, J. (2001). Identifying science understanding for functional scientific literacy. *Studies in Science Education, 36,* 1–42.

Sadler, T. D. (2009). Situated learning in science education: Socio-scientific issues as contexts for practice. *Studies in Science Education, 45,* 1–42.

Sadler, P. M., & Tai, R. H. (2001). Success in introductory college physics: The role of high school preparation. *Science Education, 85,* 111–136.

Schreiner, C., & Sjøberg, S. (2007). Science education and youth's identity construction: Two incompatible projects? In D. Corrigan, J. Dillon, & R. Gunstone (Eds.), *The re-emergence of values in science education* (pp. 231–247). Rotterdam: Sense Publishers.

Scott, P., Asoko, H., & Leach, J. (2007). Student conceptions and conceptual learning in science. In S. K. Abell & N. G. Lederman (Eds.), *Handbook of research on science education* (pp. 31–56). Mahwah: Lawrence Erlbaum.

Shanahan, M-C. (2009). Identity in science learning: Exploring the attention given to agency and structure in studies of identity. *Studies in Science Education, 45,* 43–64.

Sillitoe, P. (Ed.). (2007). *Local science vs. global science: Approaches to Indigenous knowledge in international development.* New York: Berghan Books.

Symington, D., & Tytler, R. (2004). Community leaders' views of the purposes of science in the compulsory years of schooling. *International Journal of Science Education, 26,* 1403–1418.

Tobias, S. (1990). *They're not dumb, they're different.* Tucson: Research Corporation.

Tobin, K., & McRobbie, C. (1997). Beliefs about the nature of science and the enacted science curriculum. *Science and Education, 6,* 355–371.

Tobin, K., Seiler, G., & Smith, M. W. (1999). Educating science teachers for the sociocultural diversity of urban schools. *Research in Science Education, 29,* 69–88.

von Aufschnaiter, C., Erduran, S., Osborne, J., & Simon, S. (2008). Arguing to learn and learning to argue: Case studies of how students' argumentation relates to their scientific knowledge. *Journal of Research in Science Teaching, 45,* 101–131.

Weinstein, M. (1998). Playing the paramecium: Science education from the stance of the cultural studies of science. *Educational Policy, 12,* 484–506.

Weinstein, M. (2008). Finding science in the school body: Reflections on transgressing the boundaries of science education and the social studies of science. *Science Education, 92,* 389–403.

Wood, N. B., Lawrenz, F., & Haroldson, R. (2009). A judicial presentation of evidence of a student culture of "Dealing." *Journal of Research in Science Teaching, 46,* 421–441.

Yager, R. E., & Krajcik, J. (1989). Success of students in a college physics course with and without experiencing a high school course. *Journal of Research in Science Teaching, 26,* 599–608.

Chapter 8
Japanese Elementary Rika Teachers' Professional Beliefs and Knowledge of Rika Teaching: How are they Indigenized?

Masakata Ogawa

Introduction

Information about the nature and the quality of professional beliefs and knowledge held by science teachers (experienced or novice) is crucial for science teacher educators when developing, evaluating, and revising quality science teacher education programs, whether or not they are pre-service or in-service. This is one of the reasons why research on professional beliefs and/or professional knowledge has been a strong focus in the field of science education. Now we have a rich and extensive collection of knowledge about science teachers' professional beliefs and knowledge.

The primary purpose of the research reported in this chapter is to uncover Japanese elementary Rika[1] teachers' professional beliefs and knowledge. But immediately, I realize that in writing about this research in English for non-Japanese readers I am facing a serious problem because of the very nature of elementary Rika.

I have previously shown (Ogawa 1998) that the overall objective of Rika in elementary school level consists of two mutually conflicting components. The first component is the education of "western modern science" (simply expressed as "science" below); the second component is the education of "Shizen", a Japanese traditional cosmology that embraces a specific values system. In a Rika, especially a unit on living things, we can confidently predict that the two components work simultaneously within the minds of both learners and teachers. Thus the two components are difficult to keep intact (i.e. true to their original meaning). It may well be that a learner or teacher merges the two components into a third hybrid (and idiosyncratic) meaning.

[1] Rika is the Japanese name for school science, and "elementary Rika teacher" in this study refers to "elementary teacher who identifies himself/herself to be a science major teacher" and in most cases, he/she holds lower-secondary Rika teacher's certificate as well as elementary teacher's certificate, because in Japan there is no Rika-specific certificate for elementary school.

M. Ogawa (✉)
Tokyo University of Science, Tokyo, Japan

D. Corrigan et al. (eds.), *The Professional Knowledge Base of Science Teaching*,
DOI 10.1007/978-90-481-3927-9_8, © Springer Science+Business Media B.V. 2011

The problem is that this kind of duality/heterogeneity of the objectives within one school subject has never been presupposed in developing or examining theoretical frameworks for research on science teachers' professional beliefs and knowledge in the western world, where we find a fruitful collection of research findings on professional beliefs and knowledge and/or pedagogical content knowledge among teachers in general (Shulman 1986, 1987; Grossman 1990; Fernandez-Balboa and Stiehl 1995) and science teachers in particular (Jones and Carter 2006; Gess-Newsome and Lederman 2001; Abell 2006; Appleton 2006).

Until now, unfortunately, there has been little research specifically on Japanese elementary Rika teachers' professional beliefs and knowledge, where duality/heterogeneity of an elementary Rika program is taken seriously. There are a small number of research studies on teachers of lower-secondary Rika, but the overall objectives of these studies do not contain the Shizen component (Isozaki et al. 2007). Further, other studies of Rika teachers' views on "nature of science" have paid no attention to the Shizen component, including when elementary Rika teachers were among the targets of the studies (Toda 1992; Shimizu 2002; Tanzawa et al. 2003).

Thus, this chapter reports preliminary explorations of elementary Rika teachers' professional beliefs and knowledge, something that requires a new type of theoretical framework.

How to do this? One possible way is to adapt a set of theoretical constructs developed for deciphering western science teachers' professional beliefs and knowledge separately for each of the two components (science and Shizen). Such an approach would assume that Japanese elementary Rika teachers' professional beliefs and knowledge consist of those about science and those about on Shizen.

However, this assumption may be dangerous because it excludes (a) possible differences in the nature of teaching/learning of the two components, and (b) possible amalgamation of these two components in actual teaching/learning settings. Ogawa (1986) has argued that elementary Rika teachers tend to teach Rika classes without awareness of the duality/heterogeneity of the two components. Hence, the very first step in research on elementary Rika teachers' professional beliefs and knowledge should be an exploratory study on the nature of elementary Rika teachers' ideas on elementary Rika without depending on any specific theoretical framework. When their ideas on elementary Rika and Rika teaching are elaborated, especially from the aspect of duality/heterogeneity of the two components, we can then proceed to develop a more valid theoretical framework to decipher elementary Rika teachers' professional beliefs and knowledge.

Japanese Rika and Rika Teachers: Literature Review

Extensive overviews on the Japanese school subject Rika are found in Ogawa (1998, 2001, 2002a). These include substantial information on fundamental statistics, school system, historical development of the subject, Rika teacher education system, Rika curriculum policy development processes, current problems and issues in Rika, on-going revisions in Rika, and so on. Aspects relevant to the present chapter are now very briefly outlined.

Ogawa (1986, 1998) discussed some unique characteristics of Rika compared with school science in western societies. He identified two mutually conflicting components within Rika; one relevant to science and the other relevant to Shizen (an indigenous view of the natural world that has been shared among Japanese people who embrace a certain set of indigenous values). The notion of Shizen among Japanese people is a very difficult and problematic philosophical issue, not only for Japanese educators but also for philosophers. It is usually assumed that "Shizen" can be validly translated as "Nature", but the two notions are quite different in their origins, and both have been deeply linked with their respective language systems (Kawasaki 1996) and worldviews (or cosmologies) (Ogawa 1998, 2001, 2002b; Aikenhead and Ogawa 2007).

Turning our attention to practices in Rika classes, Kawasaki (1992, 1999) argued that students' "observation" (in its scientific meaning) in Rika classes is a different kind of activity to "observation" in a Western science classroom. This difference is guided by the spirit of the Japanese term, "Kansatsu" (borrowed as the translation term of "observation"). Kawasaki insisted that scientific observation could be achieved when the observer intended to separate the target object from the observer. On the other hand, Japanese Kansatsu could be performed by an orientation in which the observer and the observed should be ultimately united into a "oneness" through the action of Kansatsu.

Aikenhead and Otsuji (2000) investigated Japanese Rika teachers' awareness of potential culture clashes within their own Rika classrooms by comparing Japanese teachers with their Canadian (Saskatchewan) counterparts. The researchers found that both groups were unaware of the many culture clashes experienced by students in the typical Rika (science) classroom. The teachers were not ready to implement culture brokering skills without some form of in-service professional development. Concerning Rika teachers' professional development, Ogawa (2002a) found that they were sceptical about the effectiveness of "officially provided" in-service or pre-service programs for professional development, However, they showed strong confidence in the effectiveness of communication with experienced Rika teachers and colleagues characterized as non-formal, daily-based, deep, apprenticeship-like, or in some sense, family-like. Similar findings were shown by Isozaki (2002).

Rika teachers' professional beliefs and practical knowledge concerning Rika teaching, especially their views about the nature of science, have been explored by Japanese researchers (Toda 1992; Shimizu 2002; Tanzawa et al. 2003). For example, Toda (1992) explored pre-service science teachers' views on the nature of science, and Shimizu (2002) investigated in-service Rika teachers' views on the nature of science and their corresponding preferences of teaching methods. Both studies revealed that most Rika teachers (including pre-service or in-service, elementary or lower secondary teachers) held classical views on (1) the nature of science and (2) scientific methods. Maeda and Sato (2004) tackled pre-service teachers' and elementary teachers' views of Rika teaching, and found that the two groups shared a common set of emphases on the objectives of Rika teaching, but with different priorities among these emphases. Isozaki et al. (2007) found that experienced lower secondary Rika teachers' pedagogical content knowledge differed considerably from those of their novice counterparts.

However, none of these studies on Rika teachers' beliefs and knowledge paid special attention to the nature of elementary Rika as consisting of the two components previously discussed (science and Shizen). While most researchers have referred to "beliefs and knowledge about Rika or Rika teaching" in their reports of their research, these researchers have actually been researching (have actually meant) "beliefs and knowledge about science". This indicates how Japanese Rika teachers, as well as Rika educators, have readily misunderstood Rika as simply "science".

Purpose and Methodology

The purpose of this study is to uncover characteristics of elementary Rika teachers' professional beliefs and knowledge, especially in terms of the "aims of elementary Rika". For that purpose, three rather independent empirical studies were conducted.

The first investigated the "overall objectives of elementary Rika" as described in Course of Study (Gakushu Shido Yoryo: GSY)[2] for elementary schools, a legal document by the Ministry of Education, Culture, Sports, Science and Technology (MEXT) which all practicing teachers and textbooks must follow. In the processes of compiling GSY, a significant number of experienced Rika teachers were intensively involved, and hence it is clearly reasonable for us to believe that their ideas on elementary Rika were reflected in the descriptions in GSY.

The second study explored leading elementary Rika teachers' ideas on the aims of elementary Rika teaching, as these were implicitly expressed in their opinion and/or practical papers published in a famous professional journal in Japan.

The third study analysed practicing teachers' comments and criticisms over an episode of an elementary Rika lesson which was taught by one of the country's leading elementary Rika teachers. Results from these three investigations were integrated into a collection of elementary Rika teachers' ideas on elementary Rika. It served as an evidence-based starting point for reflecting or developing a unique theoretical framework for deciphering Japanese elementary Rika teachers' professional beliefs and knowledge.

[2] Gakushu Shido Yoryo (GSY) is written in Japanese. Translation into English, if needed, was done by the present author.

Study I: Official Views of Elementary Rika Objectives

Elementary school education is under the legal control of the central government in Japan. School curricula and course contents for all subjects are legislated under GSY, which prescribes the overall objectives of the subject, contents, and how to deal with the contents. Teachers have no freedom to ignore it. Textbooks, though written by teams consisting of leading in-service or retired teachers, supervisors, university professors, and teacher educators in their respective subjects and published by private publishers, must be authorized by MEXT. Teachers must use the authorized textbooks in their classes. Thus, pupils of the same age all over Japan learn virtually the same set of school subjects, with the same content at the similar time allotment. Elementary Rika is not an exception. In this sense, descriptions of the overall objectives of elementary Rika found in Sect. 4, Rika, of the GSY can be regarded as appropriate material for analysis to determine MEXT's official views on the aims of elementary Rika.

In the current GSY for elementary school (1998 revised version), six elements within the overall objectives of Rika are readily identified. They are (1) to nurture pupils' problem-solving abilities, (2) to nurture feeling of loving Shizen, (3) to commune with Shizen, (4) to perform Kansatsu (observation) and Jikken (experimentation), (5) to develop an understanding of natural phenomena, and (6) to foster scientific view and ways of thinking. As Ogawa (1998) argued, these can be divided into two categories: (a) science-orientated category (elements 1, 4, 5, and 6) and (b) Shizen-oriented category (elements 2 and 3). The latter category is unique when compared with elementary science programs in other countries, where the objectives for science are all of the science form of the first category.

Since "Shizen" and "Shizen education" in the Japanese context are important notions in considerations of Rika programs (at both elementary and secondary levels), these terms need further explanation (e.g. Ogawa 1986, 1995, 1998, 2002b; Aikenhead and Ogawa 2007). In short, Shizen is a perception of the natural world surrounding human beings living in close and dense inter-relationships between humans and nature, shared with most Japanese people for thousands of years. While an English equivalent might be seen to be "nature", the meaning of Shizen is quite different to that of nature, especially in terms of recognition of existence and form of relationships between humans and nature. Shizen serves as a kind of cosmology for the Japanese, and it is a type of indigenous value system. Thus, Shizen education can serve as cosmology education at the deeper level, but in concrete classroom settings it appears as teaching and learning in indigenous ways of seeing, doing, interacting with, feeling, and feeling empathy with Shizen (or our natural world surrounding pupils). A "spirit of science", can be quite sympathetic to the idea of nature, but it is very different from the "spirit of Shizen". In this sense, the overall objectives of Japanese elementary Rika programs consist of two totally different "spirits" or "cosmologies".

Turning now to the category "Shizen education elements" of elementary Rika program, Ogawa (1998) has argued that the element "loving Shizen" has been in-

Table 8.1 Historical changes in elements in the overall objectives of elementary Rika in Gakushu Shido Yoryo

Elements	Year of promulgation									
	1891	1900	1941	1947	1958	1968	1977	1989	1998	2008
Kansatsu and Jikken	X	X	X	X		X	X	X	X	X
Knowledge of natural things & phenomena	X	X	X	X	X	X	X	X	X	X
Scientific views & ways of thinking			X	X		X		X	X	X
Problem-solving abilities								X	X	X
Scientific attitudes (Shizen Ninshiki)			X	X		X				
Relationship between science and daily life	X	X			X					
Abilities of & attitudes towards exploring Shizen				X			X			
Attitudes towards learning directly from Shizen			X		X					
Feelings of communing with Shizen			X	X	X	X	X	X	X	X
Feelings of loving Shizen	X	X	X	X	X		X	X	X	X

Respective sources of regulations are as follows:

1891: Shogakko Kyosoku Taiko (Elementary School Regulations: Mombusho (Ministry of Education) Order, No.11).

1900: Shogakko Rei Shiko Kisoku (Enforcement Regulations of Elementary School Ordinance Revised in 1900).

1941: Shogakko Rei Shiko Kisoku (Revised Enforcement Regulations of Elementary School Ordinance Revised in 1941).

1947: Gakushu Shido Yoryo (Shian) (Course of Study (tentative)). 1958: Gakushu Shido Yoryo (Course of Study).

1968: Gakushu Shido Yoryo (Kaitei) (Course of Study (revised)). 1977: Gakushu Shido Yoryo (Kaitei) (Course of Study (revised)).

1989: Gakushu Shido Yoryo (Kaitei) (Course of Study (revised)). 1998: Gakushu Shido Yoryo (Kaitei) (Course of Study (revised)).

2008: Gakushu Shido Yoryo (Kaitei) (Course of Study (revised)).

volved in the overall objectives of Rika since its very beginning as a school subject in 1891. Table 8.1 is a summary of historical changes in elements of elementary Rika overall objectives from 1891 to the present day (in 2008, the date the latest revision of the GSY has been promulgated). It clearly shows that the two major categories, "spirit of science" and "spirit of Shizen", have been vividly alive and co-existing simultaneously within elementary Rika program for more than 120 years, from its very beginning in the school curriculum.

Since the school education is extensively governed by the national government, the descriptions found in the GSY have had very strong influence on teachers' ideas about the respective subjects in the curriculum. Elementary teachers need to know these GSY descriptions because their teaching must reflect the descriptions. The "spirit of Rika" expressed in the GSY descriptions can thus be regarded as indicating teachers' ideas on elementary Rika program at any time.

Given that the elementary Rika programs are required to contribute to the overall objectives, it is reasonable to infer that Japanese elementary Rika teachers possess a set of professional beliefs and knowledge that are quite different from "elementary science" teachers in western societies. Rika teachers need to prepare for professional knowledge consisting of (1) science and science education, (2) Shizen and Shizen education, and (3) the management of possible conflicts and dilemmas caused by dealing with the first two components. In addition, teachers will be concerned with similar conflicts and dilemmas for learners and with regular class management issues, etc.

Another serious issue emerges when science and Shizen are set side by side. Elementary Rika teachers' ideas about "science" tend to be "indigenized" through continuous, deep and daily contact with the spirit of Shizen, which results in "neo-science" (Ogawa 1995). In the elements relevant to science objectives in Table 8.1, for example, we find "Kansatsu (observation) and Jikken (experimentation)". In scientific activities, observation means, of course, "scientific observation"; but in Japanese elementary Rika contexts it sometimes turns out to be another kind of "observation", discussed briefly in the section above and extensively discussed by Kawasaki (1992, 1996, 1999).

The point here is that the two different components, in terms of epistemology and cosmology, co-exist simultaneously and side by side in the same school subject, elementary Rika. In such situations, each component cannot stand by itself, or cannot be isolated from the other. Both are readily open to change under mutual influences. Ogawa (2002b) called this process an "amalgamation" of the two. Therefore, elementary Rika is not a simple subject consisting of two mutually isolated components, but an "amalgamated" or "indigenized" elementary science that emerges within the Japanese context. Elementary Rika teachers cannot escape (consciously or unconsciously) the strong influences of this duality/heterogeneity of the overall objectives of elementary Rika, if they have been (as they must) following the GSY, in which the duality/heterogeneity has existed for more than 120 years.

In summary, elementary Rika teachers have to maintain a mixed image of elementary Rika, in which the image of science-educationalized Shizen and the image of science with a Shizen flavour co-exist in an inseparable form. It is too difficult to extract "pure" science and/or "pure" Shizen within elementary Rika classes in Japan. Thus, elementary Rika teachers' ideas on Rika ought to differ from those of "pure" science, or those of "pure" Shizen. From this point of view, the theoretical dichotomy of the six elements among the overall objectives of elementary Rika, which Ogawa (1998) argued, seems to be rather simplistic. The reality of the Rika classes is expected to be much more complicated. For example, Kansatsu (observation) and Jikken (experimentation) could be categorized in both "science" and "Shizen" simultaneously.

Study II: Teachers' Ideas on Elementary Rika Objectives

In the second study, leading elementary Rika teachers' ideas on elementary Rika aims are the target of investigation. For that purpose a valid collection of data is needed in which participants freely express their respective ideas on the aims of teaching Rika. However, it is not so easy to obtain such expressions. Elementary Rika teachers must hold their respective ideas on how elementary Rika should be or what aims of elementary Rika should be, through their daily experiences of teaching. But it is very rare for them to express their own ideas on Rika aims explicitly in their writings, because (1) teachers hesitate to do so since the official "objectives of elementary Rika" are pre-defined in the GSY, and (2) there are few channels for teachers to express such ideas openly while there are many chances to express their ideas on class activities, teaching methods, and newly developed teaching materials.

The data source selected for this second study is the set of opinion and/or practical papers published in a popular monthly journal, "Rika no Kyoiku" (Science Education Monthly), edited by the Society of Japan Science Teaching (SJST). This is a journal for practicing and pre-service Rika teachers, supervisors, science educators, and Rika teacher educators. Over the last 57 years, the journal has published opinion and/or practical papers and non-academic research papers about quality teaching. More than two thousand copies of each issue have been published and distributed widely (e.g. to 1,500 members of SJST and sold in bookstores to non-member Rika teachers). Each Journal issue had a certain specific theme relevant to Rika teaching, and includes 8–10 papers with usually two or three papers among them written by leading elementary Rika teachers. Thus, about 30 papers by elementary Rika teachers appear in the journal every year.

Since the editorial team of SJST designs the specific themes in advance and invites appropriate Rika teachers to write, there is no official review system for these papers. This system has merits for the present study because (1) the authors can express their "opinions" much more freely than is the case under an official review system, and (2) they happen to "confess" their personal "tacit" ideas or beliefs in "what Rika should be" with little consciousness because they pay much more attention to the specific theme given by the editorial team. Thus, such papers are a very good data source for the present case study.

Two sets of papers from the Journal were identified for analysis. The first set consists of recent (2006–2008) papers while the second set consists of papers published earlier. Considering the revision year of the GSY (see Table 8.1), the publishing years of the target papers were set around 1966–1968, 1976–1978, 1986–1988, and in 1996–1998. Within the first set, there are 110 papers written by elementary Rika teachers, while the second set consists of 314 papers in total. Each of the target papers was carefully examined from two viewpoints: whether or not the author's personal ideas on and beliefs in the aims of elementary Rika were expressed despite the paper's specific themes, and whether or not the paper contained components not directly relevant to teaching or learning "science". The second criterion

Table 8.2 Collected excerpts (2006–2008) on elementary Rika teachers' beliefs about Rika aims

[Excerpt A1]
In the activities of Kansatsu, science teachers should make much of pupils' viewpoints and ideas. In these cases, Shizen means natural environments with tremendous amount of information, and *pupils can grasp Shizen by going into it with fully open mind and heart.* (Murayama 2008, p. 21)

[Excerpt A2]
Nurturing "problem-solving" ability among children is one of the original purposes of school education. One of the reasons for valuing the process of "problem-solving" in Rika classes is that *it contributes to the achievement of the original purpose of human development:* to nurture abilities of thinking, decision making, and an expression for solving their respective problems. (Sakita 2008, p. 14)

[Excerpt A3]
Pupils first learn Rika at 3rd grade. The teacher does not teach them how to Kansatsu (observe), but *supports them in finding out how to Kansatsu by themselves, for example, helping them think how they can uncover Shizen through much detailed and precise Kansatsu, and by discussion among themselves.* Kansatsu is not a means to teach natural sciences, but *a means for making pupils change their views of Shizen and ways of thinking of Shizen by themselves.* (Watanabe 2008, p. 44)

[Excerpt A4]
My conclusion is that we should not destroy Japanese traditions of Rika cultivated in its unique culture. *Rika is one of the school curricula that serves as "educating human" based upon Japanese original views of Shizen, views of culture, and views of human.* (Ishii 2006, p. 17)

[Excerpt A5]
Rika is originally a school subject for learning the relationships between human beings and Shizen. Learning Shizen and "*learning from Shizen*" itself are fun. And within the school education system, learning in close relationship among peers is much more fun, I believe. (Ito 2006, p. 21)

[Excerpt A6]
What I have been keeping in my mind is "educating human" or "formation of healthy individuals" through teaching the subject Rika. It is also my wish or dream. (Ueno 2006, p. 19)

reflected the fact that elementary Rika is meant to include components other than science, as discussed above. As had been predicted, only six excerpts from the first set (Table 8.2) and only four excerpts from the second set (Table 8.3) were successfully collected and used for analysis (see relevant section in the References for a listing of the papers analysed).

Several components irrelevant to science were found in the excerpts. The first was the nature of Kansatsu (Excerpt A1 and A3, Table 8.2) as one of the major components of Shizen education. The descriptions clearly indicated that the respective authors' ideas on Kansatsu lay far beyond the nature of observation in scientific investigations. The idea of Kansatsu "grasping Shizen" (Excerpt A1) is directly derived from the sense of Gyo in the Buddhism tradition, which is "a practice, especially a repeated exercise or activity, easily performed in an exact manner without trainees' criticism or judgment" (Kawasaki 1999, p. 266), but the author of Excerpt A1 himself was unaware of this.

Another example of Shizen education as a kind of Gyo was "learning from Shizen" (A5, B1). In scientific learning, it should be "learning (something) from (observing or manipulating) Shizen (natural world)". But in Shizen education,

Table 8.3 Collected excerpts (1966–1997) on elementary Rika teachers' ideas about Rika aims

[Excerpt B1]
I think that Rika is not a subject to teach outcomes of natural sciences, but a subject to nurture children *learning directly from Shizen*. And the ability to *learn from Shizen* is cultivated by nurturing scientific views or ways of thinking within themselves. The scientific views and ways of thinking are cultivated through the processes of pursuing facts, principles, and laws. (Nagai 1966, p. 27)

[Excerpt B2]
One of the aims of Rika teaching is *making children commune with Shizen* and work on natural events, thereby developing scientific abilities and attitudes within children. It is important for Rika teachers to guide pupils towards activities in outdoor fields, to *make them interact directly and commune with Shizen,* and to make them grasp Shizen as it is, find out Rika questions and solve the questions. (Tanino 1976, p. 19)

[Excerpt B3]
In each grade and each unit, what is needed is a kind of Rika instruction where pupils not only understand objective phenomena on living things, but also *deepen their emotional feelings and sympathy to these living things as being alive (and dead, too)* through treating them warm-heartedly. (Igarashi 1987, p. 34)

[Excerpt B4]
One of the aims of Rika teaching is to nurture abilities of and attitudes toward problem-solving *through direct experiences with Shizen,* and through scientific views and ways of thinking. Pupils' direct contact with natural events and phenomena make them become sensitive, taste curiosity and enjoyment of Shizen, as well as the joy of discovery. (Kikkawa 1997, p. 20)

emphasis is always laid on the aspect of learners' practice, concentrating on facing and communing with Shizen. Among other aspects of Shizen education, nurturing "affective feelings toward Shizen" (B2, B3) and "experiencing Shizen" (B4) were also found. The idea, "learning a relationship between human beings and Shizen" (A5) was irrelevant to science, but quite relevant to Shizen education.

Besides Shizen, another component of elementary Rika education is learning as a part of human development (Excerpts A2, A4 and A6). This idea is deeply related to the ultimate purpose or goal of schooling itself. The primary purpose of elementary schooling for the authors of these excerpts was to help students develop as individual humans, and all the educational activities (not only learning in school subjects including Rika, but also learning other school activities) should contribute to that specific purpose. This is neither an element of Shizen education, nor of science education. The idea that all school activities should primarily contribute to an individual student's development as a human has been a shared value among Japanese teachers, especially elementary teachers.

Thus, leading elementary Rika teachers' ideas on the aims of elementary Rika are scattered around the notions of "science education", "Shizen education" (including Kansatsu, and affection for Shizen) and "educating humans". Of course, teachers' ideas concerning "science education" were not expressed explicitly because for the authors it was assumed that their ideas concerning "science education" had already been "indigenized".

Study III: Elementary Rika Teachers' Beliefs and Knowledge-in-Action

In order to explore professional beliefs and knowledge-in-action, one typical class episode excerpted from an opinion paper written by a leading elementary Rika teacher was analysed critically by myself (with 27 years of experience as an elementary Rika teacher educator) and by elementary Rika teachers (former students of mine) to make unstructured and free comments about the episode without any suggestions or directions on my part. Out of nine teachers invited to participate, five responded voluntarily to my invitation (four with rich teaching experience, and one in his early career stage). In this section, results are presented on the basis of my own initial critical analysis, plus my critical analysis of the free comments from the five practicing Rika teachers.

Decipherment of the Episode

Table 8.4 show an episode from a 5th grade lesson extensively used in Rika. The learning activity is on "effects of light on plant growth", which is a very popular biological theme in elementary science in most countries. However, there are several different factors deeply involved in the activity.

Dilemma Between the Two Components of Rika Objectives The episode first describes the student K's inner thoughts and feelings. It is clear that she (K) realized a requirement of a relevant scientific experiment: a comparison of experimental results between light-on and light-off conditions. In this sense, most of the pupils, including K, and the Rika teacher were undoubtedly pursuing a lesson objective of the form "to perform Kansatsu (observation) and Jikken (experimentation) with their own prospectus" (quoted from the overall objectives of elementary Rika located in the umbrella of aims of "science" learning). But there is another Rika objective at work: "to nurture feeling of loving Shizen (nature)" (again from the overall objectives of elementary Rika). Prior to the lesson being described, the pupils had been taking care of their respective kidney seedlings, probably from the time of planting. During the process of taking care, a feeling of affection (or love) for their respective seedlings had emerged among the pupils. The feeling is symbolically expressed as "my" seedling. As was discussed in the analysis in Study I (the Rika objectives in GSY), the following are outcomes of the objectives: "to commune with Shizen (nature)", and "to nurture feeling of loving Shizen (nature)".

The feeling readily provokes a difficult dilemma among the pupils, derived from a conflict between two components of Rika objectives with different value orientations. That is, "respecting a plant's life" versus "investigating scientific truth". Both are among the six Rika objectives in GSY. In the episode, K's struggle with the dilemma within her mind is vividly described. However, she obtained neither resolution nor a kind of "trade-off" by herself. It seems to me that a similar struggle

Table 8.4 Excerpt from an opinion paper written by a leading elementary Rika teacher—the target episode for the Study III analysis

K (name of a girl of 5th grader) was at a loss.

All her group members believed that light was needed for their kidney bean seedlings to keep growing. They also realized correctly that they needed an experiment that compared results of how the seedlings grow in light-on and light-off conditions if they want to obtain evidence. But in the process of developing their experimental design, another issue came up in their minds: "Whose kidney bean seedlings should be set in the light-off condition?"

Four kidney bean pots, which had been well taken care of by each of four group members, were on the group table. They were sprouting several leaves and were expecting to keep thriving. All the members knew from their daily experience that plants would grow by themselves by receiving sunlight on the leaves, they were expected not to grow in the shadow. Thus, plants in the dark would die soon (sometimes such examples were seen) while they did not understand its mechanism.

Every member did not want his/her own kidney bean pot in the dark. But they realized that one of them should do this. K decided, "If someone's kidney seedling needs to be sacrificed, there is no help for her kidney… It is okay if her kidney bean is sacrificed for our group." "But, is this experiment setting such seedlings into the light-off condition really needed, because the results are not in doubt?"

She must have felt this way long before deciding her kidney seedling seedlings be sacrificed. Group discussion ended by K's proposal of sacrificing her seedling. Among other groups, however, nobody wanted to provide his/her kidney seedling for the light-off condition, because all of the pupils loved their own kidney seedlings.

The class teacher realized the pupils' feelings, and decided to provide kidney seedlings he took care of for the probable use of the light-off condition. K was actually relieved with the decision and murmured that she did not want to put her seeding to the light-off condition, either.

The pupils actually felt affection or love to their respective kidney seedlings. It comes from their "relationship" to the seedlings, because they had been taking care of it from the time it sprout. It is clearly different from the scientific case of seeds. In the experimentation of examining factors effective to seed germination (appropriate warmth, water, and air), no such special affective feelings emerged among the pupils. The action, taking care of kidney seedlings by themselves, triggered the feelings of relationship to develop and thus their affection or love for their seedlings. (Tsuyuki 2007, p. 20)

could be expected to occur among every pupil joining the lesson. Eventually, the teacher presented a resolution by providing seedlings for the light-off condition, cultivated by the teacher himself. Pupils did not need to "kill" their own seedling in the experiment. Through the whole process, the pupils faced their own feelings of affection for the seedlings. In this sense, the objective "to nurture feeling of loving Shizen (nature)" was achieved.

Although there is no description in the episode, probably the "scientific experiment" was successfully performed since the teacher provided the seedlings for light-off condition. Thus, we may say that another Rika objective was also successfully achieved in this lesson: "to perform Kansatsu (observation) and Jikken (experimentation)".

Ethical Dilemma in the Rika Lesson Another issue remains. K suffered from a kind of ethical dilemma. Within her group, she faced a decision over whose seedlings would be sacrificed in order to achieve her group's objective to perform a scientific investigation. Within her mind, she struggled with two conflicting ideas: (1) asking someone to sacrifice his/her seedling so her own seedling would be secured, and (2) proposing that her seedling be sacrificed. This ethical dilemma was temporarily resolved by the second option, where she became satisfied to think that she contributed to the group's happiness. A group's decision making is a popular topic in moral education or ethics classes in Japan. The point is that such issue can readily be integrated into the group activities, even in Rika lessons. The experienced Rika teacher in the episode resolved the situation by providing his own seedlings to every group in the class. What kind of relevant professional beliefs and knowledge were working on this specific case? It is relevant neither to science and its teaching nor to Shizen and its teaching. It is a kind of belief and knowledge necessary for elementary classroom teachers in general, not specific to Rika teachers. However, it is not negligible when thinking of the professional beliefs and knowledge domain for/of elementary Rika teachers.

The episode again suggests that elementary Rika teachers' professional beliefs and knowledge will be different from those of elementary science teachers in Western countries. Also, elementary Rika teachers need to develop unique capabilities to cope with factors other than teaching science. In this sense, we may say elementary Rika teachers' professional beliefs and knowledge in such a context should be "indigenized".

What are the differences between the nature of elementary Rika teachers' professional beliefs and knowledge for teaching indigenized science (elementary Rika in a Japanese context) and those of Western science teachers? Why and how has the "indigenization of professional beliefs and knowledge among elementary Rika teachers" occurred? Why and how do Rika teachers develop such indigenized professional knowledge? These are interesting but yet unresolved questions, indeed.

Considering Elementary Rika Teachers' Comments on the Episode

How do practicing elementary Rika teachers respond to the episode? I sent the episode (the original Japanese version) to each of the respondents, inviting their free comments on the episode. Several viewpoints were extracted from a preliminary analysis of their responses: (1) an evaluation of the lesson, (2) the two components of elementary Rika objectives (science education and Shizen education), (3) K's ethical dilemma, (4) pupils' affection (or love) for plants, and (5) other educational objectives (Table 8.5).

Evaluation of the Lesson Three of the five Rika teachers did not directly express their evaluation on whether the lesson was good or bad, nor did they express any negative comments. They seemed to share the rationale or philosophy of the lesson. The remaining two teachers, however, explicitly expressed negative impressions of the lesson. For instance, a female elementary Rika teacher with 20 years' experience

Table 8.5 Summary of the practicing elementary Rika teachers' free comments on the episode (gender, years of teaching experience)

		Y.W. (Male, 14 years)	M.M. (Male, 20 years)	C.W. (Female, 20 years)	T.I. (Male, 26 years)	S.T. (Male, 5 years)
Evaluating the lesson		Not mentioned	Not mentioned	Partially negative	Not mentioned	Negative
Two types of Rika objectives	Recognition	Yes	Yes	Yes	Yes	No
	Awareness of dilemma	Yes	Yes	Priority should have been given to learn science in Rika lessons	Difficult to achieve both objectives simultaneously in Rika lessons	No
	Pedagogical resolution to avoid the dilemma	Awareness of the need to develop alternative ways	Each pupil cultivates two seedlings	Each pupil cultivates more than two seedlings or provides seedlings cultivated by the teacher. After the experiment, the seedlings used for light-off condition should be transplanted to thrive into school garden	If priority is given to the objective, "nurturing attitude of respecting plant life," I shall do it this way. The two objectives should be treated separately because of the difficulty to achieve them simultaneously	In this specific lesson, the teacher should have made pupils cultivate plants for experiments by the whole class not by each individual pupil
K's ethical dilemma	Recognition	Not mentioned	Not mentioned	The moral dilemma is a secondary issue	Not directly mentioned	Not mentioned
	Pedagogical resolution to avoid the dilemma	Not mentioned	Not mentioned	Not mentioned	Not "each individual's seedling", but "our class' seedlings" or "group seedlings" for this experiment	Not mentioned
Pupils' affection towards plants		Usually happens. Sympathetic with their relationship with living things	Unaware of such feelings emerging among pupils in the case of plants. (while aware of the feeling in the case of animals or insects)	Not mentioned	The extent of affection for the plants is cultivated differently from the case of cultivating "by myself" and that of cultivating "by ourselves"	Not mentioned
Rika objectives and other educational objectives		In an appropriate and needed situation, priority can be given to moral education components even in Rika class	Not mentioned	Not mentioned	Not mentioned	Not mentioned

first argued that the objective of this specific lesson should have been the scientific investigation identifying environmental conditions that caused kidney seedlings to thrive. Then, she continued:

> The Rika lesson shown in the episode is apt to be accepted positively as an excellent example of a Rika lesson, but I feel that the core objective of the lesson seems to be vague. The group discussion in the lesson is not a scientific one (how to control environmental conditions for plant growth) but a moral one (whose kidney seedlings will be sacrificed?). (C.W.)

Although she expressed such a negative opinion, she also commented:

> Considering it from the viewpoint of respecting living things, the teacher should have declared to pupils, in advance, that all the seedlings shall be transplanted in the school garden after confirming the results of the experiment, in order to avoid their dying out. (C.W.)

Such comments indicated that she did not want to exclude the component "respecting living things" from the Rika lesson, but insisted that much more emphasis should have been laid upon the objective "scientific investigation" in this specific lesson. In this sense, we can say she is aware of the existence of the two components of Rika objectives.

On the other hand, a male teacher, still in early in his career (5 years experience) claimed:

> It seemed to me that the teacher wrongly mixed Rika teaching (through experimentation) the process of plant growth with moral education on making pupils take care of plants with deep affection. Of course, while I believe that both are important educational activities in elementary school level, for me the teacher lacked an appropriate "prospective" on the aim of the activity concerned.
> Since they (5th graders) already experienced an educational activity to make pupils take care of their plants during their earlier grade days, I do not feel the need for these 5th graders to be involved again in such an activity to take care of plants with deep affection. (S.T.)

He realized the two components of educational objectives, but did not consider that the objective "making pupils take care of plants with deep affection" to be relevant to one component of Rika objectives. He simply identified it as an objective in moral education. Also, though he referred to the pupils' experience of taking care of their personal plants, he thought it was not an appropriate activity in Rika classes but instead it belongs in Seikatsu-ka (Life Skills) classes for 1st and 2nd graders, because there is no Rika classes in these two grades. We can conclude that he seems to be unaware of the existence of the two components of Rika objectives.

Thus, the evaluations of the lesson were not consistent among the teachers. The main reason for this discrepancy seems to be the difference among their opinions on the extent to which emphasis should be placed on each of the two components of Rika objectives in this specific lesson.

Dilemma Between the Two Components of Rika Objectives As is shown above, the four experienced teachers (but not the one early career teacher) recognized the existence of the two components of elementary Rika objectives. And, some of them did propose ideas on a pedagogical resolution to avoid the dilemma, which suggests they may have had similar experiences in their teaching.

Mr. M.M. (male, 20 years' experience) proposed that each pupil should cultivate two seedlings as follows:

> If I were the teacher, without being at a loss, I would make each pupil cultivate two pots of seedlings from the beginning, because I used to prefer to make each pupil do experiments individually. (M.M.)

Interestingly, Mr. S.T., who realized the objective "making pupils take care of plants with deep affection" as a component to moral education but not to a Rika lesson, showed his alternative idea, which was the same as the above for the experienced teachers:

> Since the purpose of this experiment was just to compare growth of plants in light and dark conditions, the teacher knew in advance that the plants in the dark condition would not thrive further. Thus, the teacher should not have made each pupil take care of their plant, but should have made the class members, as a whole, take care of all the plants needed for the experiment, for example, as in their school garden. (S.T.)

He did not mention precisely why he believed that the alternative could resolve the dilemma.

Mr. T.I. with 26 years' experience did not show any concrete alternatives on this very specific point, but confessed as follows:

> If the lesson would aim to cultivate affection for plants in addition to the ordinary scientific objectives, I could teach it in a similar way shown in the episode. But, I feel that these two different objectives, cultivating affection towards living things and scientific knowledge and/or ways of thinking, are difficult to achieve simultaneously. Both of them seem to be incompletely achieved.
> The similar situation happens with dissections, too.
> In order to cultivate a close relationship between pupils and living things, or to cultivate love or affection for Shizen, another kind of learning setting should be planned. These two different components should be taught separately. Rika teachers continue to struggle with a dilemma or contradiction to teach two different things, if they are forced to achieve both objectives simultaneously. (T.I.)

His struggle dealing with the two different objectives simultaneously is clearly obvious. I know that he is very familiar with the issues of Shizen education and the affection towards Shizen among Japanese people. The deeper he understood the difference between science and Shizen, the more he felt difficulty in teaching the two simultaneously. In this sense, elementary Rika teachers with rich knowledge on Shizen and that on science may consider the duality/heterogeneity of the elementary Rika objectives much more seriously than teachers with an incomplete knowledge.

As Ogawa (1995) has suggested, Rika teachers should be aware that the two components can be treated in parallel as distinctively different objectives even in the same lesson. Of course, T.I.'s idea (the two different components of the Rika objectives should be treated separately in different settings) is an alternative, but even in that treatment, on the learners' side, each pupil must resolve by him/herself, the dilemma or conflict between the two different values nurtured within his/her mind.

The teachers who responded to the survey showed similar alternative solutions. This implies that they had similar experiences in their daily practice and they identified appropriate pedagogical resolutions to avoid or minimize possible dilemmas or conflicts between the two components of elementary Rika objectives. Since such pedagogical knowledge is not provided in formal pre-service or in-service Rika teacher training programs, we can conclude that it is acquired through their personal experiences or through a kind of non-formal, daily-based, deep, apprenticeship-type, or in some sense, family-type of communication with their senior Rika peers who have rich experiences (see Ogawa 2002a). It is not explicit knowledge but tacit knowledge to be shared among the community of elementary Rika teachers.

K's Ethical Dilemma Another kind of dilemma found in the episode is K's ethical dilemma. On this point, three teachers did not mention anything. They were neither aware of the issue nor concerned about it. But two teachers commented. Ms. C.W. with 20 years' experience insisted that the primary objective of this specific lesson should have been science. While a moral dilemma found in the lesson is a secondary issue for her, she suggested possible resolutions:

> The discussion that happened in the class is not a scientific one (how to control environmental conditions for plant growth) but a moral one (whose kidney seedlings should be sacrificed for all). In order to avoid this kind of moral dilemma, and to honestly think of the central objective of this class, the teacher should have made each pupil cultivate more than two seedlings, or the teacher should have prepared extra seedlings in advance. (C.W.)

The alternative she presented was the same proposed by others for resolving the dilemma of Rika objectives. But she believed that they resolved a pedagogical issue: to avoid or minimize the ethical dilemma.

Mr. T.I. with 26 years' experience mentioned the following:

> In my case, I usually try to make our pupils cultivate more kidney bean seedlings than the total number of class members; not as "my seedlings" for each individual but as "our class' seedlings" or "our group's seedlings" as a whole. I think that I have finally come to this way of teaching this lesson in order to avoid the situation shown in the episode, based on my experiences in teaching this unit for several years. I know not only Rika teachers but also non-Rika major teachers do so.
>
> If cultivating kidney seedlings as "our group's seedlings", even in four members' group, pupils can prepare for the experiment to set seedlings for the light- and dark-conditions without expressing their negative feelings towards the experiment just as the episode showed.
>
> Actually, the extent of affection for their kidney seedlings is quite different between "my seedling" and "our seedlings". (T.I.)

Strong attachment to individual plants emerges when each pupil cultivates their personal plant. This belief came from his rich teaching experience. Thus, he "invented" a way to avoid the ethical dilemma shown in the episode: Making a group, or a whole class cultivate "their" plants.

These two teachers realized that the ethical dilemma described in the episode served as a kind of impediment to pursue the elementary Rika objectives. Viewing it from the reverse side, Rika lessons, especially in the teaching unit "life and living things", cannot be free from ethical or moral issues. Rika lessons cannot work in ethical or moral vacuums. Elementary Rika teachers also need to develop professional knowledge of and professional pedagogical skills for coping with such ethical or moral issues.

Pupils' Affection Towards Plants Three teachers referred to the issue of pupils' affection towards plants

. Mr. Y.W. with 14 years' experience expressed his sympathy:

> K came to feel deep affection towards her kidney seedling, which was the object of her scientific experimentation and observation though the processes of experimentation and observation. Similar situations could readily emerge in ordinary Rika lessons. As a teacher, I feel sympathy for K's attitude toward kidney seedlings. (Y.W.)

Mr. T.I.'s quotation above also indicates an interesting story of pupils' affection towards plants. It is worth emphasizing that "the extent of affection to their kidney seedlings is quite different between 'my seedling' and 'our seedlings'". For Mr. T.I., it is one of the important professional beliefs and knowledge needed when teaching units on plants.

Mr. M.M. with 20 years' experience confessed his new finding in the episode:

> Honestly speaking, I am not aware that the issue of affection for living things appears, not only in dealing with animals or insects but also in dealing with plants. I hesitate because of my unawareness of this issue.
>
> Actually, when treating animals or insects, I have been very seriously keen about the issue of affection for life. But I haven't cared about it in the case of plants. I have not taught this specific Rika lesson with the idea that pupils can anticipate a terrible result of the experiment and may show psychological resistance or question the experiment itself from their affection towards kidney seedlings.
>
> From the episode, I realize that Rika teachers should take the issue seriously. (M.M.)

For him, affection towards plants has not previously been part of his thinking. But the present episode made him aware that it is one of the key factors to influence the lesson significantly. If he had a chance to read Mr. T. I.'s comment, "The extent of affection for their kidney seedlings is quite different between 'my seedling' and 'our seedlings'", his idea may be deepened further.

Other Educational Objectives Only one teacher, Mr. Y.W. with 14 years' experience mentioned the relationship between Rika objectives and other educational objectives.

> Among the Rika objectives, there is an objective, "nurturing respecting living things". In addition to this, as is shown in GSY, components of moral education and/or Tokubetsu Katsudo (special activities) should be treated appropriately if they emerge. So, even in the Rika lessons, moral issues or group management issues emerge, priority can be given to cope with such issues. (Y.W.)

This idea is very popular among elementary teachers. Elementary Rika teachers should serve as "elementary teachers" before serving as a "Rika teacher". Thus, they are primarily sensitive to the overall objectives of elementary education. From this point of view, elementary Rika teachers' professional beliefs and knowledge should include those of elementary teachers in general. Elementary Rika teachers must wear a kind of dual spectacles, one lens of a general elementary teacher and the other lens of a Rika teacher. In Japan, the priority has always been given to the former. Elementary Rika teachers should be excellent elementary teachers before serving as excellent Rika teachers.

Conclusions and Implications

The three empirical studies revealed certain common findings about the unique characteristics of Japanese elementary Rika, which are summarized in Fig. 8.1.

Study I found that within the "spirit" expressed in the overall objectives of elementary Rika, there have been two distinct components: science-oriented education and Shizen-oriented education. Each of these two components consists of several elements respectively. These two components stand on separate and independent epistemological and cosmological foundations. More importantly, the two cannot exist together as their original separate forms, or cannot be isolated from each other. Both of the two change under continuous mutual influences, or as Ogawa (2002b) has described it "amalgamation". In other words, elementary Rika is not a simple subject consisting of the two mutually isolated components, but it is "amalgamated" or "indigenized" elementary science in a Japanese context. Elementary Rika teachers could not escape from the strong influences of the duality/heterogeneity of the overall objectives of elementary Rika if they had been obeying and taking seriously the descriptions of the GSY. This duality/heterogeneity has been present for more than 120 years from the very beginning of elementary Rika program.

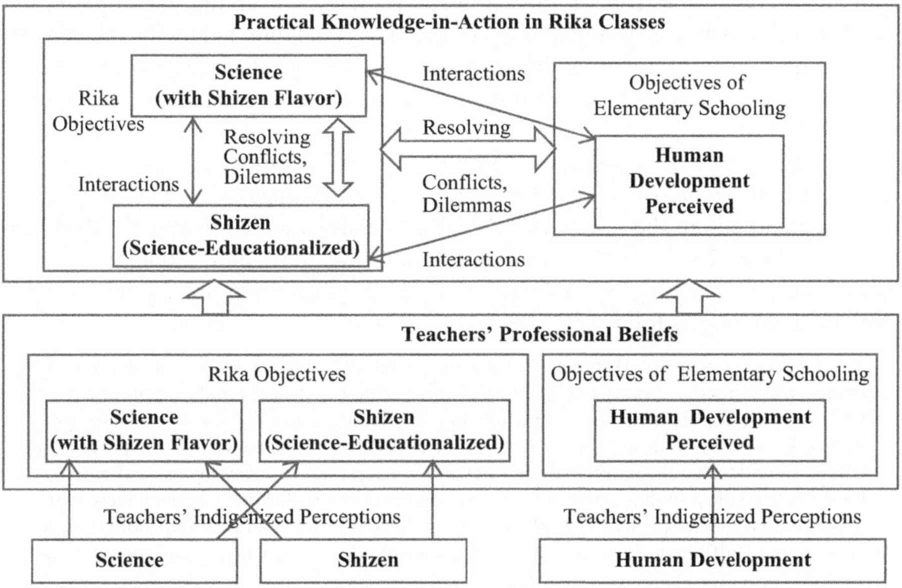

Fig. 8.1 Japanese teachers' beliefs and practical knowledge of elementary Rika teaching

Study II, the analysis of experienced elementary Rika teachers' ideas in the aims of elementary Rika, showed that in addition to the two components of elementary Rika mentioned above, another important component—"learning Rika as a part of educating human"—is present. This comes from the idea that elementary schooling as a whole should contribute to children's development as a complete human being. To become a man or woman of knowledge is not the top priority. Rather the top priority is to become a man or woman with moral and ethical norms and/or having wisdom as a human being.

The origin of this priority may date to the very beginning days of elementary schooling in Japan. In the first few years of the modern Japanese school system, Pestalozzian pedagogical methods were welcomed by educational administrators, stressing the natural gifts of the child. The child was encouraged to learn from experiences and not by memorization or by abstraction. But soon more conservative voices spoke in favour of the need for moralizing methods to implant in a child certain fixed moral ideas (Murthy 1973). And finally a German style of elementary schooling system and philosophy of Herbertian pedagogical principles (Herbert 1892; Duke 2009) was adopted, in which moral training was the chief goal for all educational endeavour (Murthy 1973). In Shogakko Rei (Elementary School Ordinance), as revised in 1890, the primary purpose of elementary school was first defined as follows:

> Primary (Elementary) Schools are designed to give children the rudiments of moral education and of civic education together with such general knowledge and skills as are necessary for life while due attention is paid to their bodily development. (Murthy 1973, p. 363)

The core was a new type of morals education, which was not the traditional Confucian morals education. Duke (2009) argued "Rather than serving the emperor, the new school system was designed to serve the state. Each unit led by the elementary school had a specific role in an integrated structure to produce loyal patriotic citizens who would contribute to a strong and independent nation in the next century" (pp. 337–338).

Although space does not allow more discussion of this historical perspective, contemporary Japanese teacher educators as well as teachers themselves, tend to be more sympathetic to the notion of "Bildung" ("cultivation/edification" (Masschelein and Ricken 2003)) in the German tradition (Westbury et al. 2000; also Chap. 3) than that of "teaching" or "education" in the English tradition. Wimmer (2003) explained Bildung as follows:

> The German concept of Bildung encompasses a highly complex web of meanings and usages which render it particularly untranslatable. Bildung denotes both the processes of learning—the development of the personality or identity—and the results of those processes. In contrast to the concept of learning or development, the concept of the process of Bildung implies that the individual goes beyond himself (sic) in a way that is neither teleological nor goalless in the course of his (sic) individual self-realisation and the concomitant advancement of the species. This process is considered to have no goal (freedom) and to have a goal (fulfilment or perfection), to be determinate (inner nature) and indeterminate (self-creation). (p. 185)

The spirit of Bildung was acceptable in Japan because certain aspects were regarded to fit or harmonize with the notion of how Japanese children should be trained in a newly established system of elementary schooling. However, it was not the idea of "cultivation of independent free citizen" but that of "cultivation of the state's good and quality subjects" under the strong ideology of nationalism at that time.

The third Study clearly showed how these three complex components are vividly present in elementary education and how they interact in complicated ways in actual class settings, causing certain kinds of dilemmas and confusions among practicing elementary Rika teachers and their students. While experienced elementary Rika teachers had already developed certain ways to resolve these tough situations (probably through their rich experiences), early career teachers may still be struggling with these challenges in their classrooms. The teachers' reactions were rather diverse and each seems to be acceptable.

Several implications follow from the empirical studies. First, research on professional beliefs and knowledge among elementary Rika teachers needs to develop an appropriate theoretical framework, something that cannot be borrowed directly from frameworks developed in western school science settings. The Japanese framework serving for real Rika classroom settings should contain at least three major components ("science", "Shizen" and "educating human [Bildung]"). Second, Japanese educators need to investigate lower and upper secondary Rika teachers' professional beliefs in and knowledge of Rika teaching as well. In lower- and upper-secondary Rika, the overall objectives in GSY have excluded the component, Shizen education. The official intention is clear: at the secondary level, Rika should deal with "science" alone. However, do secondary Rika teachers also believe that "loving Shizen" should be one of the important objectives of Rika? If so, the reality of secondary Rika classes could be complicated in a similar way as elementary Rika. Third, for teaching contexts outside of Japan, (e.g. in countries where indigenous knowledge and wisdom still function in communities), indigenization of a school science program might occur, especially at elementary level. Science teachers' beliefs about "science" might be different from western modern science itself. And fourth, a debate-provoking question can be added: Is elementary school science in western countries already "amalgamated", too?

Finally, the present study strongly implies that Japanese elementary Rika classes deal with more than Rika (i.e. more than science and Shizen). Because many science educators are interested in the efficient and effective ways to improve students' performance in learning science, they unconsciously focus on teaching/learning "science" alone and overlook the Shizen and other things taking place in Rika classes. If science educators observed a Rika class through the eyes of anthropologist, for instance, they would see more than science/Shizen being taught. They would notice a rich collection of enterprises, events, hidden rules and communication codes, etc. because students in Rika classes are encouraged to enjoy and be committed to their engagement with materials (e.g. kidney seedlings) and/or activities (Kansatsu or Jikken) presented to them. This engagement not only draws upon students' cognition (sometimes referred to "alternative frameworks" or "misconceptions"), but also upon their affective attachment to these materials and/or activities. Accordingly, teachers in elementary Rika classes are expected to attend to the development of the whole child, which requires interactions beyond Rika activities. Thus, the world of professional beliefs and knowledge-in-action among elementary Rika teachers is more holistic and encompassing than the more limited world of reductionist science educators.

References

Abell, S. K. (2006). Research on science teacher knowledge. In S. K. Abell & N. G. Lederman (Eds.), *Handbook of research on science education* (pp. 1105–1140). Mahwah: Lawrence Erlbaum Associates.

Aikenhead, G. S., & Ogawa, M. (2007). Indigenous knowledge and science revisited. *Cultural Studies of Science Education, 2,* 539–591.

Aikenhead, G. S., & Otsuji, H. (2000). Japanese and Canadian science teachers' views on science and culture. *Journal of Science Teacher Education, 11,* 277–299.

Appleton, K. (2006). Elementary science teaching. In S. K. Abell & N. G. Lederman (Eds.), *Handbook of research on science education* (pp. 493–535). Mahwah: Lawrence Erlbaum Associates.

Duke, B. (2009). *The history of modern Japanese education: Constructing the national school system, 1872–1890.* New Brunswick: Rutgers University Press.

Fernandez-Balboa, J. M., & Stiehl, J. (1995). The generic nature of pedagogical content knowledge among college professors. *Teaching and Teacher Education, 11,* 293–306.

Gess-Newsome, J., & Lederman, N. G. (2001). *Examining pedagogical content knowledge.* Dordrecht: Kluwer.

Grossman, P. L. (1990). *The making of a teacher: Teacher knowledge and teacher education.* New York: Teachers College Press.

Herbert, J. F. (1892). *The science of education.* London: Swan Sonnenschein

Isozaki, T. (2002). *Research on curriculum development of science and mathematics teacher education for promoting the twenty first century science and mathematics education.* Higashi-Hiroshima: Hiroshima University. (In Japanese).

Isozaki, T., Yoneda, N., Chujo, K., Isozaki, T., Hirano, T., & Tanzawa, T. (2007). Theoretical and empirical research on the pedagogical content knowledge in the case of lower secondary science teachers. *Journal of Science Education in Japan, 31*(4), 195–209. (In Japanese with English abstract).

Jones, M. G., & Carter, G. (2006). Science teacher attitudes and beliefs. In S. K. Abell & N. G. Lederman (Eds.), *Handbook of research on science education* (pp. 1067–1104). Mahwah: Lawrence Erlbaum Associates.

Kawasaki, K. (1992). An epistemological study on "Kansatsu" believed to be a precise equivalent for "Observation." *Bulletin of Society of Japan Science Teaching, 33*(1), 71–80. (In Japanese with English abstract).

Kawasaki, K. (1996). The concept of science in Japanese and western education. *Science & Education, 5,* 1–20.

Kawasaki, K. (1999). A deductive description of cultural diversity of "observation" in science education. *Journal of Science Education in Japan, 23*(4), 258–270.

Maeda, K., & Sato, T. (2004). Teacher-training course students' and elementary school teachers' views of science teaching. *Memoirs of the Faculty of Education, Kumamoto University. Natural Science, 53,* 51–57. (In Japanese).

Masschelein, J., & Ricken, N. (2003). Do we (still) need the concept of Bildung? *Educational Philosophy and Theory, 35*(2), 139–154.

Murthy, P. A. N. (1973). *A historical study of the role of education in the making of modern Japan*. New Delhi: Ashajanak Publications.
Ogawa, M. (1986). Toward a new rationale of science education in a non-western society. *European Journal of Science Education, 8,* 113–119.
Ogawa, M. (1995). Science education in a multi-science perspective. *Science Education, 79,* 583–593.
Ogawa, M. (1998). A cultural history of science education in Japan: An epic description. In W. W. Cobern (Ed.), *Socio-cultural perspective on science education* (pp. 139–161). Dordrecht: Kluwer.
Ogawa, M. (2001). Reform Japanese style: Voyage into an unknown and chaotic future. *Science Education, 85,* 586–606.
Ogawa, M. (2002a). How are the novice getting to be the expert?: A preliminary case study on Japanese science teachers. *Journal of the Korean Association for Research in Science Education, 22,* 1082–1102.
Ogawa, M. (2002b). *Nature of indigenous science: A stratified and amalgamated model of knowledge and cosmology*. Paper presented to the 33rd Annual Meeting of the Australasian Science Education Research Association, Townsville, Australia.
Shimizu, M. (2002). Teachers' views of the philosophy of science and the way science is taught in the classroom. *Journal of Research in Science Education* (Society of Japan Science Teaching), *42*(2), 43–50. (In Japanese with English abstract).
Shulman, L. S. (1986). Those who understand: Knowledge growth in teaching. *Educational Researcher, 15,* 4–14.
Shulman, L. S. (1987). Knowledge and teaching: Foundations of the new reform. *Harvard Educational Review, 57*(1), 1–22.
Tanzawa, T., Kumano, Y., Tsuchida, S., Katahira, K., Imamura, T., & Nagasu, N. (2003). Characteristics of Japanese views on science and technology. *Journal of Research in Science Education* (Society of Japan Science Teaching), *44*(1), 1–12.
Toda, K. (1992). Pre-service science teachers' view of science and its influence on their selection of science teaching methods. *Bulletin of Society of Japan Science Teaching, 33*(1), 59–70. (In Japanese with English abstract).
Westbury, I., Hopmann, S., & Riquarts, K. (Eds.). (2000). *Teaching as a reflective practice: The German Didaktik tradition*. Mahwah: Lawrence Erlbaum Associates.
Wimmer, M. (2003). Ruins of Bildung in a knowledge society: Commenting on the debate about the future of Bildung. *Educational Philosophy and Theory, 35*(2), 167–187.

List of the Papers Used for Study II and Study III (All the papers were written in Japanese and the excerpts referred were translated by the present author)

Igarashi, H. (1987). Teaching materials on living things and its environment in elementary school Rika. *Rika no Kyoiku, 36*(4), 34–37.
Ishii, M. (2006). Considering Rika from the origin. *Rika no Kyoiku, 55*(2), 16–17.
Ito, H. (2006). Seeking to make students experiences joy of learning, by doing science. *Rika no Kyoiku, 55*(11), 19–21.
Kikkawa, H. (1997). Learning Rika by children's viewpoints. *Rika no Kyoiku, 46*(1), 20–22.
Murayama, T. (2008). Course contents of new B-unit "Life and Earth." *Rika no Kyoiku, 57*(5), 20–23.
Nagai, M. (1966). Instruction guiding pupils toward discovery of facts, principles and laws. *Rika no Kyoiku, 15*(7), 27–29.
Sakita, T. (2008). To nurture children loving science. *Rika no Kyoiku, 57*(4), 14–16.

Tanino, Y. (1976). Improving instruction of outdoor Kansatsu in the units of living things. *Rika no Kyoiku, 25*(7), 19–22.
Tsuyuki, K. (2007). What children commune with Shizen. *Rika no Kyoiku, 56*(2), 20–22.
Ueno, K. (2006). What would Rika be in next 10 years? *Rika no Kyoiku, 55*(2), 18–19.
Watanabe, M. (2008). What fan it is to observe in detail! *Rika no Kyoiku, 57*(5), 44–46.

Chapter 9
Chinese Teachers' Views of Teaching Culturally Related Knowledge in School Science

Hongming Ma

Introduction

Cultural studies in science education have been gaining more attention in recent years. One much-debated theme in this area is teaching culturally related knowledge in modern school science. Within the context of science education, "culturally related knowledge" refers to the knowledge that can be treated as science, but has different cultural origins from that of Western Modern Science (WMS) or Euro-centric science (Aikenhead and Ogawa 2007). Different points of view about the nature of science and the relationship between science education and culture have a direct influence on people's opinions of the content and pedagogy of science teaching in school.

For those who believe that Western science is the best form of knowledge to account for particular phenomena, traditional knowledge should only be used for the purpose of enhancing understanding of Western science. Speaking from a solely Western science perspective, Matthews (1994) claimed that only WMS could meet modern challenges and therefore he assumed that adopting multi-science perspectives resulted in teaching second-best or inadequate understandings. Matthews' claim reflects a persistent ignorance and marginalisation of the potential value that a diversity of perspectives can bring to both the understanding of natural phenomena in general and of students' needs to meet modern challenges in diverse local contexts.

Starting from the premise that school science education is a cultural phenomenon, some educators and researchers suggest the introduction of a pluralistic science education which does not insist that students with diverse cultural backgrounds be enculturated in Western science. Reiss (1993) criticises the dominance of a Western view of science in school syllabi and textbooks, and points out that it is an obstacle to equal opportunities in science education in a pluralist society. Aikenhead (2000) argues that a pluralistic multi-science approach is the only way to attain the goal of "science for all".

Although remaining controversial, pluralistic science education has also attracted the academic attention of many educators and researchers from non-Western

H. Ma (✉)
Monash University (Gippsland campus), Churchill, VIC, Australia
e-mail: hongming.ma@monash.edu

D. Corrigan et al. (eds.), *The Professional Knowledge Base of Science Teaching*,
DOI 10.1007/978-90-481-3927-9_9, © Springer Science+Business Media B.V. 2011

countries. In these countries, non-Western students are the majority in the society, and yet the non-Western science tradition is often being marginalised in school science education. For example, Loo (2001) points out that while many academic literatures might have focused on the relationship between Science and Christianity, such knowledge may not be relevant in Islamic communities. Jegede (1994) proposes a "conceptual ecocultural paradigm" (p. 130) for African science educators, which gives a significant consideration to the African view of Nature, socio-cultural factors, and ways of thinking in teaching science within African society.

The majority of students in China come from a non-Western cultural background. However, the mainstream understanding of science in Chinese modern school science curricula is WMS. Given the cultural context of contemporary China, what does the debate mean to Chinese science education? Discussion concerning teaching culturally related knowledge in Chinese modern school science has been rare. An example of bringing traditional perspectives into the modern educational context comes from Hua et al. (1999). In their paper "Taoism and its implications for science education" they discuss the relationship between an ancient Chinese philosophy, Taoism, and the contemporary environmental crisis. They argue that "a conventional science curriculum is probably sacrificing the chance for teachers and learners to realise the important holistic ideology eminent in systems thinking—the whole is greater than the sum of its all parts—toward the Universe" (p. 9). In part, Hua et al. try to seek compatibility between Taoism and modern science by analysing the contemporary environmental crisis through Taoist perspectives.

The marginalisation of Chinese Native Knowledge (CNK) in science may be at the expense of both Chinese cultural reproduction in science education and a deeper understanding about the nature of science itself. However, to what extent and how CNK can be introduced into the school science classroom remains unknown. Science teachers are the ones who have a significant influence on the implementation of an intended curriculum. In fact, teachers have been described as "curriculum makers" (Clandinin and Connelly 1992). However, little is known about their opinions about the above issues. This chapter is based on a research study of Chinese science teachers' views on teaching CNK in school science. Before presenting the research findings, it is necessary to briefly introduce the cultural situation of Chinese science education. After some methodological concerns are addressed, teachers' views of how to deal with CNK in their teaching are presented. This is followed by discussions of the interplay of teachers' understanding of the nature of science and their pedagogical practice. Some implications for developing pedagogical strategies for teaching pluralistic views in science classroom are also discussed.

CNK in Chinese School Science Curricula

Before starting a brief history tour, it is necessary to clarify the use of some terms in this chapter. There would be no objection when the term "Chinese science" is used to refer to the contemporary scientific enterprise in China (mainly associated with communities of Chinese native scientists studying Eurocentric science). However, it remains controversial in academic circles as to whether some Chinese traditional knowledge can be called "Chinese science" (Needham 1956; Hart 1999). The present study recognises that not all indigenous knowledge can be called science, while there is the possibility that some ideas and practices can be treated as science (though not necessarily according to Eurocentric criteria). Given this understanding, the term "Chinese Native Knowledge", as opposed to the "imported" science, is used here to refer to ways of knowing and interacting with Nature (including achievements, practice and ideas) rooted in Chinese traditional culture in a broad sense. Where the issue of traditional knowledge as "science" is raised, especially in science education practice, the term "traditional Chinese science" is used to indicate the position that certain criteria should apply to distinguish scientific knowledge from other ways of knowing.

Traditional Chinese science education dates back to the Spring and Autumn period (770 BC–476 BC) and the content was usually a hybrid of both scientific and technological ideas and practice developed in ancient times (Wu 2002). It was not until the late 1800s that Western scientific knowledge was systematically taught in Chinese schools. Since then, Western scientific knowledge started systematically replacing traditional Chinese knowledge in school curricula at all levels.

In the early 1900s, Chinese society experienced unprecedented cultural conflict when traditional values and beliefs confronted a serious challenge from Western cultures. Since then, conflicts between "traditional and modern" and between "Chinese and Western" have been the main themes in any debate about socio-cultural reform (including educational reform). Within the fields related to science, however, Western scientific knowledge seldom confronted any substantial resistance in systematically taking the place of CNK except in the field of medicine, where there have always been debates about the legitimate status of Traditional Chinese Medicine (TCM) in science.

CNK in the current school curricula mainly serves as a link to history and values education. Many Chinese traditional science and technology achievements are mentioned in the history curriculum. For example, according to the history syllabus for the Chinese general senior high school (Ministry of Education 2002), traditional science and technology achievements in different historical periods are introduced together with other historical events that happened during the same periods. Chinese traditional science and technology achievements are also mentioned in science textbooks. Together with Chinese modern scientific achievements, they are used to acknowledge Chinese scientists' contributions (both ancient and modern) to, on the one hand, the development of science, and, on the other, to serve as materials for values education. Expressing these points in terms of the physics curriculum, Mu (2003) explained:

> Combining education with patriotism, introducing ancient China's brilliant contribution to science and technology and China's achievements in physics research after liberation can enhance students' sense of pride and confidence in our nation. It can inspire ambition, arouse love of our country, foster a sense of mission and responsibility, and elevate students' motivation of learning Physics. (p. 52)

The following is an example of the involvement of Chinese traditional scientific achievements from the senior high school chemistry textbook used by participant teachers at the time when the interviews for this study were conducted:

> The discovery of Arsenic: With regard to the discovery of Arsenic, Western historian of chemistry all believed that it was the German Albertus Magnus who first produced Arsenic in 1250 using realgar as raw material. Recently, studies conducted by our scholars have found that in fact, an ancient alchemist in our country was the first discoverer of Arsenic. According to the record of classical literature, in the year about AD 317, alchemist Ge Hong produced Arsenic using realgar, pine resin, and saltpetre. (Renmin Jiiaoyu Chubanshe Huaxueshi 2003, p. 7)

Similar examples can also be found in physics and biology textbooks. Traditional achievements included in textbooks are those that are compatible with the Western scientific knowledge system so that it can be integrated in some way as a Chinese contribution to the "contemporary" knowledge system, which is often seen as "universal" though based on Eurocentric science.

TCM is a body of knowledge that is deeply rooted in Chinese traditional culture and underpins its fundamentals with typical Chinese ways of thinking. On the one hand, some of its practice can be understood and explained using modern medical theories based on Eurocentric science, for example, the composition of herbal medicine. On the other hand, TCM contains a large amount of tacit knowledge gained through intuitive strategies and some of its practice is based on holistic way of viewing the relationship between human body and the environment, which often fails to offer analytical evidence required by modern medicine. For example, Qi is an important concept in TCM. However, there has been no consensus about what it exactly refers to—matter or spirit. Detailed information about TCM and its relationship with Chinese traditional culture is given in Ma (2008).

Although being acknowledged to a greater extent after the establishment of the People's Republic of China in 1949, knowledge about TCM which contains aspects incompatible with Western scientific knowledge system is not included in the general primary and secondary education system. Debates and controversy about whether TCM can be treated as valid and reliable scientific knowledge are still ongoing. However, TCM is taught in secondary vocational schools and specialised schools at post-secondary and postgraduate level. The Chinese modern medical system was established with reference to the Western medical system. Even so, there are specialised TCM hospitals and private clinics. TCM practitioners can also be found in many public hospitals where the practice is dominated by Western medicine. Traditional herbal medicine is still widely used in people's everyday life. It seems that there is no immediate threat to the survival of TCM in modern Chinese society, however, the lack of introduction of basic TCM ideas in general education means that the younger generation has very little chance of knowing about it and this suggests a not so optimistic future for the development of TCM.

The Purpose of the Research

The study reported here is part of a larger research project aimed at portraying the profile of the images of science in relation to Chinese culture held by Chinese secondary school science teachers. The images of science held by teachers illustrated in relation to Chinese views of nature have been reported elsewhere (Ma 2009). This chapter presents teachers' views on dealing with CNK in their teaching practice.

Methodological Issues

The study adopted a qualitative approach. Data were collected using a semi-structured in-depth interview protocol. Two different kinds of probes about dealing with CNK in science teaching were involved in the protocol: one based on teachers' real experience and the other on a hypothetical basis through a "scenario".

The first kind of probe started with the following question:

> There are some Chinese traditional scientific and technological achievements involved in science textbooks. How do you usually treat them in your teaching?

The second kind of probe was based on the following scenario:

> Xiao Lin's father is a doctor of TCM, so his views on the relationship between illness and health are influenced by his father. For example, falling ill is because the balance of Yin and Yang of the human body has been disturbed; acupuncture can make Qi running more smoothly through the human body and therefore is helpful for curing disease; etc. Xiao Lin has found these perspectives are quite different from knowledge he has learned from his science textbooks, so he feels very confused.

In discussing this scenario, teachers were asked how they would have dealt with students' confusion like Xiao Lin's and whether they thought it was necessary to introduce ideas about TCM in science teaching.

Participants comprised Chinese secondary school science teachers (n=25) from two secondary schools in the same city in the North of China. The schools were general public secondary schools. According to the government's educational policy, public schools in China all follow the same national curricular framework and have little freedom in selecting teaching materials such as textbooks. Likewise, teachers are required to follow the given syllabuses and textbooks. Although they may organise the classrooms in their own way, the examination mechanism is seen to not allow them to go beyond what is required on the official syllabuses. This is the case with all school subjects, not just science.

There were eight Biology teachers, eight Chemistry teachers and nine Physics teachers among the participants. Whenever the subject that teachers taught showed significant influences on teachers' views, comparisons were made. However, because of the size of the sample, and the exploratory nature of the research, no statistical description related to gender, age, and years of teaching was sought when interpreting the data. Teachers participated on a voluntary basis and the interviews were conducted face to face with each teacher within their school precinct (classroom, meeting room, or teacher's office). All interviews were conducted in Chinese. The audiotapes were first transcribed in Chinese for analysis, and then later translated into English for reporting.

Data analysis was based on the participants' responses to semi-constructed questions. Given the small size of the sample and the explanatory nature of the study, in analysing the participants' views, both general trends and particular opinions were seen as of the same importance in representing the complexity of human thinking (Ma 2009). As a result, the following presentation of findings gives equal consideration to both opinions held by most participants and those expressed only by one or two teachers.

Findings

Two themes are identified from the semi-structured interview. One is related to teachers' attitudes towards teaching CNK in school science and the other is about dealing with possible students' confusion as a result of teaching CNK. The findings are presented accordingly.

Attitude Towards Teaching CNK in School Science

There are some traditional scientific and technological achievements included in Chinese secondary school science textbooks. TCM, a CNK system, is not included in secondary school curricula. As a result, teachers were asked how they usually dealt with those achievements already included in textbooks and whether they thought it would be necessary to introduce knowledge about TCM in their science teaching.

Teaching CNK that Has Been Included in Current School Science Curricula: Most teachers expressed that they would mention those traditional achievements already included in textbooks in their teaching. The most frequently mentioned purpose of doing this was to nurturing a sense of pride among students. Typically, as Wei put it:

> At least you should know this Chinese stuff. You shouldn't think blindly that anything of the West is better than Chinese one. It's not right. For example, when teaching the magnetic declination, surely Shen Kuo [an ancient Chinese scientist] has to be mentioned. There's only one Chinese in all junior high school [Physics] textbooks. There's no reason not to remember him carefully. …We [the Chinese] knew it [magnetic declination] four hundred years earlier than them [the West], why not mention this?

One teacher, Lei, showed an obvious negative attitude towards mentioning traditional achievements for the purpose of nurturing a sense of pride. It was obvious that she thought those traditional achievements should belong to the past:

> This might be able to motivate students to some extent. But I feel the effect wouldn't be very big. In fact, for me, it depends on whose theory is more advanced at present. Personally, if currently my compatriots get some achievements, it will inspire me most. Nevertheless, there should be this kind of knowledge in textbooks. After all, as a Chinese, to know

> about our history is also useful. But you can't carry this history on your shoulder, showing off every day.

In talking about the traditional achievements, most teachers did not distinguish different knowledge production systems. They treated those traditional achievements as Chinese examples of a culture-free "universal" scientific knowledge system. Wen provided a possible explanation:

> Usually when mentioning, for example, Four Inventions, it is basically because of teaching magnetics, mentioning "Sinan" [a kind of compass, one of the Four Inventions in ancient China] in passing. Basically, there seems to be not much involved in, so it won't involve very contradictory points. …Our teaching materials have all been screened in advance, so those contradictory points may be rejected beforehand. This is just my guess.

Teaching TCM Which Has Not Been Included in Current School Science Curricula: Compared with native knowledge already included in science textbooks, knowledge about TCM can hardly be found in secondary school science textbooks. Therefore, teachers were asked whether they thought it would be necessary to introduce knowledge about TCM in their science teaching. Almost all teachers agreed that it was not necessary to teach TCM as a special subject at secondary school level, however, the majority of the teachers showed a positive attitude to introducing basic TCM knowledge in their teaching. Their reasons combined the concern about both reproducing traditional Chinese culture and developing valuable scientific knowledge. For example, Rui affirmed that TCM is a kind of science. He first used the word "cultural heritage" to describe TCM, however, he then quickly corrected himself using only the word "culture". This implied that for him, TCM should not be seen as something only belonging to the past, rather, it should have modern value. Here, he seemed not to pay much attention to the distinction between "culture" and "science":

> I feel that TCM itself is a kind of science. …It should be kept and developed as a kind of Chinese cultural heritage, not a kind of heritage, just a kind of culture, kind of valuable thing. If, only until someone goes to university and selects the subjects of TCM, can TCM be noticed; or only when one's own dad or mum who studies TCM, would one know about it, for TCM, that would be a kind of sadness, a waste. (Rui)

Many teachers argued that to introduce TCM in science teaching is to give students a chance to know what TCM is about and to give them a chance to consider whether to study TCM in their future. Many worried about the loss of valuable TCM knowledge due to the marginalisation of TCM in the secondary school curricula. Several teachers expressed a similar concern. Wen's view was typical:

> You can see that it [TCM] is seldom mentioned in textbooks. I remember when I was at school, it wasn't mentioned. That is why when talking about TCM, you feel that it is false or nonsense. This is partly because there is no mention in textbooks. It should be involved in. At least, if you mention a bit, students who are interested in learning more would try to do more searches. …If you don't even mention it, there would be no chance for the students.

Depending on their different knowledge about TCM and their different pedagogical beliefs, teachers' views on the possible status of TCM knowledge in the science curricula diverged. Some suggested that knowledge related to TCM and Western medicine could be put together for the purpose of making comparisons and eliminating misunderstandings. As Ping put it:

> In Physics learning, it's called "comparative learning", which can strengthen students' impression and lead to thorough understanding. Teach TCM and Western medicine in a comparative way and thoroughly make a distinction between how TCM works and how Western medicine works. This can strengthen students' impression and make them understand better.

Some teachers admitted that putting these different kinds of knowledge together might cause confusion. However, they believed that students could make their own decision about how to deal with it. Typically, as Wen and Na put it:

> It's hard to avoid confusion. ...It [the confusion] may arouse students' curiosity to work on this project and achieve something. (Wen)

> Different views and different attitudes can co-exist. You could tell students to accept all of them and then make their own judgement, if you feel hard to judge them yourself. (Na)

Other teachers insisted that differences should be avoided in designing the curricula:

> When designing teaching materials, stuff about TCM and Western medicine could be relatively separated so as not to cause much confusion. (Hai)

> You could consider not including this [controversy]. ...First, you're not a professional. Second, you [students] can't reach that level. Why seek trouble for yourself [as a teacher]? ...Don't let the contradiction arise. ...As a teacher, that is annoying. What is said in this book is different from what's in that one. How could you do? (Fei)

More specifically, Ying suggested that experience-based TCM treatments which have not been well explained should be avoided, while some herbal medicine whose composition has been worked out could be included:

> Many TCM treatments are based on experience. Even doctors seem not to be able to explain them clearly, nor can it be explained clearly through technoscience. Absolutely don't teach stuff like this. Otherwise the students would be more confused. There is, for example, some herbal medicines whose composition has been worked out. ...It actually combines TCM and Western medicine. At least, it can be explained well. (Ying)

Similarly, Meng suggested that teachers should only mention things that have obvious effect and have been universally accepted. She used the Chinese word "Xuan" to describe things which are so obscure that they often sound unbelievable and may cause confusion among students:

> There is something that's hard to be clarified and it's easy to make students feel "Xuan". I feel it's better not to mention too much about things that are easy to cause students feel "Xuan". If it is universally accepted and its effect is really obvious, it could be mentioned.

Although almost all teachers believed that TCM is scientific knowledge, the above excerpts imply that they had certain reservations in treating it as of equal status with the scientific knowledge already included in textbooks. Therefore, it is not surprising that many teachers suggested TCM knowledge be treated as extra-curricula reading material. Qian's view was widely shared. Qian also used the word "Xuan" to describe TCM. She made a comparison between Western science, which she thought of as clear and definite, and TCM, which is "Xuan". By using this word, she emphasised the "tacit" or "fuzzy" nature of TCM—relying more on "intuition" to understand which is too abstruse and more difficult for students to follow. She then concluded that TCM should be put into extra-curricula reading material:

> The difficulty is that it is not like science, Western science, the one that we call science, which is just like "one is one, two is two". TCM, I can use one word to describe, it's "Xuan". It can only be understood by intuition and can't be taught by words. For example, feeling the pulse. Everyone can feel the pulse. But you may not be able to feel the differences, while they [practitioners of TCM] can. Things like this, are hard for kids to understand and it's also hard for you to explain. So I feel it could only be treated as after-class reading materials.

Similarly, Yang suggested talking about TCM in a "casual" way and Jin thought it could be a discussion topic for after-class activities rather than for "formal" teaching:

> It could be taught as extra knowledge, extracurricular knowledge. It's not necessary to put it into textbooks. Just like a chat, tell the students what TCM is about and there is the idea of Qi. They only need to know a bit. It's not necessary to put it into textbooks. (Yang)

> I feel it could be a discussion topic for after-class activities. Encourage students to collect materials for reference and do some research. I feel it's not quite suitable to put things that you can't clearly explain in formal teaching. You can never explain it clearly and kids may form a wrong and not scientific view. I feel, at least, this is not a normal thing that a teacher should do. (Jin)

Two teachers clearly objected to the idea of involving TCM knowledge in secondary school science curricula. However, they had very different reasons for their views. Li's reason was that TCM is not particularly close to students' everyday life:

> It's not necessary. …[It will] put extra burden on students. What's more, I feel it's no use if you only teach some superficial knowledge about TCM. Curricula should be closely related to students. I feel that nowadays the relation between TCM and kids is not very close. I feel it's not necessary to offer this.

One of Qing's reasons was that TCM needs complex systematic learning while secondary school curricula should only focus on basic knowledge:

> I feel it's not necessary [to teach TCM] at secondary school level. …Secondary school just offers basic knowledge. …To teach TCM, you have to start from its root, which can't be learned in a short time. …TCM is hard to learn. [Only teaching a little bit] is no use.

Another reason Qing gave was that there has not been a complete explanation of TCM from the aspect of science, which may cause students' confusion.

> In addition, this set of theories, as just said, there has been no complete interpretation in terms of science. If it is included, it may make students easier to get confused.

He noticed that in general, TCM "contains something that is not very compatible with Western science" (Qing). This incompatibility, according to him, seemed more to do with "cultural differences". He described the current situation in China as being invaded by Western culture. This description was quite unusual among the teachers in this study:

In fact, China is now experiencing cultural invasion from the West. It [Western culture] has become the mainstream culture in China. …Western science, Western culture, especially science, has taken over almost one-hundred-percent. There's no Chinese stuff in it any more.

Putting all his reasons together, he seemed to argue that the cultural context for learning TCM at secondary school level has been lost, which may lead to students' confusion:

If you want them to learn [TCM], you have to teach them from very young, from kindergarten. …Imagine there is no basis at all and suddenly it is included, they [the students] can't afford it.

Dealing with Students' Confusion

The discussion about how to deal with students' confusion was elicited by the scenario which adopted TCM as an example. Interview questions focused on teachers' pedagogical strategies. There were mainly two different kinds of students' confusion involved in the discussion, one was related to issues on which there has not been broad consensus, the other was close to student understanding of knowledge in existing school curricula. It was obvious that teachers handled these two kinds of confusion differently.

Confusion Caused by the Differences Between Explanations of TCM and Knowledge in Science Textbooks: Teachers showed a cautious attitude in dealing with students' confusion of this kind. Most of them admitted that their knowledge about TCM was limited and tried to avoid making a definite judgement about knowledge from both sources and suggested that students needed to do some more research on related issues. Rong's concern and Ying's strategy were shared widely:

[I] really don't know much about TCM. If I knew about it, I would probably be able to find out the similarities and differences between TCM and Western medicine, or their respective theoretical bases. Then [I] would introduce them better to students. …Because you don't know much about TCM and Western medicine, you can't rashly teach students so as not to mislead students. (Rong)

In this case, you can guide them to look up literature, or interview some experts, authorities, or do some research, some detailed inquiries. Guide them, once getting a chance in the future, they can take it as a potential direction of their future development or direction of inquiry. Don't directly make a conclusion—besides, I can't make any conclusion anyway. (Ying)

Teachers' strategies to deal with students' confusion actually suggested a defence of TCM. In doing this, teachers' different beliefs about TCM made their strategies slightly different in detail. Some teachers thought that there should not be any essential differences between the two kinds of explanations. Wei claimed that confusion simply came from the marginalisation of TCM in the school curriculum:

I feel it is just because Western medicine is introduced relatively more in textbooks and TCM is introduced less, kids don't know enough about TCM. It doesn't mean that they

> [TCM and Western medicine] are very different. He [a student] felt confused just because he didn't know TCM enough. …This is just the problem of textbooks. If the textbook introduces more about TCM, the students will know. (Wei)

Lei and Yi suggested that differences may arise as a result of different wording:

> I feel when teaching students, Instead of saying that the knowledge is different, one should say that it is just different wording. (Lei)

> Is it possible that TCM has its own interpretation of "Qi", just like what you said about the human body? I really don't know. I feel what is called "Qi" in TCM may correspond to some parts of the human body, though the wording is different, it may refer to a combination of several parts, or something like that. (Yi)

Some teachers, such as Jin, suggested teachers guide students to find more shared points:

> I would take advantage of this chance to guide the kid to do more search. "Since you raised this issue, you could go to find out what exactly their shared points are."

Further, Jin argued that once the shared points were worked out, teachers would be able to make comparisons in teaching and then students' confusion would be solved naturally.

> I think, when talking about health, or "Qi", once it can be clearly pointed out that Qi refers to which part of the human body, it can be included in the teaching as an example. When teaching anatomy, "what TCM says 'the circulation of Qi and blood' refers to such and such". Then, I feel, confusion like Xiao Lin's would be solved naturally.

There were also some other teachers who emphasised the differences more. These teachers were more inclined to explain students' confusion in terms of ways of thinking. Typically, as Ning put it:

> His [the kid in the scenario] confusion may be because that he has already known something about TCM and he's been familiar with ways of thinking of TCM. Probably what is being taught in textbooks is something related to Western medicine which requires him to change his ways of thinking and he can't do it suddenly. …I feel that [the teacher] should let him know that to understand things, one should approach them from multiple aspects, not just from on aspect. TCM understands the phenomenon from one aspect, Western medicine from another.

Yun also identified that relevant knowledge in Biology textbooks is within the "knowledge framework of Western medicine" and distinguished TCM as belonging to "different knowledge system". Similar to Ning's view, Yun suggested that "all aspects can be used for reference":

> The stuff taught in Biology class, if distinguishing in terms of TCM and Western medicine, I feel that it belongs to the knowledge framework of Western medicine. They are knowledge of different systems and shouldn't be mixed up. …Shouldn't simply say which is true and which is false, because TCM has its theories and so does Western medicine. As for issues like this, all aspects can be used for reference.

Some teachers moved beyond just talking about TCM and made a generalisation to the pedagogical strategy of how to deal with controversial issues. They especially emphasised the importance of holding a dialectical, flexible and tolerant view. For example, Yang thought that being flexible was important to learning everything and, to understand the world, one should approach it from two sides:

> I feel that the learning of knowledge—science, anything, material world, including the understanding of TCM and Western medicine—needs flexibility. Flexibility is especially important. Maybe for people who insist on the pursuit of truth, truth is truth, one is one, two is two. I don't think so. You must be flexible. I mean no matter how you understand, one should understand the world from two sides.

Qian admitted that contradictions exist. Rather than focusing on the conflict, she suggested that shared points and true points of each be found:

> I would tell him that currently there really exists the contradiction. Since the contradiction exists, I would tell him my personal view that the contradiction is objective reality. Although there is contradiction, each of them has its true stuff and they just approach the issue from different aspects. Ask him to look for shared points or for each of their true stuff.

Na showed her tolerant attitude in a more obvious way. Instead of suggesting a tentative and critical attitude, she intended to accept all opinions before a judgement could be made:

> I would suggest that he accept both. These two views, two attitudes, can co-exist. Because presently they can't be explained, all of them can be accepted. …When there is no way to tell which is true and which is false, you could accept them all. As time goes by, if gradually it can be worked out which one is false, then you should reject the false one.

Confusion Caused by Students' Alternative Conceptions: During the interview, teachers identified another kind of students' confusion which they showed more confidence in dealing with. This kind of confusion is usually caused by students' lack of understanding or misunderstanding of knowledge in school curricula. A relatively simple source of this kind of confusion is exercises in school work. To deal with this kind of confusion, Lei's strategies were widely shared:

> Most students ask questions to do with the exercises. …Not every exercise has standard answer. I feel some answers are wrong and then I would tell the students my ideas, say, "this is how I answer this question". In this case, students could also provide their answers. Sometimes a consensus can be reached, while sometimes each one just keeps their own views.

Another source of this kind of confusion is students' misunderstanding of scientific concepts in textbooks. Teachers showed more authority and confidence in dealing with this kind of confusion. As shown in the following excerpt:

> Their [students'] views are different from those in the textbooks. In most cases, it is because the students don't understand. …In most cases, I feel it is them [the students] who are wrong. …I would help them to find out where their problems lie, according to the textbooks. I would try to unify students' views with those in the textbooks, finding out why it is explained in this way in the textbook and where their mistakes lie. …Because we [teachers] must learn the knowledge in the textbooks first, we're sure to be able to explain well to students. I haven't experienced the situation when knowledge in the textbooks can't be explained. Because what we are teaching is very basic, it all can be well explained. (Ping)

It seemed that teachers' confidence came from their trust in the reliability of knowledge in school curricula. No matter what subject they taught, teachers showed the same view that controversial issues are not likely to be included in school science textbooks. The above excerpt is from a physics teacher's point of view. Similar opinions can also be found from biology and chemistry teachers. For example:

> At the level of senior high school, it [divergence of views] is seldom an issue. More may be found at university level—for the same phenomenon, there are different theories and different people put forward different ideas. But at senior high school, I feel there's less. …Most of them are definite conclusions. (Hui, Biology teacher)

> I feel that for school science, there are fewer situations like this [which one is right is not clear]. Not like TCM and Western medicine, which are two different theory systems. In chemistry, there are fewer situations like this. (Ling, Chemistry teacher)

This view may explain why teachers adopted different strategies for different kinds of students' confusion. For example, Lei explicitly compared these two kinds of confusion and distinguished different strategies that she would adopt. The example she gave for "definite conclusion" was "the Earth runs around the Sun", while TCM belonged to that for which no consensus has been reached:

> Some definite conclusions are definitely true. For example, the Earth runs around the Sun, which is definitely true. If a student said "teacher, the Sun runs around the Earth", surely you would tell him "that's wrong". Things like this, the true is true, the false is false. It can be explained clearly. As for TCM and Western medicine, if in my classroom, I feel that I shouldn't tell them if it's true or false.

For the "definite conclusions", she showed more confidence in her ability to "convert students" while for those on which no consensus has been reached, she was more cautious in terms of imposing her own views:

> You know that there has been no universally accepted theory, so you can't explain it. You can't impose your view on students who may have their own views. Don't impose. …I feel if there is universally accepted view, with my disposition, I would impose this view on students. I would tell them "it's just like this. This is discovered by scientific research. You all should know this."

Summary of Findings

Most teachers showed positive attitudes towards introducing in their teaching Chinese traditional achievements already included in textbooks. There was very little agreement with ideas that TCM should be made a special subject at secondary school level, though the majority of the teachers showed a positive attitude to introducing basic TCM knowledge in their teaching. Given that teachers in Chinese schools do not have much freedom in selecting topics, not being included in officially selected textbooks means the impact of TCM remains limited in general public schools.

The findings also identified teachers' different strategies in dealing with different types of confusion. Teachers showed a cautious attitude in dealing with students' confusion caused by the differences between explanations of TCM and knowledge in science textbooks. Believing that (Eurocentric) scientific concepts presented in textbooks are rather certain than tentative, teachers often adopted more authoritative strategies in dealing with students' alternative understanding of these concepts. The findings that different strategies were adopted by teachers for different types of student confusion raise the question as to how to develop teachers' pedagogical knowledge for teaching pluralistic views in school science.

Developing Pedagogical Knowledge for Teaching Pluralistic Views in School Science

The existence of culturally different perspectives in science means that science teaching inevitably involves intercultural encounters as in many other cross-cultural social activities. To avoid an uneasy epistemological relativism in science, Harding (1991, 1998) proposes "strong objectivity" which requires scientists to take account of perspectives from alternative paradigms, especially from those groups marginalised by orthodox science communities. From a Chinese traditional point of view, differences need not be restrictive; they can also be productive. This means a co-existence of possible different perspectives towards the same phenomenon in school science curricula. The relationship between different perspectives is not of one replacing the other, but as alternative frameworks, or even in complementary ways.

Teaching different perspectives can be a big challenge for teachers since they have to switch their thoughts between these different perspectives during their teaching process. Because different perspectives may be situated in different discourses which hold "certain concepts, viewpoints, and values at the expense of others" (Gee 1996, p. 132), the change of one single perspective may involve switching from the whole ideology of one discourse to another. Therefore, learning to deal with cultural differences is crucial for the teaching of pluralistic sciences.

Hofstede (2001) describes the ability of managing cultural differences as "intercultural competence". He summarises three phases for the acquisition of intercultural competence: awareness, knowledge, and skills. The following discussion is in relation to these three phases.

Awareness

Groenfeldt (2003) put awareness of cultural values and cultural distinctiveness as the starting point to reclaim a cultural identity. According to Hofstede (2001), "intercultural contact does not automatically breed mutual understanding. Rather, it confirms the groups involved in their own identities and prejudices" (p. 424). Therefore, he points out that the awareness phase "teaches participants to perceive people in their cultural context and to dig up the unconscious knowledge of their own mental programs" (p. 428).

Teachers' attitudes to and opinions of dealing with CNK in school science reflect their different understanding of the nature of science. In turn, it is the teachers' beliefs about the nature of science that determines how science is portrayed in the classroom. Awareness of the relationship between culture and science, between Euro-centric science and CNK, can help teachers to form a more culture-oriented understanding of science educational practice.

In general, the findings suggest that there are inclinations to see science as both universal and as multicultural; as well as inclinations to see scientific knowledge as both tentative and as certain. Knowledge included in existing textbooks is seen by the teachers as certain and they showed greater confidence in dealing with it. This kind of knowledge is mainly Euro-centric science. Some CNK that is compatible with a Euro-centric view also belongs to this category and it is often seen as Chinese contribution to a "universal" knowledge system (e.g., the discovery of arsenic as mentioned earlier in this chapter). While, for some teachers, TCM represents different ways of thinking and can be seen as "legitimate" science though "tentative" in nature.

The core debate around native knowledge is related to the problem of demarcation which itself has caused a great deal of controversy. The lack of consensus about what counts as science may make decision-making difficult in relation to whether or not the school science curriculum should include certain native knowledge. However, the controversy itself may make native knowledge valuable as culture-specific material for discussions about the nature of science—putting different knowledge systems together in the discussion about the nature of science may facilitate the recognition of the culture-based nature of both native knowledge and Euro-centric science.

Relevant Knowledge

The next phase is learning relevant knowledge. This phase is actually a process of identifying the similarities and differences of ways of thinking, ways of valuing, and worldviews between different cultures. Relevant studies have shown that there are both compatible and incompatible points between native knowledge and Euro-centric sciences (Aikenhead 2006; Stephens 2001; Aikenhead and Ogawa 2007; Wu 2002; Wang and Jin 2004). Different knowledge systems may be based on different worldview presumptions (attitudes towards Nature, ways of thinking, reasoning, feeling, valuing and believing) and to understand each of them may involve switching from one whole "Gestalt" to another fundamentally different one.

The idea of pluralistic sciences in modern school science requires the identification of the different paradigms adopted by different sciences. To do this, the view that science should be a unified entity of knowledge should be questioned. Instead of seeking a unified standard for all sciences, a separate criterion may be more appropriate for each topic with a specific purpose. In terms of teaching science in the Chinese cultural context, both Chinese ways of knowing and Western ways of knowing are involved, as has been shown in this study. Re-evaluating CNK may involve the consideration of ideas which are both compatible and incompatible with the Euro-centric ones. Thus, teachers have to make sense first of both of them before consciously dealing with differences and possible conflicts between them.

It is crucial to acknowledge that the "relevant knowledge" does not only involve cultural knowledge that can be identified as "science". The relevant knowledge should include the cultural knowledge as a whole. There are two reasons for this. Firstly, one may argue that it is almost impossible to distinguish "independent science" from the cultural knowledge in question. McKinley (2007) observes that there has been an effort among science educators to distinguish between "indigenous knowledge"(IK) and "traditional ecological knowledge" (TEK): "one might see TEK as the 'science' of IK—that IK which is validated through scientific criteria" (p. 205). However, she argues:

> Because indigenous knowledges are deeply integrated with locality (nature) and experience, along with many other characteristics, IK is difficult to separate from TEK from an indigenous point of view. In fact, the extraction of the knowledge from the knower undermines the very basic of IK. (p. 205)

The other reason is that one cannot gain deeper understanding of science without understanding the underlying cultural beliefs and values. In terms of learning WMS, some Chinese scholars argue that Chinese people cannot completely understand science if they only focus on the utilitarian aspect of science without developing the spirit of science at the same time (Zhang 2003). Li (2004) criticises the "lack of gene (or collective unconsciousness) of the spirit of science" (p. 70) among Chinese people. He also argues that the spirit of science had been so deeply rooted in the Western culture that it had already become a part of the "gene" of Western people. What he meant by the "spirit of science" mainly refers to "rationalism" and "empiricism", which he thought were not supported by Chinese traditional culture. From a multicultural perspective, Li's argument is problematic in that he only recognises the spirit of Western science and ignores the possible different spirit of Chinese native science. From the same multicultural perspective, Li is right that one cannot understand Western science thoroughly without understanding Western culture. The same is true when understanding CNK. One can gain only some superficial understanding of CNK without understanding Chinese culture.

Skills

In addition to awareness and knowledge, some skills must be developed in order to manage practical problems. Controversial issues may arise during the process of presenting different views. Teachers have to learn how to deal with them. Many researchers and educators approach this issue from different aspects.

Aikenhead (1996, 2000) describes the activity of consciously moving back and forth between different discourses as cultural border crossing. Aikenhead (2000) identifies several functions of a teacher in helping students to cross cultural borders successfully.

> This help can come from a teacher (a culture broker) who identifies the cultural borders to be crossed, who guides students back and forth across those borders, who gets students to make sense out of cultural conflicts that might arise, and who motivates students by drawing upon the impact western science and technology have on their life worlds (not upon the contribution western science and technology have made to a monoculture determined by a privileged class). (p. 246)

Reiss (1993) suggests three models to teach controversial issues: advocacy, affirmative neutrality, and procedural neutrality. Advocacy is where "the teacher argues for the position she/he holds" (p. 53). Because the teacher is usually more powerful in a classroom context, her/his loyalty to a particular value may turn the negotiation of different ideologies and paradigms into the demonstration of one dominant one. Affirmative neutrality is where "the teacher presents to her/his students as many sides of a controversy as possible, without, at least initially, indicating which she/he personally supports" (p. 53). This model tries to seek a more balanced presentation of different issues. However, as Reiss points out, "what would constitute a balanced presentation, for many issues, is itself controversial" (p. 54). The third model, procedural neutrality, requires the teacher to act as a facilitator, who elicits different points of view from the students and controversial issues from resource material without revealing her/his own position. This model requires careful consideration of the suitable resource material to avoid it turning into affirmative neutrality or advocacy. More detailed in practical strategies, Snively (1995) suggests a five-step process for exploring both Western science and indigenous science. Her approach emphasises reflecting on different perspectives and evaluating the process itself.

There are also other factors that may influence which model or approach that teachers may choose in their classroom practice. Different teachers may personally value different models. However, the choice may also depend on the society's common values. A discourse of school science classroom forms historically within any socio-cultural context. The beliefs and action of teachers both shape and are shaped by the discourse at the same time. Philippou and Christou (1998) point out that beliefs and attitudes of teachers are related to their actions, but are also influenced by experience and reflection on the actions. Therefore, both teachers' beliefs and their interaction with the classroom discourse, within which they find themselves, are crucial for the successful intercultural exchange in school science classrooms.

In terms of teaching a pluralistic science curriculum in the Chinese cultural context, likewise, the teachers' classroom practice is influenced by both their personal values and the society's common values. The latter may be more important in the context of Chinese culture, in which the teachers often emphasise collective values. However, further exploration is needed of Chinese teachers' attitudes towards the different approaches of dealing with controversy and different points of views as well as how the Chinese cultural values in the society may shape teachers' behaviour in teaching controversial issues in the classroom.

Conclusion

A three-phase strategy for developing teachers' pedagogical knowledge of teaching pluralistic views in school science is proposed. The three phases include becoming aware of the cultural distinctiveness of science, obtaining relevant knowledge and developing teaching skills. TCM could be used as a specific example for discussion about pluralistic sciences. One of the challenges for teachers is obtaining relevant knowledge, as TCM itself is a complicated body of knowledge that requires years of study to understand both its principles and practice. Materials particularly for the purpose of promoting pluralistic views in school science need to be developed. It is also worth noting that promoting a pluralistic science curriculum and emphasising learning cultures as part of learning science does not mean that one should integrate different values into one's worldview. The interplay between an individual's worldview and his or her understanding of science is far more complicated and is beyond the purpose of this chapter. One of the consequences of a pluralistic science curriculum is to enable science teachers to make a conscious effort to consider different perspectives, to develop a richer understanding of the nature of science and to make decisions accordingly as to how to deal with them.

References

Aikenhead, G. S. (1996). Science education: Border crossing into the subculture of science. *Studies in Science Education, 27,* 1–52.

Aikenhead, G. S. (2000). Renegotiating the culture of school science. In R. Millar, J. Leach, & J. Osborne (Eds.), *Improving science education: The contributions of research* (pp. 245–264). Buckingham: Open University Press.

Aikenhead, G. S. (2006). *Science education for everyday life: Evidence-based practice.* New York: Teachers College Press.

Aikenhead, G. S., & Ogawa, M. (2007). Indigenous knowledge and science revisited. *Cultural Studies of Science Education, 2,* 539–620.

Clandinin, J., & Connelly, F. M. (1992). Teacher as curriculum maker. In P. Jackson, (Ed.), *Handbook of research on curriculum (a project of the American Educational Research Association)* (pp. 363–401). New York: Macmillan.

Gee, J. P. (1996). *Social linguistics and literacies: Ideology in discourses* (2nd ed.). London: RoutledgeFalmer.

Groenfeldt, D. (2003). The future of indigenous values: Cultural relativism in the face of economic development. *Future, 35,* 917–929.

Harding, S. (1991). *Whose science? Whose knowledge?* Milton Keynes: Open University Press.

Harding, S. (1998). *Is science multicultural?—Postcolonialisms, feminisms, and epistemologies.* Bloomington: Indiana University Press.

Hart, R. (1999). On the problem of Chinese science. In M. Biagioli (Ed.), *The science studies reader* (pp. 189–201). New York: Routledge.

Hofstede, G. (2001). *Culture's consequences*. (2nd ed.). London: Sage Publications.

Hua, H. P., Chang, C. Y., & McRaven M. M. (1999, March). *Taoism and its implications for science education.* Paper presented at the annual meeting of the National Association for Research in Science Teaching, Boston, MA.

Jegede, O. (1994). African cultural perspectives and the teaching of science. In J. Solomon & G. Aikenhead (Eds.), *STS education: International perspectives on reform* (pp. 120–130). New York: Teachers College Press.

Li, X. M. (2004). *Zhongguo xiandai kexue sichao* [The trend of thought on science in modern China]. Beijing: Kexue Chubanshe.

Loo, S. P. (2001). Islam, science and science education: Conflict or concord? *Studies in Science Education, 36,* 45–78.

Ma, H. (2008). *Chinese Secondary School Science Teachers' Images of Science—The Role of Chinese Culture*. Unpublished PhD dissertation. Monash University.

Ma, H. (2009). Chinese secondary school science teachers' understanding of the nature of science—Emerging from their views of Nature. *Research in Science Education, 39*, 701–724.

Matthews, M. R. (1994). *Science teaching: The role of history and philosophy of science.* New York: Routledge.

McKinley, E. (2007). Postcolonialism, indigenous students, and science education. In S. K. Abell & N. G. Lederman (Eds.), *Handbook of research on science education* (pp. 199–226). Mahwah: Lawrence Erlbaum Associates.

Ministry of Education (People's Republic of China). (2002). *Quanrizhi putong gaoji zhongxue lishi jiaoxue dagang* [History syllabus for the general senior high school]. Beijing: Renmin Jiaoyu Chubanshe.

Mu, X. Y. (2003). Wuli kecheng jiegou yu gongneng biange [The reform of the structure and function of Physics curricula]. In B. Q. Liao, X. Y. Mu, F. Chen, X. J. Wang, & Y. L. Zhao (Eds.), *Zhongxue wuli kecheng gaige de mubiao yu shishi* [The aim and practice of the reform of secondary school Physics curricula] (pp. 31–59). Beijing: Gaodeng Jiaoyu Chunbanshe.

Needham, J. (1956). *Science and civilisation in China* (Vol. 2: History of scientific thought). Cambridge: Cambridge University Press.

Philippou, G. N., & Christou, C. (1998). *Beliefs, teacher education and the history of mathematics.* Paper presented at the 22nd Conference of the International Group for the Psychology of Mathematics Education, Stellenbosh, South Africa.

Reiss, M. J. (1993). *Science education for a pluralist society*. Buckingham: Open University Press.

Renmin Jiiaoyu Chubanshe Huaxueshi. (2003). *Huaxue* [Chemistry] (Vol. 2). Beijing: Renmin Jiaoyu Chubanshe.

Snively, G. (1995). Bridging traditional science and western science in the multicultural classroom. In G. Snively & A. MacKinnon (Eds.), *Thinking globally about mathematics and science education* (pp. 53–75). Vancouver: University of British Columbia, Research and Development Group.

Stephens, S. (2001). *Handbook for culturally responsive science curriculum.* Fairbanks: Alaska Native Knowledge Network.

Wang, Q., & Jin, F. (2004). *Zhongguo jishu sixiangshi lun* [On the history of technological thinking in China]. Beijing: Kexue Chubanshe.

Wu, C. (2002). *Gudai zhongguo kexue fanxing* [Paradigms of science in ancient China]. Beijing: Zhonghua shuju.

Zhang, H. X. (2003). *Kexue jiujing shi shenme* [What really is science?]. Beijing: Jiaoyu Kexue Chubanshe.

Chapter 10
Teaching Secondary Science in Rural and Remote Schools: Exploring the Critical Role of a Professional Learning Community

Debra Panizzon

Introduction

Educational literature consistently emphasises the need for teachers to adapt their pedagogical knowledge and practices to cater for student diversity as a means of enhancing student learning, engagement and interest. This focus becomes even more critical in subjects like science and mathematics where there is evidence of a global decline in student enrolment and participation in these subjects at the secondary school and tertiary levels of education (Fensham 2007; Schreiner and Sjøberg 2007). The cumulative effect of this trend over the last decade in many OECD countries is a decrease in the number of students pursuing and entering careers involving science, technology, engineering and mathematics (STEM) (OECD Global Science Forum 2006; Tytler et al. 2008). Explaining aspects of this decline in student interest, Aikenhead (Chap. 7) discusses a number of key components around quality teaching as it relates to science. However, if our teachers are to meet the needs of their students and cater for diversity then they too must be provided with an opportunity to grow professionally by being immersed in an environment that is collegial, supportive and challenging (Bascia and Hargreaves 2000). Specifically, they need to engage in a professional learning community with a shared interest around science education (Loughran 2010).

While access to such a community is more likely in urban areas, it becomes a great deal more problematic for science teachers located in schools in rural and remote contexts in particular countries. Key reasons identified in the literature include: (1) fewer science teachers being positioned in specific schools or in the broader community, (2) extensive distances between schools often restricting the frequency of face-to-face contact, and (3) lower rates of teacher retention resulting in a high turnover of science teachers in these schools. This is certainly the case in Australia, with the confounding factor that the majority of science teachers in rural and remote schools have either just graduated or are in the early stages of their

D. Panizzon (✉)
Flinders Centre for Science Education in the 21st Century, Flinders University, Adelaide, SA, Australia
e-mail: debra.panizzon@flinders.edu.au

D. Corrigan et al. (eds.), *The Professional Knowledge Base of Science Teaching*,
DOI 10.1007/978-90-481-3927-9_10, © Springer Science+Business Media B.V. 2011

careers (Lyons et al. 2006). In the absence of a collegial network comprising experienced science teachers there is often little opportunity for these new teachers to continue learning about pedagogy while developing discipline content knowledge in areas of science that lie outside their specialisation (Sharplin 2002). Not surprisingly, the majority feel professionally isolated (Roberts 2005).

So why focus on rural and remote contexts in a book about pedagogy? Alarmingly, there is evidence to suggest that student learning, achievement and engagement in these schools in some countries is different to students attending urban schools (Panizzon 2011). For example, an analysis of the Programme for International Student Assessment (PISA) 2006 data for scientific literacy highlights significant differences between the achievement of students located in rural and/or remote schools compared with those in urban locations for each of Australia, Canada, Korea and New Zealand (Canadian Council on Learning (CCL) 2006; Panizzon 2011). However, this pattern is not apparent for students in similar geographical locations in the UK, USA and Denmark (OECD 2006; Panizzon in press). While there is likely to be a number of factors contributing to these differences, the quality of science teaching available in these schools is clearly a key variable (Hattie 2003). As such, inclusion of rural and remote contexts in relation to pedagogy is appropriate and valuable, but must be considered in relation to particular countries.

This chapter explores aspects of secondary science teacher pedagogical practice pertinent to working in rural and remote Australian schools. These schools provide a unique context when compared to many other Western countries. With this Australian rural and remote context described, a range of pedagogical aspects aimed at enhancing both student learning in science and teacher engagement is discussed with a central focus around the important role of a professional learning community and how the lack of access to this critical resource hinders teacher growth and scholarship. The discussion presented here is based on research, and is influenced by the author's years of experience in preparing and supporting preservice, graduate and experienced science teachers working in rural and remote schools. Interestingly, although a few syntheses exist that highlight the challenges and issues around teaching science in these geographical locations (Holloway 2002; Panizzon 2009; Oliver 2007) there is relatively little research that discusses specifics around pedagogy. Hence, this chapter provides an important contribution to the literature.

Rural and Remote Australian Contexts

Australia is a Western and highly industrialised country sharing many commonalities with, for example, USA, UK, Canada, and Germany. Australia has a current population of approximately 22 million (ABS 2009). Even though the country's surface area is equivalent to USA and it has a much smaller population than that country, it is one of the most highly urbanised countries in the world (State of the Environment Advisory Council 1996). Most of its population is located around the coastal fringe, so between large cities there are great expanses located across the continent that contain relatively few towns of varying size. Students in many of these small towns attend local schools for their primary education (i.e. 5–11 years of age) but often travel significant distances to attend secondary schools located in larger centres.

Given this physical isolation it is not surprising that there is a strong connection between schools in remote and rural areas of Australia and the communities in which they reside. A sense of this emerges from the following situational analyses (using fictional names) written by preservice secondary science teachers for science curriculum units designed and taught while undertaking teaching placements in rural schools.

> The students who attend Blackberry High School are mainly from a lower socioeconomic group. Resources in the school are limited so there is a constant struggle for teachers to gain the equipment and texts needed for teaching science in particular. The community is a close-knit farming area that represents both low and higher socio-economic groups and there is a clear distinction between these groups in the town. There is a small Indigenous community within the general population.

> The community offers a great deal of support to the school with projects and fundraisers and other endeavours to raise funds for resources for the school. Much of the livestock and feed is donated annually to the school by local parents and farmers, thus lowering the amount of funds needed for the maintenance of the agriculture plot and therefore freeing more of the science/agriculture funds to purchase resources for the students and expand the school farm. This is a huge assistance to a relatively small school that may not receive sufficient funds otherwise and allows the gradual collection of textbooks and other laboratory equipment to ensure that these are available to all students and teachers whenever they are needed.

> Northwest High School is a semi-rural coastal school with a 25% turnover in students. The school has 1200 students with approximately 70 members on the teaching staff. There are 7 science laboratories with students in junior years having 7 lessons per fortnight. Junior science classes are graded (steamed), with between 20–30 students per class. However, academic classes are larger. In general there is a lack of high achieving students with a large proportion of students lacking motivation.

These descriptions provide an insight into what Cornbleth (1990) refers to as the sociocultural contexts of rural and remote communities in Australia. It is the connectivity between the school and the local community that underpins much of the social and economic networking that occurs within a town. Importantly, the degree of connectivity has changed over the last century due to urbanisation (DeYoung and Lawrence 1995). Traditionally, most of these communities were based around agriculture with the expectation that children would inherit farms and continue the family farming business. However, increasing urbanisation since the early 1900s, powered by major shifts in government economic and political policy, has left many of these rural and remote towns with reduced populations and struggling to survive, resulting in the closure of many related services such as banks and government departments (Squires 2003).

Complicating this urbanisation issue further is the need for students from rural and remote communities to translocate to major cities in order to access tertiary education. Once qualified for their professions, many are often unable to return to their original communities because of limited career opportunities. In other words, in order to return home they must accept a position for which they are usually overqualified. This conundrum may lead to social norms operating in particular localities that conflict with the broader educational aspirations pertaining to the population as a whole (Howley et al. 1996). For example, rural parents perceiving the loss of their children from working the family farm as their livelihood may hold lower academic aspirations than urban parents who perceive a university degree as a means of enhancing the future employment, travel and economic opportunities of their children. As a result, teachers in rural and remote schools attempting to balance the educational standards of the nation may be placed in the centre of these conflicting priorities. On one hand they will try to encourage as many students as possible to complete secondary schooling and strive for a tertiary education qualification to bring Australia in line with other Western countries (OECD 2009) while on the other hand feeling responsible for meeting the needs and expectations of the local community (Bush 2005).

There are further compounding factors beyond these sociocultural aspects. Most of the Australian research exploring the quality of education in rural and remote schools alludes to the difficulties in attracting and retaining experienced teachers to rural and remote schools (Yarrow et al. 1998); the inadequacy of resources in those schools (Roberts 2005); feelings of isolation experienced by many teachers in the schools (Herrington and Herrington 2001); the lack of professional development opportunities in close proximity to their schools for these teachers (Vinson 2002); insufficient preservice preparation of teachers (Boylan 2003); and the lack of availability of experienced casual staff (Roberts 2005). In considering science specifically, a major national study undertaken by Lyons et al. (2006) comparing Australian secondary science teachers in rural/remote schools with those in urban schools found all these factors to be significantly different across geographical locations ($p<0.001$).

Clearly, there are a range of critical issues impacting on the quality of education, including of course science education, being received by many Australian students in rural and remote schools.

Enhancing Pedagogy by Learning Among Colleagues

In this chapter pedagogical knowledge and expertise is considered as a learning process that combines theory and practice over time (Loughran 2010) as teachers "collectively question ineffective teaching routines, examine new conceptions of teaching and learning, find generative means to acknowledge and respond to difference and conflict, and engage in supporting one another's professional growth" (Little 2003, p. 914). Viewed in this way, building knowledge is about science teachers working together to form a professional learning community that supports, challenges and extends their thinking so as to ultimately impact on their students' learning and engagement. Clearly, this facet is essential regardless of geographical location. However, this becomes even more critical in rural and remote Australian schools given the unique characteristics of this context and consequent difficulties outlined earlier.

Professional Learning Communities

Research in the area of effective schooling (e.g. Reynolds et al. 1996; Sammons et al. 1995) identifies a number of contributory factors around teachers including the need for purposeful teaching, establishment of high expectations, positive student reinforcement, and use of appropriate mechanisms for monitoring student progress. While these relate specifically to individual teachers and their classrooms (Hattie 2003), there is another level of factors necessary if science teachers are to collectively meet the needs of cohorts of students each year. In other words, we know that outstanding individual teachers will make a difference to the students in their classrooms but what about the other science classes in the school?

A major Australian study entitled An Exceptional Schooling Outcomes Project (ÆSOP) was designed to identify and analyse junior secondary (12–16 years of age) schooling processes that generated high value-adding for low, middle and high achieving students in urban, rural and remote government schools in New South Wales (Panizzon et al. 2007). Emerging from this study was the critical role played by subject departments (i.e. science teachers as a collective) around enhancing student learning. Although the pedagogies and practices used by individual teachers in their classrooms were important, it was the ability of the science teachers to work as a *team* that had the major impact on value adding for students in these schools. Specifically, it was only by thinking and operating collectively that a *culture* was created that embodied a shared vision and "camaraderie" reflected by collegiality, support for new and inexperienced staff, enculturation of new staff, use of a range of approaches to cater for student differences, sharing of expertise and resources with others, and the implementation of common assessment practices to ensure a high level of consistency across the science team (Panizzon et al. 2007). Within these more collegial departments, informal discussions around teaching and learning were a normal part of the day, with experienced teachers in particular areas of science (e.g. physics) supporting and nurturing other staff in developing the discipline knowledge necessary for their own scientific learning. This connectivity was strengthened further by most of the science teachers in these schools participating in different types of professional development and learning outside their school. Consequent access to a broader networks of science teachers provided access to greater diversity, expertise and experience, thereby enhancing further the quality of the learning for the teachers in the ÆSOP schools.

These findings highlight the central role of professional learning communities (both within a school and beyond the school) in not only supporting teachers in their work but also in ensuring that they continue to grow and develop as professionals. According to Westheimer (2008, p. 759), learning communities for teachers should aim to achieve six interconnected goals:

1. Improve teacher practice so students learn better
2. Make ideas matter to both teachers and students by creating a culture of intellectual inquiry
3. Develop teacher learning about leadership and school management
4. Promote teacher learning among novice teachers
5. Reduce alienation as a precondition for teacher learning
6. Pursue social justice and democracy

Achieving these goals in Australian rural and remote contexts is difficult due to the (i) severe lack of experienced science teachers and predominance of graduates or early career teachers in these schools (Lyons et al. 2006), and (ii) lack of a critical mass of science teachers either in the school or in the local community (with 1–2 schools present in the area). Subsequently, the majority of our new science teachers are working in an environment where they have minimal access to teachers with the necessary discipline and pedagogical content knowledge (i.e. specific pedagogies relevant to the teaching of physics, chemistry or biology) to support their ongoing learning. This quickly leads to strong feelings of isolation and inadequacy (Roberts 2005). Compounding these feelings further is that many of these new teachers will have to assume leadership roles within their schools, such as coordinating the science department because they are the only qualified science teacher available. Ultimately, this means they have less time to concentrate on their teaching or engage in professional dialogue (even if it were present) while being expected to overseer the direction of science in the school.

Placing these science teachers in a situation where they are unable to access a professional learning community in their immediate environment is detrimental to their on-going development, as demonstrated by the high levels of teacher burnout and attrition rates in these schools (Hudson and Hudson 2008; Lyons et al. 2006). The outcome in Australia is that most science teachers will remain in rural and remote schools for only between 1 and 3 years, after which time they will transfer to urban or coastal schools (Roberts 2005). This is of concern given that the literature around teacher scholarship indicates that it takes about "five years to proceed from the novice stage of development to the advanced beginner stage to the competent stage of development" in teaching (Berliner 2000, p. 360). Hence, there is also an unfortunate irony here—the pedagogical expertise developed around working in these contexts is then consistently lost from these schools again with the science teacher to be replaced by a new graduate.

An additional component related to teacher expertise is actual discipline knowledge. Recent national data indicate that many secondary science teachers in Australia are or will at some stage be teaching outside their specific discipline area (Harris et al. 2005). This study found that 43% of senior (final years of schooling) physics teachers lack a major in physics, with one in four not having studied physics beyond first-year university. Chemistry fares better with only 25% lacking a major in chemistry. In contrast, senior biology teachers are the most highly qualified with 86% having a major in biology. Lyons et al. (2006) suggest that science teachers in rural and remote schools are more than three times as likely as those in urban area schools to be teaching a science subject for which they are not qualified (e.g. biology teachers teaching senior physics classes).

This discussion identifies the important role of a teacher learning community for science teachers in rural and remote schools who are usually graduates with limited pedagogical expertise and content discipline knowledge in a specific area of study. The way in which this lack of access to professional support impacts the science classroom is explored in greater detail below.

Pedagogical Knowledge and Practice

Given the importance of pedagogy, it is surprising that little research exists addressing this aspect in relation to science in rural and remote environments. Fortunately, there is some evidence that accentuates the contribution made by teachers in these schools in developing and implementing practices that meet the needs of students in these locations.

> Many so-called "innovations" being championed today were born of necessity long ago in the rural schoolhouse...cooperative learning, multi-grade classrooms, intimate links between school and community, interdisciplinary studies, peer tutoring, block scheduling, the community as a focus of study, older students teaching younger ones...all characterize rural and small school practices (Stern 1994, p. 1).

Interestingly, the strategies and practices mentioned here are equally relevant and applicable in urban schools, particularly those in lower socioeconomic areas (Calabrese 2007). The difference in rural and remote contexts is that science teachers are usually implementing a number of these strategies at any point in time within their classes. While this is challenging on its own, the lack of support, pedagogical expertise and mentorship provided by access to a learning community merely heightens the difficulties confronting science teachers in these schools.

Small Senior Classes: Most schools located in rural and remote areas have smaller student populations when compared to urban schools and this impacts the breadth of subjects available in any year as schools try to remain economically viable (Russon et al. 2001; Stern 1994). In the case of mathematics and science, these subjects are usually maintained even with a small clientele given the negative message sent to the community if these are removed from the curriculum. Dealing with small numbers of students in a classroom, particularly at a senior level, places a high degree of responsibility on the science teacher to engage students and ensure that they are being academically extended. For example, as observed frequently in rural and remote schools, Year 12 physics may involve a class of two or three students only (Lyons et al. 2006). While this might at first appear to be "easy teaching", consider how does the teacher move the students through the syllabus or curriculum so that they construct (i) their knowledge and understanding of physics concepts, (ii) hone their thinking skills so that they are able to critique, analyse and justify their perspectives, and (iii) develop and practice other communication skills when there is very likely not a diversity of opinion available in the classroom? In other words, such a situation reflects the potential lack of a critical mass of ideas to initiate and generate the interactive dialogue necessary in a group situation (Walsh and Golins 1976).

It is important to think about this scenario in terms of teacher knowledge and pedagogy. Critical to this environment is the ability of the science teacher to generate open-ended and challenging questions that illicit student engagement (Goodrum 2004; Loughran 2010). Given the possible lack of diversified points of view, it might be that the teacher must also play "devil's advocate" by challenging students' thinking and encouraging them to justify their scientific understandings. In addition, the teacher needs to plan lessons carefully, structuring experiences and activities just as required with larger cohorts of students if teaching time is to be maximised (Soliman 1999). Otherwise, there is a tendency to be easily deviated from the core purpose with the view that it is easy to catch up because of the small group of students. What is paramount here is that students are provided with equivalent opportunities to build up their conceptual scientific understandings and skills regardless of the number involved or the school location. While an experienced teacher of larger physics classes will have access to a "tool bag" of strategies to motivate and extend students, this becomes difficult for the graduate who is essentially "flying blind" with limited pedagogical expertise and little clear idea about the academic expectations required.

Multi-Grade Teaching: One commonly applied strategy for dealing with small numbers of students in science and mathematics classes is to adopt some form of multi-grade or multi-age teaching. For example, within the senior school it may incorporate classrooms comprising students of different year levels who are taught independently for most of the time with some overlap occurring occasionally (e.g. combining Years 11 and 12 physics students). Alternatively, in the junior secondary school it may comprise classrooms with students from three or four different age levels who are taught as a homogenous group with particular students being extended due to their aptitude and ability not age (Lloyd 1999).

In considering the senior secondary physics classroom as an example, there are clear advantages and disadvantages in combining Years 11 and 12. An obvious advantage for the students is the growth or development of physics concepts that occurs over the 2-year period. Accessibility to Year 12 work by Year 11 students enables them to understand the relevance and meaning of the physics they are currently studying. Such a situation also allows academically capable Year 11 students to progress beyond where the teacher may have pitched the work in a conventional classroom while giving Year 12 students who are struggling conceptually the opportunity of reviewing the more difficult concepts with the younger students. Careful planning and structuring by teachers encourages communal teaching for particular concepts along with shared investigative activities.

However, there are disadvantages. For such a teaching situation to enhance student learning, it requires that the physics teacher plans, programs and organises the lesson sequences carefully so that each group of students meets the curriculum requirements for the specific year level. As with smaller class sizes, being able to establish workable routines in a multi-grade classroom that addresses the needs of students requires a high degree of teacher competency (Marland 1993) that is usually beyond the realm of the majority of graduate teachers working in rural and remote contexts. Without access to experienced science practitioners many flounder in being able to maximise the learning opportunities for their students resulting in lower achievement (e.g. Thomson and De Bortoli 2008).

Establishing and Maintaining Academic Standards: Linked to smaller classes and multi-grade classrooms is the concern that many science teachers share anecdotally about the need to ensure that academic standards are being attained. There are two major issues here. Firstly, senior students often forget that they are not competing for grades and further educational opportunities with the three or four students in their physics or chemistry class but with the cohort of Year 12 students across the state. (With the exception of the state of Queensland, in Australia, all end-of-secondary schooling grades and selection for further education programs derive from external examinations at the state level). While teachers can try to motivate students to strive for higher achievement it is ultimately up to the student. Secondly, many science teachers raise concerns about their ability to maintain specific standards in classrooms if they are not provided with the opportunity to engage and discuss aspects of curriculum and assessment with colleagues, especially experienced senior science teachers.

Attaining and maintaining educational standards becomes even more complicated when a clash of academic aspirations exists between the expectations of the science teacher and those of the parents and broader community. The following quote from a teacher in a rural Australian school exemplifies this conflict (cited in Lake 2007, p. 3)

> Yeah some kids come in and say oh I don't do that at home because Dad says I should be doing chores, school work is for school and stuff like that. A lot of families living out here have lived here all of their lives and there's generations that have lived here and never lived anywhere else but here and I just think they don't know what it is like outside of our community so they lack the life experiences that some of the other children who come in, maybe the transient population like teachers and those children bring into this school…The kids are limited here to what their parent's value if they don't value reading and writing, maths and science. That sort of stuff is not valued by kids either.

Without opportunities to regularly engage in academic discourse within their specific field of science, share assessment tasks, and participate in cross-marking experiences, secondary science teachers find it difficult to ensure that their academic standards are being maintained and kept in line with their professional colleagues (Lyons et al. 2006).

Using Rural Contexts to Provide Meaningful Experiences for Students: Russon et al. (2001) allude to the different world views held by students living in rural and remote locations, particularly those with agricultural backgrounds, and these students' preference for more practical and "hands-on" experiences. Similarly, Oliver (2007) argues that if rural science teachers are to meet the needs of their students they:

> …must use a frame of reference that consciously builds a curriculum with a cooperative inclusion of community, the unique student and school needs found in that community, and the inimitable capabilities of the teachers found in those schools (p. 364).

Looking at the evidence available in Australia, Lake (2007) found that the majority of teachers in remote schools tended to deliver science through the internet rather than through "direct environmental experiences" (p. 4), thereby missing opportunities to make science meaningful to their students by relating it to the contexts in which the students live.

The distinction between student need and implementation exemplified by these research perspectives is unnecessary in Australia given that most of the science curricula available encourage teachers to select contexts that relate to the real world of their particular students as a means of teaching the concepts, understandings and skills of science. For example, Fensham (Chap. 17) refers to water as a context for teaching about properties, chemical composition and structure while providing an opportunity for exploring environmental issues around water usage and sustainability. This type of approach is possible in Australia but the literature suggests that the majority of secondary science teachers do not understand the notion of teaching via a context and so tend to remain tied to teaching within the traditional disciplines of science (Tytler et al. 2008).

Clearly, teaching using contexts that are relevant to students is sound pedagogical practice regardless of geographical location. However, it becomes more critical in rural and remote schools because of the close links between the school and community. Again, the difficulty becomes helping to inform current teachers about teaching through contexts and supporting them in building up the confidence necessary to move away from traditional approaches to teaching science. A critical step in this change of culture is that teacher educators prepare our graduate science teachers for utilising local rural and remote contexts so that they become agents for change within these schools (Oliver 2007).

Engaging Indigenous Students: Any discussion of teacher pedagogy and knowledge for rural and remote Australian schools is incomplete without consideration of Indigenous students because of the higher proportion of the national population of these students attending schools in these locations. This is a particular area of concern because Australian Indigenous students do not achieve as highly as non-Indigenous students, including in science and mathematics. For example, PISA in 2006 identified a full proficiency level representing two and half years of schooling difference between the results of Indigenous and non-Indigenous 15-year-old students for science (Thomson and De Bortoli 2008).

It is evident from the literature that science teachers are aware of their lack of expertise and skill in catering for the diversity of student needs in their classroom, particularly in relation to Indigenous students (Lyons et al. 2006). However, a review of the science education literature provides little direction about ways of enhancing student engagement and helping support achievement in science. This situation differs markedly from parts of Canada, for example, where Aikenhead has undertaken considerable research around Indigenous knowledge in science and technology (2001, 2006). Importantly, when a broader review of educational literature is undertaken there is a predominance of studies identifying the issues and challenges around Indigenous education in Australia while a dearth exists around research evidence that helps inform teacher pedagogical knowledge and practice in this specific area (Panizzon 2009). This remains a critical area of need in the Australian rural and remote context.

Nurturing the Building of Teacher Knowledge

The discussions presented in this chapter highlight that if secondary science teachers are to enjoy teaching in rural and remote schools, better engage their students, and not merely survive the experience, they require access to a network of colleagues. This community comprising teachers with a range of experiences, expertise and discipline knowledge creates a forum for sharing ideas and resources, with mentors providing examples of strategies, activities or tasks that work in the science classroom. With time, graduates and early career teachers can access the discipline knowledge, breadth, and depth of teacher knowledge for *just in time learning* (Panizzon et al. 2009) thereby facilitating their transition from graduate to novice to competent teacher (Berliner 2000). In addition, such a community also supports teachers on a personal level given that preservice students undertaking practicums in rural and remote schools often raise major concerns about aspects of the personal and social challenges they face working in such close-knit environment (Sharplin 2002).

As articulated here, establishing these networks in this context is difficult given the tyranny of distance. While there is often a significant emphasis for these networks to be available online, teachers still enjoy face-to-face contact given that learning is viewed as a social activity. The reason for this is that initiating these networks requires the building of relationships, which is always complex in that it requires personal commitment, engagement, and a degree of negotiation by those involved in the relationship (Corrigan 2004). Given that face-to-face meetings are difficult, a compromise is necessary with an obvious way forward being through multimedia technologies.

In creating an electronic professional learning community, it is important to consider the purpose of the interaction, as this will highlight the most appropriate media to use. For example, email, blogs, wikis, and instant messaging are useful for exchanging ideas or resources once a rapport between teachers is established. However, ichat, skype, or webex allow teachers to interact in real time with images available, thereby providing an equivalent face-to-face experience (Sobel-Lojeski and Reilly 2008). As with any meeting, establishing and maintaining these networks requires an individual with the motivation, time, and technological expertise to coordinate the process. This may be an opportunity for the graduate science teacher to take on an active role and make a substantive contribution to the group given that they are more likely to be up-to-speed with these forms of media.

Use of technologies in this manner also paves the way for engagement with members of the broader community or expertise that may be located at a distance.

For example, Rennie (Chap. 2) discusses the *Scientists in Schools* programme and the opportunity for teachers and their students to interact in many different ways with a range of scientists. While there may be scientists working in rural and remote communities, engaging with them face to face may not be possible. But use of these electronic technologies makes many new interactions possible not only locally but nationally and internationally.

A further mechanism for facilitating the sharing of expertise and resources while initiating the beginnings of a professional network is through a school clustering approach (Panizzon et al. 2009). This allows a qualified and experienced physics teachers in one rural school to teach students in other rural or remote schools either face to face (using residential intensive periods) or via the range of technologies mentioned earlier. While there is some literature from the UK available around clustering, finding appropriate models and processes for implementation in schools that suit the unique Australian rural and remote context are currently not readily available.

Conclusion

Regardless of geographical location, access to a professional learning community provides science teachers with the opportunity to develop and build upon their specific discipline knowledge (i.e. physics, chemistry, biology) while engaging with innovative practices in teaching and learning (i.e. use of electronic assessment tasks). In contrast to professional development that "consists of a patchwork of course, curriculum, and programs and may do little to enhance teachers' content knowledge or techniques and skills needed to teach science effectively" (National Research Council (NRC) 2001, p. 33), interaction with a community allows science teachers to learn in a meaningful and dynamic manner (Panizzon et al. 2009). As argued in this chapter, accessing such a community becomes critical for secondary science teachers in rural and remote Australian schools who already feel professionally isolated given the tyranny of distance. The main issue for these teachers is that while the implemented pedagogies are equally pertinent to urban schools (e.g. small senior classes), rural teachers are often dealing with a number of these strategies simultaneously with varying degrees of support, mentorship, and expertise available. Furthermore, the majority of these teachers are either graduates or in the early stages of their careers. Clearly, universities begin the journey of building teacher knowledge through their preservice teacher education programs. Ensuring that this process of learning continues in a constructive, informative, and engaging manner is essential not only for the teachers but also for the communities in which they reside.

Acknowledgements Appreciation is extended to Professor John Pegg, Ms Teresa Wright, and Dr Bruce Cameron for their insights and views about teaching in rural and remote schools that significantly helped in the formulation of this chapter.

References

Aikenhead, G. S. (2001). Integrating western and aboriginal sciences: Cross-cultural science teaching. *Research in Science Education, 31,* 337–335.

Aikenhead, G. S. (2006). *Science and technology education from different cultural perspectives.* A keynote presented to the 12th Symposium of the International Organization for Science and Technology Education (IOSTE), Batu Ferringghi, Penang, Malaysia, July 30–August 4.

Australian Bureau of Statistics. (2009). Population clock. http://www.abs.gov.au/ausstats/abs%40.nsf/94713ad445ff1425ca25682000192af2/1647509ef7e25faaca2568a900154b63? OpenDocument. Accessed 20 Nov 2009.

Bascia, N., & Hargreaves, A. (2000). Teaching and leading on the sharp edge of change. In N. Bascia & A. Hagreaves (Eds.), *The sharp edge of educational change, teaching, leading and the realities of reform* (pp. 3–26). New York: Routledge Falmer.

Berliner, D. (2000). A personal response to those who bash teacher education. *Journal of Teacher Education, 51,* 358–371.

Boylan, C. (2003). *Putting rural into preservice teacher education.* Paper presented at the joint Australian Association for Research in Education/New Zealand Association for Research in Education Conference, Auckland. Resource document. http://www.aare.edu.au/03pap/alpha.htm. Accessed 20 Oct 2006.

Bush, W. S. (2005). Improving research on mathematics learning and teaching in rural contexts. *Journal of Research in Rural Education, 20*(8). Resource document. http://www.umaine.edu.jrre/20-8.pdf. Accessed 3 March 2009.

Calabrese, B. A. (2007). Science learning in urban settings. In S. Abell & N. Lederman (Eds.), *Handbook of research on science education* (pp. 319–343). Mahwah: Lawrence Erlbaum Associates.

Canadian Council on Learning. (2006). Lessons in learning. Resource document. www.ccl-cca.ca/CCL/Reports/LessonsInLearning/LiL1March2006.htm. Accessed 8 Sep 2008.

Cornbleth, C. (1990). Reforming curriculum reform. *Education Action, 1*(2), 33–43.

Corrigan, D. (2004). Understanding mentoring. In D. Corrigan & J. Loughran (Eds.), *Snapshots of mentoring: Vignettes of practice* (pp. 5–11). Clayton, Melbourne: Monash Print Services.

DeYoung, A. J., & Lawrence, B. K. (1995). On hoosiers, Yankees and mountaineers. *Phi Delta Kappan, 77*(2), 104–122.

Fensham, P. (2007). Values in the measurement of students' science achievement in TIMSS and PISA. In D. Corigan, J. Dillon, & R. Gunstone (Eds.), *The re-emergence of values in science education* (pp. 215–229). Rotterdam: Sense.

Goodrum, D. (2004). Teaching strategies for science classrooms. In G. Venville & V. Dawson (Eds.), *The art of teaching science* (pp. 54–72). Crows Nest: Allen & Unwin.

Harris, K-L., Jensz, F., & Baldwin, G. (2005). *Who's teaching science? Meeting the demand for qualified science teachers in Australian secondary schools.* Report prepared for the Australian Council of Deans of Science. Melbourne: ACDS.

Hattie, J. (2003). *Teachers make a difference: What is the research evidence?* Paper delivered to the Australian Council for Educational Research Conference, in October. Resource document. http://www.acer.edu.au/workhops/documents/Teachers_Make_a_Difference_Hattie.pdf. Accessed 20 July 2006.

Herrington, A., & Herrington, J. (2001). Web-based strategies for professional induction in rural, regional and remote areas. Resource document. http://www.aare.edu.au/01pap/her01711.htm. Accessed 8 July 2006.

Holloway, D. L. (2002). Using research to ensure quality teaching in rural schools. *Journal of Research in Rural Education, 17*(3), 138–153.

Howley, C. B., Harmon, H. L., & Leopold, G. D. (1996). Rural scholars or bright rednecks? Aspirations for a sense of place among rural youth in Appalachia. *Journal of Research in Rural Education, 12*(3), 113–118.

Hudson, P., & Hudson, S. (2008). Changing preservice teachers' attitudes for teaching in rural schools. *Australian Journal of Teacher Education, 33*(4), 67–77.

Lake, D. (2007). Science education: Innovation in rural and remote Queensland schools. Educational Research for Policy and Practice. Resource document. http://www.springerlink.com/content/85275k332112842q/. Accessed 8 Jan 2008.

Little, J. W. (2003). Inside teacher community: Representations of classroom practice. *Teachers College Record, 105,* 913–945.

Lloyd, L. (1999). Multi-age classes and high ability students. *Review of Educational Research, 69,* 187–212.

Loughran, J. (2010). *What expert teachers do: Teachers' professional knowledge of classroom practice.* Sydney & London: Allen & Unwin, Routledge.

Lyons, T., Cooksey, R., Panizzon, D., Parnell, A., & Pegg, J. (2006). *Science, ICT and mathematics education in rural and regional Australia: Report from the SiMERR National Survey.* Canberra: Department of Education, Science and Training.

Marland, P. (1993). *Preparing teachers for multi-grade classrooms: Some questions and answers.* Paper presented at the Australian Association for Research in Education (AARE) Conference in Fremantle, Perth. Resource document. http://www.aare.edu.au/93pap/marlp93147.txt. Accessed 25 Apr 2009.

National Research Council (NRC). (2001). *Educating teachers of science, mathematics, and technology: New practices for the new millennium.* Washington: National Academy Press.

Oliver, J. S. (2007). Rural science education. In S. Abell & N. Lederman (Eds.), *Handbook of research on science education* (pp. 345–368). Mahwah: Lawrence Erlbaum Associates.

Organisation for Economic Cooperation and Development (OECD). (2006). OECD Programme for International Student Assessment (PISA) data. Resource document. http://pisa2006.acer.edu.au/interactive_results.php. Accessed 8 July 2008.

Organisation for Economic Cooperation and Development (OECD) Global Science Forum. (2006). Evolution of student interest in science and technology studies: Policy report. Resource document. http://www.oecd.org/dataoecd/16/30/36645825.pdf. Accessed 3 Sep 2006.

Organisation for Economic Cooperation and Development (OECD). (2009). Highlights from education at a glance 2009. Resource document. http://www.google.com.au/search?hl=en&source=hp&q=highlights+from+education+at+a+glance&meta=&btnG=Google+Search. Accessed 21 July 2009.

Panizzon, D. (2011). Science education in rural areas: Exploring the issues, challenges and future directions. In S. Richie (Ed.), *The world of science education: Handbook of research in Australasia* (pp. 137–162). Netherlands: Sense.

Panizzon, D. (in press). Science education in rural settings: Exploring the 'state of play' internationally. In B. Fraser, K. Tobin, & C. McRobbie (Eds.), *Second international handbook of science education.* Dordrecht: Springer (in press)

Panizzon, D., Barnes, G., & Pegg, J. (2007). *An exceptional schooling outcomes project for science.* Flaxton, Queensland: Post Pressed.

Panizzon, D., Ward, M., & Westwell, M. (2009). *Project report: School to work mathematics and science innovation project.* Adelaide: Future South Australian Certificate of Education Committee.

Reynolds, D., Bollen, R., Creemers, B., Hopkins, D., Stoll, L., & Lagerwij, N. (1996). *Making good schools: Linking school effectiveness and school improvement.* London: Routledge.

Roberts, P. (2005). *Staffing an empty schoolhouse: Attracting and retaining teachers in rural, remote and isolated communities.* Sydney: NSW Teachers' Federation.

Russon, G., Lynde, P., & Horn, J. (2001). The relationship between the drivers of educational reform and the Rural Systemic Initiatives in science, mathematics, and technology education program of the National Science Foundation. Resource document. http://www.wmich.edu/evalctr/rsi/relationship_between_drivers.html. Accessed 15 Feb 2009.

Sammons, P., Hillman, J., & Mortimore, P. (1995). *Key characteristics of effective schools: A review of school effectiveness research.* London: International School Effectiveness and Improvement Centre, University of London.

Schreiner, C., & Sjøberg, S. (2007). Science education and youth's identity construction—two incompatible projects? In D. Corrigan, J. Dillon, & R. Gunstone (Eds.), *The re-emergence of values in the science curriculum* (pp. 231–248). Rotterdam: Sense.

Sharplin, E. (2002). Rural retreat or outback hell: Expectations of rural and remote teaching. *Issues in Educational Research, 12,* 1–13.

Sobel-Lojeski, K., & Reilly, R. R. (2008). *Uniting the virtual workforce: Transforming leadership and innovation in the globally integrated enterprise.* Hoboken: Wiley.

Soliman, I. (1999). Introduction to university teaching series: Teaching small groups. Resource document. http://www.une.edu.au/tlc/staff/publications/small-groups.pdf/. Accessed 1 June 2009.

Squires, D. (2003). Responding to isolation and educational disadvantage. *Education in Rural Australia, 13*(1), 24–40.

State of the Environment Advisory Council (SOEAC). (1996). *Australia: State of the environment.* An independent report presented to the Commonwealth Minister for the Environment. Melbourne: CSIRO.

Stern, J. (1994). *The condition of education in rural schools.* Washington: US Department of Education, Office of Educational Research and Improvement.

Thomson, S., & De Bortoli, L. (2008). Exploring scientific literacy: How Australia measures up. Resource document. http://www.acer.edu.au/news/2007_PISA.html. Accessed 20 Dec 2007.

Tytler, R., Osborne, J., Williams, G., Tytler, K., & Clark, J. C. (2008). Opening up pathways: Engagement in STEM across the primary-secondary school transition. Resource document. http://www.dest.gov.au/sectors/career_development/publications_resources/profiles/Opening_Up_Pathways.htm. Accessed 4 July 2008.

Vinson, A. (2002). Inquiry into the provision of public education in New South Wales First Report September 2002. Resource document. http://www.pub-edinquiry.org/reports/final_reports/03/. Accessed 20 July 2006.

Walsh, V., & Golins, G. L. (1976). *The exploration of the outward bound process.* Denver: Colorado Outward Bound School.

Westheimer, J. (2008). Learning among colleagues. In M. Cochran-Smith, S. Feiman-Nemser, & D. J. McIntyre (Eds.), *Handbook or research on teacher education: Enduring questions in changing contexts. Third edition* (pp. 756–783). New York: Routledge.

Yarrow, A., Ballantyne, R., Hansford, B., Herschell, P., & Millwater, J. (1998). Teaching in rural and remote schools: A literature review. *Teaching and Teacher Education, 15,* 1–13.

Chapter 11
Argumentation in the Teaching of Science

Maria Evagorou and Justin Dillon

Introduction

Argumentation is a fundamental discourse of science, a part of the practice of science for evaluating, refining and establishing new theories, and is therefore considered a core element of the scientific enterprise (Duschl 1990; Osborne et al. 2004), one which can engage students in the social practices of science (Driver et al. 1996; Zembal-Saul 2009). Argumentation has been the emphasis of many studies during the last decades (Bell 2004; Brem and Rips 2000; Erduran et al. 2004; Jimenez-Aleixandre et al. 2000; Sandoval 2003; Sandoval and Reiser 2004). General questions about argumentation and specifically how students and adults argue have been addressed: for example, studies of learning in both formal and informal contexts suggest that students and adults have difficulties arguing, and tend to focus on single pieces of evidence, or only choose the evidence that best supports their initial claim (Erduran et al. 2004; Jimenez-Aleixandre et al. 2000; Kuhn and Reiser 2005; Kuhn 1993; Sandoval and Reiser 2004; Osborne et al. 2004). According to other studies, some students are able to provide better arguments after engaging with specially designed instruction whilst others fail to improve their arguments (e.g. Clark and Sampson 2006; Osborne et al. 2004). A question that still remains unanswered however is "Why do some students engage in argumentation whilst others do not, and what is the teacher's role in this process?"

Toulmin (1972) suggested that there are four interconnected dimensions explaining why some students are successful in argumentation: a pedagogical dimension—this refers to the teacher's style and instructional practices; a cognitive and metacognitive dimension—this describes students' cognitive abilities; a communicative dimension—among students and teacher; and a social dimension—the interactions that take place in the classroom. In this chapter we will be exploring some of these dimensions that focus on the teacher through a review of the existing literature, and data from two different classrooms.

M. Evagorou (✉)
University of Nicosia, Nicosia, Cyprus
e-mail: evagorou.m@unic.ac.cy

D. Corrigan et al. (eds.), *The Professional Knowledge Base of Science Teaching*,
DOI 10.1007/978-90-481-3927-9_11,

More specifically, the purpose of this chapter is to provide an overview of argumentation, explain the importance of argumentation in the teaching of science, and review studies that explore how pre- and in-service teachers approach argumentation in their teaching. In the second part of the chapter we explore some of these issues through two case studies of teachers implementing the same curriculum in their classes, and conclude with suggestions for further studies in this field.

Defining Argumentation

Recently in science education there has been an emphasis on scientific literacy and meaningful science learning (Duschl et al. 2007). Teachers are encouraged to present science as a human endeavour, with a stress on both the diversity and limitations of scientific methods. According to this perspective, students should be able to understand how science is related to "daily activities, personal problems, social issues, or global concerns" (Aikenhead 2006, p. 2) and be able to critically evaluate information presented to them. Learning science in the classroom in this particular way involves "children entering a new community of discourse" (Driver et al. 1994, p. 11) in which they understand that science concepts do not only emerge from sensory experience but rather are formulated "using the conceptual foundations of scientific discourses" (Matthews 1994, p. 2) and cannot be acquired independently of social interaction. A recent U.S. science education reform document states that:

> Learners who understand can use and apply novel ideas in diverse contexts, drawing connections among multiple representations of a given concept. They appreciate the foundations of knowledge and consider warrants for knowledge claims. Accomplished learners know when to ask a question, how to challenge claims, where to go to learn more, and they are aware of their own ideas and how these change over time. (Duschl et al. 2007, p. 19)

This call for an emphasis in science education on understanding and evaluating the evidence and claims of science is associated with the shift from studying science as exploration and experiment to studying science as argument and explanation (Duschl et al. 2007). For instance, Duschl (1990) argues that if we do not present science as a process of revision and substitution of knowledge claims, we run two risks. Firstly, we may foster in students the perception that "scientific knowledge growth is governed by the addition of new ideas, facts and theories to old ones" (p. 54) and, secondly we may portray science as an activity in which scientists always agree. Hence the emphasis is on teaching argument and explanation, which can contribute to students' appreciation of both the power and the limitations of scientific knowledge claims. In this way argumentation is seen as a process of science, "a social process, where co-operating individuals try to adjust their intentions and interpretations by verbally presenting a rationale of their actions" (Patronis et al. 1999, pp. 747–748). It is also part of the practice of science for evaluating, refining and establishing new theories (Duschl 1990) and is therefore considered a core element of the scientific enterprise.

Argumentation is also central to people's ability in solving problems, making judgements and decisions and formulating ideas and beliefs (Kuhn 1991), and is essentially an approach to thinking and reasoning. According to Kuhn (2005), thinking is the process that enables us to make informed choices between conflicting claims, and a recognition of this perspective may lead a person to value thinking. Usually, when learners are constructing arguments, they need to evaluate alternative perspectives and opinions and select a solution that is supported by evidence and explanation (Cho and Jonassen 2002). Hence, argumentation is an important skill for everyday life since we are usually faced with situations in which we have to evaluate alternative solutions or scenarios and, based on evidence, decide on a course of action. The ability to challenge the authority of science and identify alternative solutions—a skill associated with argumentation—can potentially help people move towards more informed decisions in their everyday lives.

Even though we see argumentation also as a reasoning skill, the main theoretical underpinning for teaching science as argument within the context of school science is that students must develop an understanding of the scientific enterprise, such as the aims and purposes of scientific work (Driver et al. 1996). Driver et al. (1996) argue that understanding the scientific enterprise is necessary because it helps students to develop an appreciation of the power and limitations of scientific knowledge claims. They support that this appreciation is necessary in order to understand, evaluate and use the products of science and technology and it can help students to view science as an epistemological and social process in which knowledge claims are generated, adapted, reorganized, and, at times, abandoned (Lawson 2003). Specifically in science, arguments are commonly constructed to explain a phenomenon or to explain a theory or a new discovery, and argumentation is also seen as part of the process of knowledge construction in science and "can be defined as the connection between claims and data through justifications or the evaluation of knowledge claims in light of evidence, either theoretical or empirical" (Jimenez-Aleixandre and Erduran 2008, p. 13). Hence, in our work we support that argumentation is a thinking skill, and a process of knowledge construction, but is also a fundamental discourse of science, and is therefore considered a core element of the scientific enterprise (Duschl 1990; Osborne et al. 2004), and a component of scientific literacy (Duschl et al. 2007).

Teaching Science as Argument

In recent years there has been a shift from the view that teaching science is about the teacher having good subject matter knowledge to other aspects of teaching as well (Kind 2009). Teaching is a complex ability for which transformation "of teacher knowledge from diverse domains such as subject matter knowledge, general pedagogical knowledge, and knowledge of context" (Avraamidou and Zembal-Saul 2005, pp. 967–968) is required. Shulman (1987) stated that an important knowledge base for teaching consisted of a teacher's: content knowledge, pedagogical content knowledge (PCK), curriculum knowledge, general pedagogy, knowledge of learners and their characteristics, knowledge of educational contexts and educational purposes, ends and values. According to Shulman (1987) PCK is "the capacity of a teacher to transform the content knowledge he or she possesses into forms that are pedagogically powerful and yet adaptive to the variations in ability and backgrounds presented by the students" (p. 15), and curriculum knowledge is the use of the materials, textbooks, software and lab equipment to help him or her in the transformation of the subject knowledge.

Hence for teaching science as argument, or engaging students in argumentation, the teacher's subject matter knowledge, the teacher's knowledge and understanding of argumentation and the ability to transform this knowledge and present it to students are important aspects. In the next section we discuss teachers' knowledge and understanding of argumentation, and the ability to transform this knowledge of argumentation and present it to the students.

Teacher's Knowledge and Understanding of Argumentation

One of the first studies that looked at teachers' argumentation was that of Zembal-Saul and colleagues (2002) in which they explored how pre-service teachers themselves engaged in argumentation and developed arguments. The pre-service teachers (all of whom had a biology qualification) participated in a professional development program and engaged in argumentation by discussing evidence regarding the natural selection as presented in the *Galapagos Finches* software. The results suggest that the pre-service teachers could link data with claims but they did not talk about the validity of the evidence; their arguments were not complex and they did not consider alternative explanations; they did not offer justifications (perhaps because they thought that the connection between the evidence and the claim was clear and there was no need for justification); and they had a hypothesis and only collected data to support that specific hypothesis, ignoring data that contradicted that hypothesis. According to Zembal-Saul et al. (2002), it was not clear if the pre-service teachers understood that they were constructing arguments as a way to help them advance their knowledge of natural selection, a finding paralleled in studies involving students who failed to see the process of discussion and argument construction as a way to advance their knowledge. Findings from this study suggest that pre-service teachers themselves have problems developing coherent arguments, and have difficulties engaging in argumentation similar to the ones that secondary school students have. A similar study by Sadler (2006) with pre-service secondary school teachers revealed that his students were able to construct arguments, and most of them to evaluate given arguments, only after a special instruction focusing on the structure of arguments. In a more recent study Beyer and Davis (2008) focused on a different aspect of a teacher's understanding of argumentation, they investigated a beginning elementary school teacher's understanding and enactment of argumentation. One of the main findings of this case study was that the teacher held two different views of argumentation, making it difficult to explicitly and consistently focus on teaching argumentation in her science class.

Even though these studies involve pre-service or beginning, and not in-service teachers, we claim that even in-service teachers find it difficult to engage in argumentation and produce good quality arguments. This claim can be supported by Kuhn's (1993) findings with adult participants with high qualifications that suggest that even adults face the same difficulties in constructing arguments as younger children if they are not taught how to argue. Consequently, if teachers are not familiar with the structure of an argument we do not expect them to be able to construct good arguments, or hold an understanding of what argumentation is. Consequently, two issues arise for the review of this literature: (a) teachers themselves should be able to construct arguments, and (b) teachers' understanding of argumentation is an important aspect of their knowledge.

Examining Teachers' Argumentation Practices

Implementing argumentation in the classroom goes beyond implementing a new curriculum. In argumentation, the teacher may need to shift away from the role as authority, towards the role of a facilitator, and a fundamental pedagogical shift may be necessary in teaching argumentation effectively (Simon et al. 2006; Zohar 2008). More specifically, when it comes to argumentation, teaching requires a different classroom culture and a different discourse, since argumentation entails shifting teaching from what we know to *how we know what we know* and *why we believe what we know* (Duschl 2008), and a more fundamental shift on teachers' pedagogy is required in order to teach science as argument.

A curriculum development and research initiative that has explored argumentation and teacher professional development was the IDeas Evidence and Argumentation in Science project at King's College London (Osborne et al. 2004). Even though the main emphasis of the project was to investigate students' argumentation in science lessons, teachers' development of practices in argumentation was also explored. More specifically, 12 teachers attended 6 half-day workshops during one year, with the purpose of developing material and strategies to support teaching argumentation (Simon et al. 2006). The workshops served as a forum for teachers to discuss the activities they implemented in their class and share their teaching experiences with argumentation, and members of the research team visited the teachers to observe their lessons. According to Simon et al. (2006) two-thirds of the teachers changed their practices after the workshops and the analysis of the teachers' first and last lessons showed that the new emphasis was on:

> …the importance of talking and listening to others, conveying the meaning of argument through modeling and exemplification, positioning oneself within an argument and justifying that position using evidence, constructing and evaluating arguments, exercising counter-argument and debate, and reflecting upon the nature of argumentation. (Simon et al. 2006, p. 255)

Even though this study has identified some of the instructional strategies that teachers who are successful in argumentation use, it only focused on training the teachers to use a specific approach to teaching argumentation without necessarily looking deeper into the teachers' understanding of what argumentation is, their argumentation skills, or their transformation of subject matter knowledge into teaching.

The work of Avraamidou and Zembal-Saul (2005), and Zembal-Saul's (2009) into teaching science as argument, provide a deeper exploration of a model of teacher professional development in terms of preparing teachers to teach science as argument. More specifically, Avraamidou and Zembal-Saul (2005) explored a first-year elementary teacher's practices for giving priority to evidence in her teaching. Based on the findings of their study, PCK in evidence-based explanations involves:

> …understandings about the central role of evidence in the construction of explanations and providing students with opportunities to engage in a variety of assignments and tasks that support the transformations of data from observations to evidence and from evidence to explanations. (p. 980)

The teacher's PCK, in this case, consisted of: providing students with opportunities to collect evidence; providing students with opportunities to record and represent evidence; and, providing students with opportunities to construct evidence-based explanations. Furthermore, the findings suggest that the teacher was able to teach argumentation because as a pre-service teacher she engaged in learning science as arguments, and developed her understanding of argumentation and her argumentation skills. So, Avraamidou and Zembal-Saul recommend that pre- and in-service teachers engage in similar learning activities, looking at science as argument in order to be able to teach science as an argument.

Zembal-Saul (2009) has recently described a pre-service teacher program for elementary school teachers in which the emphasis is on teaching science as argument and supporting students during a methods course to teach argumentation. More specifically, in the science methods course the pre-service teachers experienced what it means to learn science as argument, they observed their mentors teaching using practices consistent with evidence-based explanations, and worked with video cases showing argumentation lessons in real classrooms. Through these activities, the pre-service teachers learned about the structure of the argument, which helped them think about the importance of scientific argumentation. They understood the importance of classroom discourse in negotiating what children meant and students' talk helped them monitor and assess their students' thinking. Furthermore, Zembal-Saul (2009) talks about a continuum for teaching science as argument followed by the teachers and how they changed from teaching science as activity-based, to teaching science as investigations-based, then evidence-based and then argument-based. Hence, one of the important outcomes is that these teachers were able to move away from teaching science as activity based, to teaching science as argument. Even though Zembal-Saul provides a coherent framework for teaching science as argument for in-service teachers there is no evidence of how the teachers transferred these practices to their future careers as science teachers, an area that is still largely unexplored.

The Case of James and Heather

The studies described in the previous section provide an insight into teachers' instructional practices when teaching science as an argument, but do not map those instructional practices that lead to students' success or failures in argumentation. In a recent study by the first author addressing issues pertaining to the implementation of a specially designed learning environment by two different teachers (see Evagorou 2009; Evagorou and Osborne 2007), even though the emphasis of the research was on the process of students' argumentation (Evagorou and Osborne 2008), it was evident that the way the teacher chose to enact the curriculum in their classroom was an important factor in the success of the students. In this section we present data from the two teachers in order to discuss the different instructional practices and the effect they might have on students' argumentation.

The Learning Environment and the Two Teachers

The design of Argue-WISE was based on socio-cultural theories of learning (Vygotsky 1978; Rogoff 2003) which argue that peer interaction, the teacher, the curriculum and dialogue are all important aspects of learning. The purpose of Argue-WISE is to engage students in the construction of arguments, as defined by Toulmin (1958). The theoretical framework that informed the design of Argue-WISE was discussed with both teachers before the implementation. The two teachers were working in two different schools in the UK, one in the suburbs of London, and one in the south of England, and were both teaching 12–13-year-old students. Even though the learning environment and the theory behind its design and argumentation were discussed in detail, they were not trained on how to teach argumentation. Hence, each of the teachers enacted the learning environment based on their understanding of what was discussed with the researcher, their understanding of argumentation and its importance in the teaching of science, and their usual instructional practices.

The analysis of conversations during the whole-class discussions suggests that the two different teachers, James and Heather, used different instructional practices during the enactment of the curriculum. Heather not only supported but also encouraged discussion in the classroom, her questions were facilitating the dialogue, providing positive feedback to the students, whilst at the same time helping them to built on each others' ideas and understand the structure of an argument. In addition, as shown in Table 11.1, Heather used most of the time for paired discussions and group work.

Furthermore, Heather tried to model argumentation by discussing examples of evidence, and their validity, and how these should support claims, and provided time for the students to discuss their ideas both during paired interactions and whole-classroom discussions. The episode below presents an interaction during the whole-classroom discussion in Lesson 4.

Table 11.1 How Heather implemented Argue-WISE in her class

Lesson	Description of lesson
Lesson 1: What is the problem?	• Introduction to WISE and Argue-WISE [5 min] • Students worked in pairs on the following: – Introduction to the problem – Stated their opinion – Went through a number of activities to help them understand the ecology of the red and the grey squirrel, and to understand how these two sub-species differ – Scaffolded with the use of prompt windows
Lesson 2: The red squirrel population: is it dropping?	• Students worked in pairs on the following: – Investigated the decrease in the red squirrel population and the causes of the change in the numbers of the population – Used SenseMaker, an argument construction tool to scaffold students to collect evidence – Studied historical data sets informing them about the population of the red squirrel before the introduction of the grey, a map comparing the population in 1940 and 1998, and several internal and external links providing information and reasons for the reduction of the red and the survival of the grey
Lesson 3: How can we save the red squirrel?	• Students worked in pairs on the following: – Learned about ways to maintain the red squirrel population – Read information from a BBC website presenting real stories of how people in Scotland acted in order to save the red squirrel, an audio interview with a representative from the UK Forestry Commission, and comments from members of the public about how they are against the grey squirrel as it invades their gardens
Lesson 4: Share your argument	• Completed their final argument in pairs and submitted it within Argue-WISE [25 min] • Presentation of their argument during a whole-classroom discussion [25 min]

Heather: All right guys, if you can just save what you have done and stop typing and we are just going to try to have a discussion. We want to have a proper debate, an argument about the squirrels. So, just to start us of then, who would like to give us an opening bit of what they think?

Student: [inaudible]

Heather: So what is the evidence for that?

Student: I got (.) [looking at the screen], oh yeah. In Ireland in the 1700 a forest was destroyed and the red squirrels that lived there became extinct.

Heather: Right so you have historical data. So we have someone here with data about forest destruction as the reason that the red squirrels became extinct or endangered. So we would like to hear someone else and what they think. For or against. You can argue.

Student N: There is some evidence for forest destruction being responsible but the grey squirrels are more responsible for the red dying.
Heather: Right, explain.
Student N: Because they carry squirrel pox, they eat all their food [inaudible] and they take red squirrels' forests.
Heather: OK, who wants to respond to that?
Joshua: There are other factors as well like road accidents, food shortage and forest destruction. So humans play a big part in the loss of a lot of red squirrels as well as the grey.

In the sequence above, a representative episode from the whole-classroom discussion, Heather initiated the discussion by asking the students to express their thoughts and then she maintained the dialogue based on students' responses and not on a predetermined agenda. For example, when the first student offered his opinion, Heather asked for supporting evidence and then she labelled the evidence offered by the student as "historical data". In addition, it is also important to note that she coordinated the discussion, without interfering with what students said. For example, at some point, after a student offered an argument, Heather said: *OK, who wants to respond to that?* This kind of discussion, one that is characterized by a constant questioning of what the students' are saying, was described as Initiation-Response-Initiation by Cazden (1988). In classrooms this style of interaction is usually present when the teacher does not know an answer to a student's question, or is not expecting a specific response to a question.

James, on the other hand, followed a pedagogy of transmission, talking most of the time, trying to "impose" knowledge on students. As shown in Table 11.2, James spent most of the time explaining the activities or presenting information.

Even during whole-classroom discussions his questions were closed and he commented on students' responses without trying to link to previous comments and without providing positive feedback. There was little evidence of modelling argumentation, defining argumentation or explaining the rationale of the activities. The episode below is representative of James' interactions with his class:

James: [...] It is this argument construction tool that we have not done much with. It appears in a couple of other places. [...] On the screen you can see some arguments, some claims that have already been started. And if you click on there then you can add something on that claim. [students talking] Shshshsh. Listen, listen, listen. There is that sentence which we could add to. Can I recommend that you really add very short claims. We should kill grey squirrels to save the red. So that is our claim. If I want to add some evidence to back up that and so if you click on new evidence you can add in. Would someone want to tell me one thing that would support killing grey squirrels? No one? We lets say there are very few red squirrels. So that might be a little bit of evidence to support that claim there. So you can make new claims ok so you could add a new one in here. Would someone want to say something else about

Table 11.2 How James implemented Argue-WISE in his class

Lesson	Description of Lesson
Lesson 1: What is the problem?	• The teacher introduced the lesson [5 min] • The teacher asked the students to log-in [5 min] • The teacher read through Activity 1 [10 min] • The students went through a number of activities to help them understand the ecology of the red and the grey squirrel, and to understand how these two sub-species differ, scaffolded with the use of prompt windows [12 min] • The teacher interrupted the pairs to present Activity 1 again and then Activity 2 [19 min] • The students worked in pairs [5 min]
Lesson 2: The red squirrel population: is it dropping? And How can we save the red?	• The teacher reminded the class of what they had been doing and introduced Activity 2 that asks students to investigate the decrease in the red squirrel population and the causes of the change in the numbers of the population [10 min] • The students worked in pairs [15 min] • The teacher introduced Activity 3 and explained to the students how to use the headphones to listen to the radio program [6 min] • The students worked in pairs—used SenseMaker [30 min] • Whole-classroom discussion, summarized the work [10 min]
Lesson 3: Share your argument	• The teacher introduced Activity [4 min] • The teacher read Activity 4 [4 min] • Group work—completed their argument and submitted it within Argue-WISE [17 min] • The teacher presented some arguments prepared by the pairs during a whole-classroom discussion [10 min] • Group work—discussion in on-line board [23 min] • Closing discussion [3 min]

this matter of red and grey squirrels? I want someone to make another claim.

Students: Leave them alone.

James: OK, so which one shall we leave alone?

Students: All of them.

James: OK, leave all the squirrels alone. So there is a (.) new claim there. We might want some evidence underneath. So lets use that tool and what I want you to do is make sure that you have gone through all Activity 2 including building your argument and evidence.

It is important to note here that through the presentation of the argument construction tool, James attempted to present the structure of an argument. More specifically, by explaining the two different sentences (*we should* and *we should not kill the grey*) he explained what a claim is. However, the talk that followed was confusing since he asked students to provide evidence to support a claim, and then he asked for additional claims. When one student offered a claim (*leave all squirrels alone*), James did not explain how a claim is different from evidence, failing in that way to present a model of an argument.

The Impact of Teaching on Students' Argumentation

It is evident from Table 11.1 and 11.2 above that Heather's students had much more time than James's students to work in pairs, and collaboratively write their arguments. This finding further supports the observation that James was focusing on transmitting the information rather than allowing time for the students to study, understand, organize the evidence, whilst Heather was different. A study (Zohar 2004) states that the majority of teachers hold a view of learning as a process of transmission as opposed to one which sees knowledge as something which must be constructed by the individual. According to Zohar (2008), teachers who viewed teaching as transmission believed that teaching thinking or presenting problems that required students' independent thinking was believed to be inappropriate because it brought frustration and confusion. These teachers lowered the cognitive demands of the task by spoon-feeding the students, or chose to teach higher-order skills only in classes with high ability students (Page 1990). On the other hand, teachers who viewed teaching thinking through the lens of a pedagogy of knowledge construction put the students at the centre of the activities. The profiles described by Zohar (2008) match the profiles of James and Heather. James, on the one hand, stated from the beginning of our interaction that the learning environment might be difficult for his class, and he tried to structure it in his own way, mostly by presenting all the information to his students instead of providing the time for them to read and understand the problems. Even though he knew that the students would follow a set of specially designed activities, he prepared lesson plans that did not follow the activities, but instead were built around the activities. Heather, on the other hand, allowed time for interaction and knowledge construction, and she scaffolded whole-classroom discussions with meaningful questions and comments that helped model the structure of an argument.

Looking at the interrelationship of teachers' instructional practices, and students' achievements in argumentation as measured by the Erduran et al. (2004) levels of argumentation, Heather's students were more successful than James' students in constructing sophisticated arguments, and providing alternative solutions for the socio-scientific issue, as shown in Table 11.3.

Hence, the claim put forward in this section is that students' performance in argumentation is directly related to their teacher's instructional practices, even though various other conditions might have also have had an effect on students' performance in argumentation (see Evagorou and Osborne 2008). Our findings also suggest that adequate learning activities designed to engage students in argumentation are not a sufficient condition—teachers' instructional practices are also important.

Table 11.3 Improvement of arguments for the two classes

	Heather's class (14 pairs)	James' class (13 pairs)
Improvement of levels	11	7
No improvement	3	6

Discussion

According to Zembal-Saul et al. (2002) and Avraamidou and Zembal-Saul (2005), studies of how teachers understand and teach argumentation are just beginning to explore issues of teachers' enactment of argumentation in their classrooms, and consequently, not much is known. Findings from McNeill and Krajcik (2008) support that important instructional practices that can help students construct better arguments include discussing the rationale of using evidence-based explanations in science. Even though McNeill and Krajcik's hypothesis was that modelling or argumentation would help students improve in argumentation, their study showed that this approach was not significant in helping students. However, other studies suggest that modelling what a good argument is, or providing examples of good arguments, can help students to improve their argumentation (Crawford 2000). Neither of the teachers in our study explicitly modelled argumentation but both of them referred to the use of evidence in support of a claim to describe what an argument should consist of. Heather insisted on using evidence to support claims and also on choosing the right evidence. Hence, the data support the McNeill and Krajcik finding that modelling the structure argumentation does not necessarily help students improve their arguments.

Even though research in how teachers enact argumentation in their classes is still an area that needs further exploration, we can assume that our knowledge from the field of teacher professional development, and teachers' practices can inform our efforts to understand how teachers teach science as an argument. For example, studies have shown that teachers' beliefs, prior experience as students, values and conceptions have an effect on their classroom practice, and those with more inquiry-oriented experiences as learners have more chance of incorporating inquiry in the class (for example, Eick and Reed 2002; Mellado 1998). Furthermore, Mellado (1998) claims that "in the complexity of the real classroom teachers construct simplified models with which they are comfortable and that they find non-conflictive and permit them to act" (p. 200). Such was the case with James and the way in which he decided to enact the learning environment so as to be consistent with the usual ways in which he was teaching science, a model of teaching however that was not consistent with the theoretical underpinnings supporting argumentation.

Crawford (2007), in a recent study with teachers, has concluded that:

> A prospective teacher's personal view of teaching science [...] is a strong predictor of a prospective teacher's actual practice of teaching science. (p. 637)

Hence an important aspect of preparing teachers to teach science as argument may be to provide them with experiences that link theory to practice, using mentors to support teachers—so mentors also need to be trained (Darling-Hammond 2006; Crawford 2007), and pre-service teachers must themselves engage in argumentation as learners (Sadler 2006; Zembal-Saul 2009). A new line of research in argumentation is already focusing on how to support pre- and in-service teachers to change their view about argumentation, improve their knowledge of the topic, and enable them to successfully teach science as argument.

References

Aikenhead, G. S. (2006). *Science education for everyday life: Evidence-based practice.* New York: Teachers College Press.

Avraamidou, L., & Zembal-Saul, C. (2005). Giving priority to evidence in science teaching: A first year elementary teacher's specialized practices and knowledge. *Journal of Research in Science Teaching, 42*(9), 965–986.

Bell, P. (2004). Promoting students' argument construction and collaborative debate in the science classroom. In M. Linn, E. Davis, & P. Bell (Eds.), *Internet environments for science education* (pp. 115–143). Hillsdale: Lawrence Erlbaum.

Beyer, C., & Davis, E. (2008). Fostering second graders' scientific explanations: A beginner elementary teacher's knowledge, beliefs and practice. *The Journal of the Learning Sciences, 17,* 381–414.

Brem, S. K., & Rips, L. J. (2000). Explanation and evidence in informal argument. *Cognitive Science, 24,* 573–604.

Cazden, C. B. (1988). *Classroom discourse.* Portsmouth: Heinemann.

Cho, K., & Jonassen, D. (2002). The effects of argumentation scaffolds on argumentation and problem solving. *Educational Technology Research and Development, 50*(3), 1042–1069.

Clark, D., & Sampson, V. (2006). *Characteristics of students' argumentation practices when supported by online personally seeded discussions.* Paper presented at the National Association of Research in Science Teaching, April 3–6, San Francisco.

Crawford, B. A. (2000). Embracing the essence of inquiry: New roles for science teachers. *Journal of Research in Science Teaching, 37*(9), 916–937.

Crawford, B. A. (2007). Learning to teach science as inquiry in the rough and tumble of practice. *Journal of Research in Science Teaching, 44*(4), 613–642.

Darling-Hammond, L. (2006). Constructing 21st century teacher education. *Journal of Teacher Education, 57*(3), 300–314.

Driver, R., Asoko, H., Leach, J., Mortimer, E., & Scott, P. (1994). Constructing scientific knowledge in the classroom. *Educational Researcher, 23,* 5–12.

Driver, R., Leach, J., Millar, R., & Scott, P. (1996). *Young people's images of science.* Buckingham: Open University Press.

Duschl, R. A. (1990). *Restructuring science education: The importance of theories and their development.* New York: Teachers College Press.

Duschl, R. A. (2008). Quality argumentation and epistemic criteria. In S. Erduran & M. Jimenez-Aleixandre (Eds.), *Argumentation in science education: Perspectives from classroom-based research.* Dordrecht: Springer.

Duschl, R. A., Schweingruber, H. A., & Shouse, A. W. (Eds.). (2007). *Taking science to school: Learning and teaching science in Grades K-8.* Washington: The National Academies Press.

Eick, C., & Reed, C. (2002). What makes an inquiry-oriented science teacher? The influence of learning histories on student teacher role identity and practice. *Science Education, 86,* 401–416.

Erduran, S., Simon, S., & Osborne, J. (2004). TAPing into argumentation: Developments in the application of Toulmin's Argument Pattern for studying science discourse. *Science Education, 88*(6), 915–933.

Evagorou, M. (2009). *Argue-WISE: Exploring young students' features of argumentation within a socio-scientific issue when they engage with an on-line learning environment.* PhD Thesis, King's College London, UK.

Evagorou, M., & Osborne, J. (2007). Argue-WISE: Using technology to support argumentation in science. *School Science Review, 89,* 103–110.

Evagorou, M., & Osborne, J. (2008). *Identifying features of young students' construction of arguments in the science classroom.* New York: Annual American Educational Research Association.

Jimenez-Aleixandre, M. P., & Erduran, S. (2008). Argumentation in science education: An overview. In S. Erduran & M. Jimenez-Aleixandre (Eds.), *Argumentation in science education: Perspectives from classroom-based research* (pp. 3–27). Dordrecht: Springer.

Jimenez-Aleixandre, M. P., Rodriguez, A. B., & Duschl, A. R. (2000). 'Doing the Lesson' or 'Doing Science': Argument in high school genetics. *Science Education, 84*(6), 757–792.

Kind, V. (2009). Pedagogical content knowledge in science education: Perspectives and potential progress. *Studies in Science Education, 45*(2), 169–204.

Kuhn, D. (1991). *The skills of argument.* London: Cambridge.

Kuhn, D. (1993). Science as argument: Implications for teaching and learning scientific thinking. *Science Education, 77*(3), 319–337.

Kuhn, D. (2005). *Education for thinking.* Cambridge: Harvard University Press.

Kuhn, L., & Reiser, B. (2005). *Students constructing and defending evidence-based scientific explanations.* Paper presented at the annual meeting of National Association of Research in Science Teaching, Dallas, Texas.

Lawson, A. E. (2003). The nature and development of hypothetico-predictive argumentation with implications for science teaching. *International Journal of Science Education, 25*(11), 1387–1408.

Matthews, M. R. (1994). *Science teaching: The role of history and philosophy of science.* London: Routledge.

Mellado, V. (1998). The classroom practice of preservice teachers and their conceptions of teaching and learning science. *Science Education, 88*(2), 197–241.

McNeill, K. L., & Krajcik, J. (2008). Scientific explanations: Characterizing and evaluating the effects of teachers' instructional practices on student learning. *Journal of Research in Science Teaching, 45*(1), 53–78.

Osborne, J., Erduran, S., & Simon, S. (2004). Enhancing the quality of argumentation in school science. *Journal of Research in Science Teaching, 41*(10), 994–1020.

Page, R. N. (1990). The lower track curriculum in a college-preparatory high school. *Curriculum Inquiry, 20*(3), 249–281.

Patronis, T., Potari, D., & Spiliotopoulou, V. (1999). Students' argumentation in decision-making on a socio-scientific issue: Implications for teaching. *International Journal of Science Education, 21*(7), 745–754.

Rogoff, B. (2003). *The cultural nature of human development.* Oxford: Oxford University Press.

Sadler, T. D. (2006). Promoting discourse and argumentation in science teacher education. *Journal of Science Teacher Education, 17,* 323–346.

Sandoval, W. A. (2003). Conceptual and epistemic aspects of students' scientific explanations. *The Journal of the Learning Sciences, 12*(1), 5–51.

Sandoval, W. A., & Reiser, B. (2004). Explanation-driven inquiry: Integrating conceptual and epistemic scaffolds for scientific inquiry. *Science Education, 88*(3), 345–372.

Shulman, L. S. (1987). Knowledge and teaching: Foundations of the new reform. *Harvard Education Review, 57,* 1–22.

Simon, S., Erduran, S., & Osborne, J. (2006). Learning to teach argumentation: Research and development in the science classroom. *International Journal of Science Education, 28*(2–3), 235–260.

Toulmin, S. (1958). *The uses of argument.* Cambridge: Cambridge University Press.

Toulmin, S. (1972). *Human understanding* (Vol. 1). Princeton: Princeton University Press.

Vygotsky, L. S. (1978). *Mind in society: The development of higher psychological processes.* Cambridge: Harvard University Press.

Zembal-Saul, C. (2009). Learning to teach elementary school science as argument. *Science Education, 93,* 687–719.

Zembal-Saul, C., Munford, D., Crawford, B. A., Friedrichsen, P., & Land, P. (2002). Scaffolding pre-service science teachers' evidence-based arguments during an investigation of natural selection. *Research in Science Education, 32,* 437–463.

Zohar, A. (2004). *Higher order thinking in science classroom: Students' learning and teachers' professional development.* Dordrecht: Springer.

Zohar, A. (2008). Science teacher education and professional development in argumentation. In S. Erduran & M. Jimenez-Aleixandre (Eds.), *Argumentation in science education: Perspectives from classroom-based research.* Dordrecht: Springer.

Chapter 12
Assessment Literacy: What Science Teachers Need to Know and Be Able to Do

Sandra K. Abell and Marcelle A. Siegel

What we hope for in our science classrooms is that students learn science with understanding. What does it take to achieve such a goal? According to Donovan and Bransford (2005), this sort of learning is supported in classroom environments that strive to be learner-centered, knowledge-centered, assessment-centered, and community-centered. In this chapter, we focus on the assessment-centered classroom. In an assessment-centered learning environment, teachers employ formative assessments to support learning, help students recognize and make improvements to their thinking, and inform instruction.

> Once the knowledge to be learned is well defined, assessment is required to monitor student progress (in mastering concepts as well as factual information), to understand where students are in the developmental path from informal to formal thinking, and to design instruction that is responsive to student progress. (Donovan and Bransford, p. 16)

That is, assessment-centered environments are critical for supporting learner-centered and knowledge-centered environments to facilitate student learning with understanding. Although K-12 and college science classrooms have made great strides in changing science instruction in the recent past, change in assessment practices has lagged behind (National Research Council (NRC) 2003; Shepard 2000). This is despite the fact that assessment plays a prominent role in science education reform policy documents around the world (e.g., in the United States, American Association for the Advancement of Science (AAAS) 1993; NRC 1996). What are challenges to achieving the assessment-centered classroom?

Historically, as views of learning have changed, concepts of teaching and assessment have shifted concomitantly. Shepard (2000) presented a framework to illustrate this shift. The dominant learning paradigms of the twentieth century included

Note: This material is based in part upon work supported by the U.S. National Science Foundation under Grant No. 0553929. Any opinions, findings, and conclusions or recommendations expressed in this material are those of the author(s) and do not necessarily reflect the views of the National Science Foundation.

M. A. Siegel (✉)
University of Missouri, Columbia, USA
e-mail: siegelm@missouri.edu

D. Corrigan et al. (eds.), *The Professional Knowledge Base of Science Teaching*,
DOI 10.1007/978-90-481-3927-9_12, © Springer Science+Business Media B.V. 2011

behaviorist and associationist theories. Learning was seen as stimulus and response, with motivation based on positive reinforcement of little steps. Thorndike and colleagues provided an associated foundation for scientific measurement using objective tests—the predominant feature of American achievement testing ever since (Shepard 2000). Rather than focusing on how teachers gather evidence about student conceptual understanding and use that evidence to enhance student understanding, achievement testing focused on measuring the number of bits of knowledge students had accumulated. The cognitive revolution brought new ways of thinking about learning, which Shepard labeled the social-constructivist framework, drawn from cognitive, constructivist, and sociocultural theories of the twenty-first century. This framework holds that learning is socially and culturally constructed; learning is more than associations and recall, and includes higher-order thinking, such as problem solving and discourse practices. Classroom assessment, based on this view of learning, addresses learning processes as well as outcomes, and includes explicit expectations, challenging tasks, and student responsibility (Shepard 2000). Classroom assessment based on this view of learning provides useful evidence for teachers to adapt their instruction and for students to improve their learning.

In addition to historical shifts in views of learning, another factor influencing the enactment of assessment-centered classroom environments is the teacher. As a matter of fact, numerous studies show that the classroom teacher is the most important factor in student achievement (NRC 2001). Thus, in creating an assessment-centered classroom, what a science teacher knows and is able to do—teacher assessment literacy—will affect how assessment is planned and implemented and how assessment data are used. The purpose of this chapter is to define a framework for science teacher assessment literacy, grounded in both theoretical and empirical perspectives. We focus exclusively on teacher knowledge for classroom-based assessment, (as opposed to knowledge related to large-scale and standardized assessment), because it is the knowledge that science teachers use on a regular basis.

Theoretical Foundations of Assessment Literacy

For over 30 years, researchers have hypothesized that the knowledge teachers need for teaching is specialized knowledge not possessed by the subject matter specialist (Shulman 1986). According to Shulman, teachers know how to transform their subject matter knowledge into viable instruction by tapping into their pedagogical content knowledge (PCK). Magnusson et al. (1999) theorized that PCK includes teacher knowledge of assessment in science as well as knowledge of learners, curriculum, and instruction. They conceptualized knowledge of assessment in science as including two dimensions: (1) knowledge of what to assess, and (2) knowledge of how to assess in science classrooms. The first dimension "refers to teachers' knowledge of the aspects of students' learning that are important to assess within a particular unit of study" (p. 108) and is closely linked to instructional goals. The second dimension of knowledge of assessment in science "refers to teachers' knowledge of the ways that might be employed to assess the specific aspects of student learning that are important to a particular unit of study" (p. 109). This view of two dimensions of knowledge of assessment in science provides a foundation for understanding what science teachers need to know about assessment to be effective.

A second theoretical foundation that contributes to our view of science teacher assessment literacy comes from the work of a U.S. National Research Council Committee on the Foundations of Assessment (Pellegrino et al. 2001). They defined assessment as a process of reasoning from evidence, portrayed as the *assessment triangle*. The three vertices of the triangle represent the key elements that underlie the assessment process: "a model of student *cognition* and learning in the domain, a set of beliefs about the kinds of *observations* that will provide evidence of students' competencies, and an *interpretation* process for making sense of the evidence" (p. 44, emphasis in original). This framework provides additional ideas for what science teachers need to know and be able to do to implement assessment effectively and use assessment evidence in their teaching.

These theoretical views of assessment processes and teacher knowledge for assessment provide a starting point for understanding science teacher assessment literacy. However, empirical findings about teacher assessment knowledge can verify the frameworks and extend our understanding of assessment literacy. In the following sections, we review the research on science teacher assessment literacy, including significant findings from our own research programs. We then generate a model of science teacher assessment literacy, and demonstrate how the model might function in science teacher education. At the end of this chapter, we describe the implications for science teacher educators and make recommendations for future research on assessment literacy in science.

Empirical Foundations of Assessment Literacy

Although there is a rich literature on classroom assessment and science learning (Bell 2007; Black 1998), the literature on teacher assessment literacy is limited. In a review of the research on science teacher knowledge, Abell (2007) found only a few studies in which researchers attempted to examine science teacher knowledge of assessment knowledge directly. Some researchers looked at science teachers' decisions to use particular assessment strategies (Duffee and Aikenhead 1992; Pine et al. 2001), while others explored the ways in which teachers interpret classroom assessment data (Kokkotas et al. 1998; Sanders 1993). Researchers also uncovered contradictions between science teachers' assessment beliefs and classroom practices (Bol and Strage 1996; Briscoe 1993; Morrison and Lederman 2003).

In another set of studies, researchers examined the types of assessment tasks that teachers employ in the classroom. Although science teacher use of assessment tasks with a wider range of formats (e.g., concept maps, portfolios, interviews, observational methods, and self, peer, and group assessment) is expanding (Bell 2007), researchers have found that teachers are not proficient at implementing a variety of classroom assessment tasks (Mertler 1999; Mertler and Campbell 2005). In one study of elementary and middle school mathematics and science curriculum in the United States, a majority of the 600 teachers surveyed infrequently used items with extended student response and justification. "Middle grades teachers are significantly more likely than elementary teachers to ask students to explain or justify their answers (1–3 times/month)" (CCSSO and WCER 2000). In a study of four middle school science teachers' informal formative assessment, Ruiz-Primo and Furtak (2007) found that the teacher whose students learned the most was the one who held the most discussions, asked the most conceptual questions, and used the most different ways of adapting instruction based on information gained from assessment.

Researchers also have examined the needs of prospective and new teachers in developing assessment literacy. Stiggins (2002) asserted that prospective teachers in teacher education programs do not have enough opportunities to learn about approaches to assessment that impact student learning positively. Prospective teachers also have a limited arsenal of assessment strategies that they select from to monitor learning (Yilmaz-Tuzun 2008). Doran et al. (1994) found that teachers commonly relied on either teacher-constructed or curriculum-related objective tests for the majority of their assessment, regardless of their level of experience. However, teacher education can help teachers learn about assessment. When teachers engaged in a professional development program to increase their assessment expertise, they learned about tools to use, ways to employ them, and how to respond to students' learning needs (Gearhart et al. 2006). However, teachers needed higher-quality assessment tools to be offered in the curricula they used and additional resources that could guide their interpretation of growth in student understanding over time (Gearhart et al. 2006).

This research base, although limited, demonstrates the need to understand science teacher assessment literacy more deeply so that we can design teacher preparation and professional development programs to meet their needs. To extend the existing research literature, each of the authors has engaged in research related to science teacher assessment literacy. In the following sections, we briefly describe our research and how it adds to the knowledge base.

Studying Assessment Literacy

In a study of prospective teachers in a secondary science methods course that emphasized assessment literacy, we (Siegel and colleagues) examined the purposes of assessment and the types of activities that count as assessment as identified by prospective teachers. This study was unique in that it examined the development of prospective teachers' assessment literacy by comparing their ideas about assessment (theoretical realm) to how they generated assessment tasks to use in the classroom (practical realm). We identified 20 activities that prospective teachers viewed as assessment strategies, such as flow charts and journaling, and how the teachers perceived the benefits and drawbacks of the strategies. The prospective secondary teachers saw many disadvantages to traditional tests, including that they were unmotivating to students and often measured reading ability instead of content goals.

Results also showed that prospective teachers held several advanced ideas about the effective use of assessment tasks:

> Participants indicated that students should be assessed by multiple methods, that assessment should be used to provide feedback to students to help them learn and practice new material, and to serve as a motivating factor. Feedback, noted several participants, must be specific, useful, and immediate. Prospective teachers believed that assessments should be used daily to improve student performance and students should be involved in the assessment process. (Wissehr and Siegel 2008)

When planning inquiry-based science units, however, the prospective teachers did not plan assessment tasks to the extent they had reflected on their assessment ideas. Despite claiming to prefer more reform-based types of assessments, when writing units, the students often reverted to traditional assessments, such as worksheets and written scientific reports. Similarly, the purpose for assessment differed between their written reflections and their lesson plans. For all participants, their reflections focused on improving learning, such as gauging prior knowledge, while their plans focused on assigning grades (Wissehr and Siegel 2008).

In a related study of two sections of a secondary science methods course, we examined prospective teachers' understanding and implementation of equitable assessment. With increasing diversity in classrooms, U.S. teachers are not equipped to reach all students, especially English language learners (Johnson 2006; Lee et al. 2007). Equitable assessment includes assessment tasks that are fair and that support learning for all types of learners regardless of their language ability or cultural/ethnic/racial background (Siegel 2007; Siegel et al. 2008). In our research, we examined prospective teachers' journals, written teaching philosophies, and inquiry-based science units to see how they conceptualized equitable assessment before and during instruction. Not surprisingly, teachers lacked awareness and understanding of equitable assessment prior to methods course instruction. Participants did not see the importance of considering students' language and cultural backgrounds when designing assessments. Understanding developed as teachers took part in discussions and reflected on their views of assessment, needs of the learner, and equitable assessment strategies (Siegel and Wissehr 2009). Prospective teachers' views of English language learners' abilities changed, and they began to see assessments as learning tools, and not merely grading devices (Siegel and Wissehr 2009). Yet, similar to the findings in our initial study, the assessment tasks used in their science units indicated that prospective teachers have a long way to go in effectively implementing equitable assessment. The results demonstrate the need to emphasize developing prospective teachers' awareness and use of equitable assessment practices so that they are better able to meet the needs of diverse learners (Siegel and Wissehr 2009).

Studying Science Teacher PCK for Assessment

We (Abell and colleagues) are engaged in a study of the development of knowledge for teaching of teachers in a post-baccalaureate science teacher preparation program (Friedrichsen et al. 2009; Lannin et al. 2008). Our work is grounded in the Magnusson et al. (1999) model of science teacher knowledge, in which PCK for science assessment is a critical component of the knowledge base for science teaching. In the project, we collect data from teachers using the lesson preparation method (van der Valk and Broekman 1999) when they enter the program and two years later after their first year as a classroom teacher. We ask the teachers to plan two consecutive lessons around a science topic from the science discipline they are planning to teach, followed by a 1-hour, semi-structured interview with each participant (Patton 2002). The participants are asked to describe the process they used in designing the lessons, and we probe for information about the various PCK components, including PCK for assessment. We also collect data four times over two years using a field-based observation cycle consisting of lesson planning, pre-observation interview, and two days of classroom observation followed by stimulated recall interviews. Each interview protocol includes questions about teacher knowledge of assessment, such as: "How will you know if students are getting it or not getting it? What is a specific example (of a homework or test item)? What will you do with the information you gain?"

To structure our analysis, we coded the data by the Magnusson et al. components of PCK, including assessment PCK. To begin our analysis of PCK for assessment, we used the two categories identified by Magnusson et al. (1999): (a) knowledge of what aspects of science to assess (e.g., science content, including the nature of science and conceptual understanding), and (b) knowledge of assessment strategies in science. As we coded, we looked for participant comments about assessment, including evidence of these categories of teacher knowledge. We soon found that the teachers' comments about assessment went beyond these two categories, which led to the creation of four new assessment PCK categories: (a) assessment philosophy, (b) purpose of assessment, (c) assessment consequences, and (d) assessment challenges. In Table 12.1, we present the coding dictionary that we developed during data analysis. It captures the various types of PCK for assessment we found in the data. Thus, through our data analysis, we have expanded our understanding of teacher knowledge of assessment.

Our research on the development of PCK for science teaching, including assessment PCK, is ongoing. The coding dictionary is a first step in data analysis that will be followed up by other analysis techniques guided by individual research questions. For example, one doctoral student examined the relationship of the development of PCK for assessment and PCK for learning among a group of teachers for his dissertation. The coding dictionary is a useful tool for analyzing data in such a study. We believe it is also useful in helping to define science teacher assessment literacy. In the next section, we present a model for thinking about teacher assessment knowledge that resulted in part from this work.

A Model for Science Teacher Assessment Literacy

Based on both of our (Siegel's and Abell's) research programs about understanding science teacher knowledge of assessment, and supported by the theoretical frameworks of Magnusson et al. 1999 and Pellegrino et al. 2001, we have developed a

Table 12.1 Coding dictionary for PCK for science assessment

Code/Sub-code	Definition
What to assess	Participant describes the science content or other areas assessed (e.g., science processes, nature of science, conceptual understanding, terminology, attitudes, lab skills).
Assessment strategies • Formal • Informal	Participant describes the design and/or use of an informal or formal assessment strategy that could be used or was used in the classroom to assess a particular science concept/process/nature of science idea. • Participant refers to assessments that are typically announced to students as such and which are scored and graded. Tests and final presentations are examples of formal assessments. Participant identifies specific items from the assessment. • Participant refers to assessment tasks embedded in instruction which might not be scored or graded. Examples include exit slips and science lab notebooks.
Philosophy of assessment	Overarching principles, beliefs, and values that guide assessment. For example, participant refers to the importance or need for different assessment types (such as multiple assessments are important in science; assessment should lead to learning science better).
Purpose of assessment • Diagnostic • Formative • Summative	Participant refers to the reason for using an assessment specific to science. • Participant refers to what students know when they come in, e.g., assessing prior knowledge/misconceptions. • Participant refers to informing/evaluating instruction, helping students self monitor, aiding student learning. • Participant refers to grading, feedback to parents, accountability.
Assessment consequences • Student reactions • Teacher actions	Participant refers to what happens as a result of data collected about science learning. • Participant talks about how students react to assessment, or what students learn from an assessment task (e.g., which science concepts students learn or that they still do not understand). • Participant talks about the actions specific to science learning that the teacher takes as a result of assessment information (e.g., giving a counter example, changing instruction, deciding where to begin or how much to include, including more opportunities for practice).
Assessment challenges	Participant discusses tensions or struggles inherent in assessment (e.g., difficulty in assessing specific concepts).

model of science teacher assessment literacy (see Fig. 12.1). This model attempts to capture the various types of assessment knowledge and skills that we believe teachers need to create an assessment-centered learning environment. At the center of the model is the teacher's view of learning, which undergirds the values, principles, and knowledge of assessment needed to teach (as per Shepard 2000). A teacher's view of learning in general relates to a core set of values and principles about science learning and assessment that guide assessment decision making. These values and principles interact with four categories of science teacher knowledge of assessment—assessment purposes, what to assess, assessment strategies, and assessment interpretation and resulting actions—which also interact with each other in practice. In the following sections, we describe each component of the model in more detail.

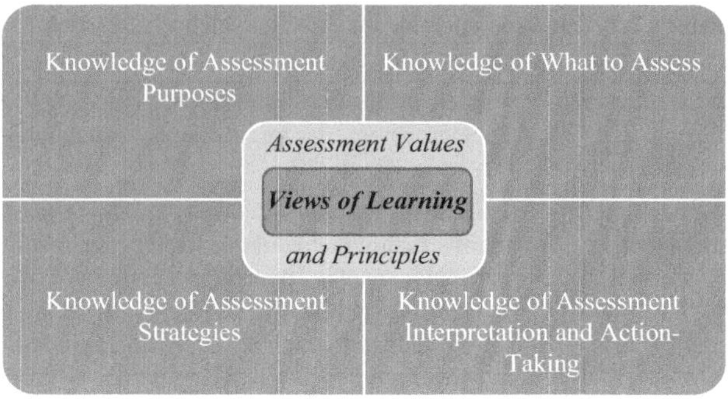

Fig. 12.1 A model for science teacher assessment literacy

The Core: Views of Learning and Assessment Values and Principles

Assessment values and principles are the overarching ideas and beliefs that guide assessment decisions in the science classroom. They are grounded in a teacher's view of student learning (Pellegrino et al. 2001; Shepard 2000) as well as in her views of what works best in assessment practice. For example, if a teacher views science learning as sense-making, then she might value particular ideas about assessment over others, such as assessment needs to go beyond multiple choice items to include opportunities for students to apply knowledge to solving problems. These values and beliefs would then influence why, what, and when she assesses; the types of assessment tasks she selects; and what she does with the assessment information (the outer ring of the model).

As science teacher educators and researchers, we value particular principles of assessment. These principles of effective assessment are based on our own research and on our review of the literature (Abell and Volkmann 2006; Banta et al. 1996; Siegel et al. 2008). These principles include:

1. Assessment is a process through which teachers can learn. Multiple types of assessment that target different kinds of knowledge, and various levels of thinking at different points in time provide a more complete view of student learning.
2. Assessment is a process from which students should learn. Assessment should be supportive, yet challenging (Siegel 2007). Assessment tasks need to challenge students to use higher level thinking as portrayed in modern views of learning. Assessment also needs to be *for* learning, and not just measure learning (Black and Wiliam 1998; Siegel 2007). Just as quality instruction includes scaffolding to help students progress, quality assessment also includes scaffolding (Siegel et al. 2008).
3. Assessment should help students be metacognitive about their developing knowledge and skills in order to self-regulate their learning. Research has shown that quality assessment tasks alone are insufficient to affect learning; how assessment tasks are introduced to and used by students, how assessment evidence

is interpreted by teachers, and how instruction is adapted are keys to fostering learning (Black and Wiliam 1998; Gearhart et al. 2006; Pellegrino et al. 2001; Siegel et al. 2006).

4. Assessment tasks need to be equitable for all learners. While it is not reasonable to expect one teacher to reduce all potential bias, assessment designers need to think about the diverse learners in their classroom, and whether the assessment task will privilege one group over another. For example, teachers need to consider if a written assessment item is comprehensible to an English Language Learner, or if it will cause undue reading skills? (See Siegel et al. 2008 for ways of reducing bias.)

We propose that these assessment values and principles, built on social-constructivist views of learning, form the core of teacher assessment literacy. Surrounding these core beliefs are four types of teacher knowledge and skills for carrying out assessment in the science classroom that interact with each other in practice.

Knowledge of Purposes of Assessment

This category of assessment literacy relates to why a teacher chooses to assess students. Some assessment purposes include:

- Helping instructors understand students' incoming ideas
- Helping instructors gauge student progress toward achieving course goals
- Providing data for instructors on which to base instructional decisions
- Assisting students in developing and applying knowledge and practicing skills
- Informing students about what they know and do not know so they can take action on their learning
- Providing evidence of student learning as compared with instructional goals for the purposes of grading

We can group these purposes into four main types of assessment that teachers need to know about:

- *Diagnostic assessment:* Diagnostic assessment occurs at the beginning of the course or unit of study. These assessments provide data about students' existing knowledge and beliefs about teaching and learning science in the classroom, and provide guidance for instruction.
- *Formative assessment:* Formative assessment occurs during instruction. Such assessment provides feedback to teachers and to students as they engage in the learning process. This feedback helps teachers think about the kinds of instructional interventions that need to occur.
- *Summative assessment:* Summative assessment provides documentation of student learning at particular points in time, most typically at the end of a unit of study or the end of a course. Summative assessment information is often the basis of course grades.

- *Metacognitive assessment:* Metacognitive assessment helps students become aware of and monitor their own learning. This kind of assessment can be combined with the previous three. It is key to helping students regulate their own learning.

These purposes/types of assessment are linked to the views of learning and assessment values that one holds. For example, if a teacher believes that teaching is telling, that students need to learn the facts, and that students are empty vessels waiting to be filled, then they will employ summative assessment of factual knowledge rather than formative assessment of the progress of student understanding. (Abell and Volkmann (2006) described in greater detail how views of assessment are influenced by principles of how people learn.)

Knowledge of What to Assess

Science teachers need to know what to assess in their science classrooms. What to assess is related to curricular goals and to values of what is important to learn and how learning occurs. For example, if a teacher believes that it is important for students to learn scientific ways of thinking in addition to science concepts, then assessment tasks will include opportunities for students to demonstrate their science process skills. If the science curriculum contains nature of science goals, then teachers will need to consider assessing student learning of nature of science concepts. In these ways, knowledge of what to assess is grounded in core values and principles of assessment and linked to assessment purposes.

Knowledge of Assessment Strategies

According to Magnusson et al. (1999), knowledge of assessment strategies refers to the ways a teacher assesses student learning in a particular unit of study. Science teachers need to know about formal assessment strategies used in summative assessment (e.g., how to design constructed response test items) and informal assessment strategies used in formative assessment (e.g., how to use minute papers/quizzes to gauge student learning at the end of a lesson). These general strategies might work for teaching/assessing any number of science topics. Teachers also need knowledge of topic-specific assessment tasks, such as a formative assessment probe (Keeley 2008) for diagnosing student ideas about density. Although not discussed by Magnusson et al., knowledge of assessment strategies also includes teacher knowledge of response strategies, such as learner-centered methods of grading (e.g., criterion-referenced scoring rubrics), effective and efficient forms of feedback (e.g., focusing on the answer to one question; self and peer evaluation), and ways to facilitate student use of feedback (through metacognition).

Knowledge of assessment strategies is linked to other parts of the assessment literacy model (see Fig. 12.1). Specifically, the types of assessment tasks that a science teacher designs and implements reflect the assessment values the teacher holds and his/her knowledge of assessment, as well as her knowledge of assessment purposes and what is important to assess. For example, if a teacher values developing scientific understanding over learning the facts of science, she will use different assessment strategies.

Knowledge of Assessment Interpretation and Action-Taking

Pellegrino et al. (2001) identified the interpretation process for making sense of assessment evidence as one vertex in the assessment triangle. We believe that science teachers need to know not only what, when, how, and why to asses, but also what do to with the assessment data. Abell and her colleagues (Lannin et al. 2008) found that science teacher knowledge included their ideas about how students would respond to assessment tasks as well as how teachers might act on assessment information. Thus, we believe that a critical component of assessment literacy is what teachers know about interpreting and acting upon assessment data. Teachers know that they can use assessment data to assign grades. More sophisticated assessment literacy includes knowing how to use assessment data to help students learn or using evidence from assessment to modify one's plans for instruction. For example, when a teacher uses an informal questioning assessment strategy in the science classroom when teaching about density, he/she learns what misconceptions students have, if another example is needed, or if students are ready to move to a new concept. Science teachers have used well-researched diagnostic assessments, such as the Diagnoser units (Hunt and Minstrell 1994), which show how a student understands a concept and recommends next steps. Each diagnostic unit contains resources for a teacher: learning goals, description of conceptions (facets), elicitation questions to uncover students' prior ideas, developmental lessons, diagnoser question sets to assess and provide counterexamples and targeted feedback, and prescriptive activities (pointers to instructional activities based on the diagnosis).

An Example of Assessment Literacy in Action

In order to illustrate our model, we present a case of assessment literacy in action in a secondary science teacher preparation program. Amelia is a composite of students we have worked with in our science teacher preparation program. Let us imagine that Amelia recently took a course about teaching science in the secondary school. In this course, the instructor asked the students to create a learning cycle unit of instruction for grade 10 biology students. Amelia decided to create a unit on the topic of mitosis. In the unit, she was required to include examples of diagnostic, formative, and summative assessment. What follows is what Amelia included related to assessment in the first draft of her unit plan, and her instructor's first round of feedback.

Amelia: For my diagnostic assessment I am going to use an assessment probe to find out students' incoming ideas about mitosis. I will ask: "How do you think your body produces new skin cells when you get a cut? How is it that the new cells are just like the old ones?"

Instructor: Amelia, I like how you asked students to think about a real life situation, that of getting a cut, instead of asking for a definition of mitosis or the cell cycle. The open-ended question will provide insight into your students' preconceptions. Another strategy that might save you time (by providing less diversity of student responses) is to give the students some choices based on the right answer and on common misconceptions. For example, Do living things grow because cells get larger? Do living things grow because cells divide? Do living things grow because cells get larger and because cells divide? That will also require less writing for your English language learners.

Amelia: For the formative assessment, students will keep a notebook from the labs and the lectures.

Instructor: Amelia, a science notebook can be a powerful formative assessment tool. Think about a specific writing task that you will ask students to carry out in each phase of your learning cycle unit. For example, during the Explore stage, when you ask the students to put the cartoons of different stages of mitosis in order, you could ask students to write a short description below each of the drawings of what they see happening. Think about what you will do with the information you gain from the science notebooks. How will you modify your instruction based on what students are learning? How can the notebooks help students track their own learning?

Amelia: In my summative assessment, I plan to give a test with the following item included: Put the following steps of mitosis in order: metaphase, telophase, prophase, anaphase.

Instructor: Amelia, in looking at your learning goals, I do not see that knowing the names and order of the stages of mitosis is included. Do you think this is a good assessment item given your learning goals? What information about students' understanding of the purpose and process of mitosis will you gain with this assessment item? Also, I noticed that most of your test items are testing lower level comprehension, and not tapping into students' higher order thinking. You might try to include at least one essay item in which students have to apply their knowledge to solving a problem (For example: You are a scientist and you are developing a drug that will inhibit the growth of cancer cells. Use your imagination and your knowledge of the cell cycle to describe how the drug would target and prevent cancer.) For your English language learners, you can try to simplify the language or add visuals while retaining the cognitive complexity of the item.

In this example, we see a dialogue between a novice with little science assessment literacy and an expert with a great deal. The expert is aware of the model for assessment literacy in Fig. 12.1, values certain assessment principles, and tries to help the student build her assessment literacy in a number of ways. First, by structuring the assignment to include three different assessment purposes, the instructor illustrates the principle of the importance of teacher learning through multiple assessments during a unit of instruction. The instructor also prompts Amelia to think about what she will do with what she learned from the assessment task. Second, the instructor emphasizes that what to assess should be connected to the teacher's learning goals and to potential student misconceptions. Including formative assessment ideas illustrates how students can learn through the assessment event itself. Third, the instructor suggests topic-specific strategies that would fit into Amelia's unit at different points in instruction, encouraging her also to think about including metacognitive strategies within the science notebooks. Fourth, the instructor uses the principle of equitable assessment when the instructor suggests that Amelia modify her assessment task to reduce the language factor while keeping the cognitive challenge high.

Assessment literacy, like other forms of teacher knowledge, does not develop overnight. Instead, teachers build their assessment knowledge over time and with different inputs from formal coursework, reflection on instruction, coaching from others, etc. We believe that science educators can use the assessment literacy model (Fig. 12.1) to help prospective and practicing teachers continue to build their science assessment literacy. In the following section, we suggest some specific implications of the assessment literacy model for science teacher education.

Implications for Science Teacher Education

Our model of assessment literacy suggests that teacher education programs should:

- Portray a view of science learning in which the learner actively constructs his/her understanding, and demonstrate how this view of learning is directly represented in views and strategies of assessment. This can be modeled by how science methods instructors assess their prospective teachers and communicated through assignments that set expectations for how to assess their students.
- Explicitly address the four principles for assessment: (1) Assessment is a process through which teachers should learn; (2) Assessment is a process from which students should learn; (3) Assessment should help students be metacognitive; (4) Assessment tasks need to be equitable for all learners. The vignette of Amelia and her instructor demonstrates one way to accomplish this.
- Provide opportunities to enhance teachers' knowledge of: assessment purposes, what to assess, assessment strategies, and assessment interpretation and resulting actions. Linking assessment ideas to other course topics, rather than adding assessment as a separate topic, may help build these knowledge types.
- Help teachers synthesize their philosophy with specific guiding principles and ways of understanding and acting in a classroom. A science teacher education program that emphasizes only the theoretical or practical side of assessment, but not both, would be deficient. For example, teachers need to understand the connections between learning goals and assessment practices.

Use of the assessment literacy model for teacher education needs to go beyond faculty awareness or a mere description in the course syllabi. One of the reasons assessment skills are difficult for teachers to develop is that they have so little experience with innovative assessment practices in their own schooling. Science teacher education programs tend to focus on content knowledge or inquiry teaching strategies, but often fail to connect these views and strategies with related assessment practices. Thus, to improve assessment literacy will take a major effort to weave a coherent philosophy of learning together with principles for assessment, while emphasizing specific assessment resources, knowledge, and practices. Providing opportunities for practice with interpreting and taking action on assessment evidence is also essential. As with any teacher education reform, connecting the theory, the methods, and the classroom is key.

Next Steps

In addition to influencing science teacher education, we hope that the assessment literacy model presented in this chapter can support researchers in investigating this under-studied topic. Different forms of comprehensive professional development aimed at improving science teacher assessment literacy could be tested. Examining assessment literacy in action is also needed. What interactions occur among the types of assessment knowledge in the model? Do strengths in one area of knowledge affect practice differently than other areas of knowledge? For example, how do improvements in science teacher assessment literacy affect outcomes for student learning?

While all facets of assessment literacy require more study, we recommend four areas of emphasis for future studies. First, learning how science teachers interpret assessment information and make instructional decisions is particularly needed. Science education researchers can find guidance about interpreting assessment information from mathematics education research (e.g., Fennema et al. 1996). Another area in great need of attention is program development and research on equitable assessment. Does a teacher with equitable assessment practices help students learn more than other teachers? Which students benefit and how? Third, the development of assessment literacy over years of teaching is another major research area we would like to explore. How can the assessment literacy model help explain teacher change over time? Finally, the relationship between assessment literacy, assessment practices in the classroom, and student learning is essential to examine.

Conclusion

In this chapter, we examined the theoretical and empirical foundations for assessment literacy, proposed a model for assessment literacy, described a case of the model in action, and suggested implications of the framework for teacher education and research efforts. Our model of assessment literacy is grounded in our own research, and extends two existing theoretical frameworks: the Magnusson et al. (1999) model of science teacher knowledge, and the Pellegrino et al. (2001) model of assessment. Our model extends Magnusson et al. by adding the categories of assessment purposes and assessment interpretation/action to their knowledge of what to assess and assessment strategies. Furthermore, our assessment knowledge categories surround a core of assessment values and principles, which we believe are parallel to what Magnusson et al. called orientations to teaching science. Pellegrino et al. represented assessment as a triangle, with models of cognition as one leg of the triangle. Our model moves views of learning to the center to demonstrate that all the other kinds of assessment knowledge are guided by how one views learning and the purposes of schooling, and which principles one values to guide assessment. Although the model in Fig. 12.1 suggests separate compartments for types of teacher assessment knowledge, our description of the model and our example of assessment literacy in action demonstrate that teachers connect the assessment knowledge types in their assessment practices. Thus, our model of science assessment literacy, based on our research with science teachers, encompasses both what a science teacher knows and does in the classroom. As a field, science education has sufficient evidence to show that how a student learns is directly affected by teachers' assessment practices. Yet science educators have a long road ahead to ensure that all science teachers are assessment literate. Classrooms will be radically changed when all students have opportunities to experience quality, equitable assessment systems that drive, support, and monitor their learning.

References

Abell, S. K. (2007). Research on science teacher knowledge. In S. K. Abell & N. G. Lederman (Eds), *Handbook of research on science education* (pp. 1105–1149). Mahwah: Lawrence Erlbaum.

Abell, S. K., & Volkmann, M. J. (2006). *Seamless assessment in science: A guidebook for elementary and middle school teachers*. Portsmouth: Heinemann.

American Association for the Advancement of Science. (1993). *Benchmarks for science literacy: Project 2061*. New York: Oxford University Press.

Banta, T. W., Lund, J. P., Black, K. E., & Oblander, F. W. (1996). *Assessment in practice: Putting principles to work on college campuses*. San Francisco: Jossey Bass.

Bell, B. (2007). Classroom assessment of science learning. In S. K. Abell & N. G. Lederman (Eds.), *Handbook of research on science education* (pp. 1105–1149). Mahwah: Lawrence Erlbaum.

Black, P. (1998). Assessment by teachers and the improvement of student learning. In B. Fraser & K. Tobin (Eds.), *International handbook of science education* (pp. 811–822). London: Kluwer.

Black, P., & Wiliam, D. (1998). Assessment and classroom learning. *Assessment in Education, 5*(1), 7–74.

Bol, L., & Strage, A. (1996). The contradiction between teachers' instructional goals and their assessment practices in high school biology courses. *Science Education, 80,* 145–163.

Briscoe, C. (1993). Using cognitive referents in making sense of teaching: A chemistry teacher's struggle to change assessment practices. *Journal of Research in Science Teaching, 30,* 971–987.

CCSSO (Council of Chief State School Officers), & WCER (Wisconsin Center for Educational research). (2000). *Using data on enacted curriculum in mathematics and science: Sample results from a study of classroom practices and subject content.* Washington: CCSSO.

Donovan, M. S., & Bransford, J. D. (Eds.). (2005). *How students learn: History, mathematics, and science in the classroom.* Washington: National Academies Press.

Doran, R. L., Lawrenz, F., & Helgeson, S. (1994). Research on assessment in science. In D. L. Gabel (Ed.), *Handbook of research on science teaching and learning* (pp. 388–442). New York: Macmillan.

Duffee, L., & Aikenhead, G. (1992). Curriculum change, student evaluation, and teacher practical knowledge. *Science Education, 76,* 493–506.

Fennema, E., Carpenter, T. P., Franke, M. L., Levi, L., Jacobs, V., & Empson, S. (1996). Longitudinal study of learning to use children's thinking in mathematics instruction. *Journal for Research in Mathematics Education, 27*(4), 403–434.

Friedrichsen, P. J., Abell, S. K., Pareja, E. M., Brown, P. L., Lankford, D. M., & Volkmann, M. J. (2009). Does teaching experience matter? Examining biology teachers' prior knowledge for teaching in an alternative certification program. *Journal of Research in Science Teaching, 46,* 357–383.

Gearhart, M., Nagashima, S., Pfotenhauer, J., Clark, S., Schwab, C., Vendlinski, T., Osmundson, E., Herman, J., & Bernbaum, D. J. (2006). Developing expertise with classroom assessment in K-12 science: Learning to interpret student work. *Educational Assessment, 11*(3–4), 237–263.

Hunt, E. B., & Minstrell, J. (1994). A cognitive approach to the teaching of physics. In K. McGilly (Ed.), *Classroom lessons: integrating cognitive theory and classroom practice.* Cambridge: MIT Press/Bradford Books.

Johnson, C. C. (2006). Enabling all students to learn science. *The Science Teacher, 73*(2), 32–37.

Keeley, P. (2008). *Science formative assessments: 75 practical strategies for linking assessment, instruction, and learning.* Thousand Oaks: Corwin and NSTA.

Kokkotas, P., Vlachos, I., & Koulaidis, V. (1998). Teaching the topics of the particulate nature of matter in prospective teachers' training courses. *International Journal of Science Education, 20,* 291–303.

Lannin, J., Abell, S., Arbaugh, F., Chval, K., Friedrichsen, P., & Volkmann, M. (2008). *Researching teacher knowledge: Further delineating the PCK construct for science and mathematics education.* Paper presented at the annual meeting of the American Educational Research Association, March, New York.

Lee, O., Luykx, A., Buxton, C., & Shaver, A. (2007). The challenge of altering elementary school teachers' beliefs and practices regarding linguistic and cultural diversity in science instruction. *Journal of Research in Science Teaching, 44,* 1269–1291.

Magnusson, S., Krajcik, J., & Borko, H. (1999). Nature, sources and development of pedagogical content knowledge for science teaching. In J. Gess-Newsome & N. G. Lederman (Eds.), *Examining pedagogical content knowledge* (pp. 95–132). Dordrecht: Kluwer.

Mertler, C. A. (1999). Assessing student performance: A descriptive study of the classroom assessment practices of Ohio teachers. *Education, 120,* 285–296.

Mertler, C. A., & Campbell, C. (2005). *Measuring teachers' knowledge and application of classroom assessment concepts: Development of the Assessment Literacy Inventory.* Paper presented at the annual meeting of the American Educational Research Association, April, Montreal, Quebec, Canada.

Morrison, J. A., & Lederman, N. G. (2003). Science teachers' diagnosis and understanding of students' preconceptions. *Science Education, 87,* 849–867.

National Research Council. (1996). *National Science Education Standards.* Washington: National Academy Press.

National Research Council. (2001). *Educating teachers of science, mathematics, and technology: New practices for the new millennium.* Washington: National Academy Press.

National Research Council. (2003). *Evaluating and improving undergraduate teaching in science, technology, engineering, and mathematics.* Washington: National Academies Press.

Patton, M. Q. (2002). *Qualitative research and evaluation methods.* Thousand Oaks: Sage.

Pellegrino, J. W., Chudowsky, N., & Glaser, R. (Eds.). (2001). *Knowing what students know: The science and design of educational assessment.* Washington: National Academy Press.

Pine, K., Messer, D., & St. John, K. (2001). Children's misconceptions in primary science: A survey of teachers' views. *Research in Science & Technological Education, 19,* 79–96.

Ruiz-Primo, M. A., & Furtak, E. M. (2007). Exploring teachers' informal formative assessment practices and students' understanding in the context of scientific inquiry. *Journal of Research in Science Teaching, 44,* 57–84

Sanders, M. (1993). Erroneous ideas about respiration: The teacher factor. *Journal of Research in Science Teaching, 30,* 919–934.

Shepard, L. A. (2000). The role of assessment in a learning culture. *Educational Researcher, 29*(7), 4–14.

Shulman, L. S. (1986). Those who understand: Knowledge growth in teaching. *Educational Researcher, 15*(2), 4–14.

Siegel, M. A. (2007). Striving for equitable classroom assessments for linguistic minorities: Strategies for and effects of revising life science items. *Journal of Research in Science Teaching, 44*(6), 864–881.

Siegel, M. A., & Wissehr, C. (2009). *Assessment of all learners: Developing expertise in equitable assessment with preservice teachers.* Paper presented at the annual meeting of the American Educational Research Association, April, San Diego.

Siegel, M. A., Hynds, P., Siciliano, M., Nagle, B. (2006). Using rubrics to foster meaningful learning. In P. Simmons (Ed.), *Assessment in science: Practical experiences and education research.* Arlington: NSTA.

Siegel, M. A., Wissehr, C., & Halverson, K. (2008). Sounds like success: A framework for equitable assessment. *The Science Teacher, 75*(3), 43–46.

Stiggins, R. J. (2002). Assessment crisis: The absence of assessment FOR learning [Electronic version]. Kappan. http://www.pdkintl.org/kappan/k0206sti.htm. Accessed 22 December 2007.

van der Valk, A. E., & Broekman, H. G. B. (1999) The lesson preparation method: A way to investigate preservice teachers' pedagogical content knowledge. *European Journal of Teacher Education, 21*(2), 11–22.

Wissehr, C., & Siegel, M. A. (2008). Unlocking assessment secrets: What are preservice teachers' views of assessment? Paper presented at the annual meeting of the Associations for Science Teacher Education, January, St. Louis.

Yilmaz-Tuzun, O. (2008). Preservice teachers' beliefs about science teaching. *Journal of Science Teacher Education, 19*(2), 183–204.

Chapter 13
Supporting Technological Thinking: Block Play in Early Childhood Education

Jill Robbins, Beverley Jane and Jacinta Bartlett

Introduction

The need for research into preschool and primary school science, engineering and technology education was identified as a major priority in the Australian Government's 2006 *Science Engineering and Technology Skills Audit Summary Report* (DEST 2006). While a significant amount of research exists about children's thinking in science, few studies have focussed on young children's thinking in design technology, and hardly any have been conducted in the context of block play. This lack of research about children's learning through block play is somewhat surprising given that resources such as wooden blocks and construction sets (for example, *Lego*, *Duplo* and *Brio*), which frequently require children to work like engineers, are used daily in most Australian early childhood centres. In this chapter we draw on a study located in two Victorian (Australia) preschools that aimed to make a contribution within this research space.

Research on Technological Thinking in the Early Years

In 2009 Australia adopted an Early Years Learning Framework, *Belonging, Being and Becoming*, a document which is designed to support and enhance young children's learning from birth to five years of age, and across their transition to school (Department of Education, Employment and Workplace Relations (DEEWR) 2009). Within this framework, technology is described as "the diverse range of products that make up the designed world. These products extend beyond artefacts designed and developed by people and include processes, systems, services and environments" (p. 46). Educators are encouraged to support children to create and construct, engage in critical thinking, enquiry processes, experiment, solve problems, transfer knowledge from one situation to the next, and to "demonstrate an increas-

J. Robbins (✉)
Monash University (Peninsula campus), Frankston, VIC, Australia
e-mail: jill.robbins@monash.edu

D. Corrigan et al. (eds.), *The Professional Knowledge Base of Science Teaching*,
DOI 10.1007/978-90-481-3927-9_13,

ing knowledge of, and respect for natural and constructed environments" (p. 29)—all of which are intrinsic to technological thinking. This current framework goes beyond earlier versions, where technology education for young children received little attention. In contrast, in England, technology education has been recognised as important since 1990, when a national curriculum for design and technology was introduced (Benson 2008; Siraj-Blatchford and MacLeod-Brudenell 1999). Building on this national curriculum, technology education for three-to-five-year-olds became mandatory upon the introduction of the *Early Years Learning Goals* in 2000 (QCA 1999).

While government-funded research has been conducted in England into primary and foundation stage technology (Benson 2008), in Australia and New Zealand a relatively small amount of research exists on technology within primary school contexts (see, for example, Cowie and Moreland 2007; Fleer 2000, 2008; Mawson 2005; Moreland et al. 2001; Webster et al. 2006) and even less has been conducted in technology within *early childhood* settings in these countries (Jane and Robbins 2004). Few examples of research into block play can be found, with the exception of limited studies considering issues such as block play and mathematics (see for example, Gura 1992; Wolfgang et al. 2001) and children's interactions in block play (Cohen and Uhry 2007; Struss and Stremmel 2004). Our study aimed to make a contribution in this space, by focussing specifically on how young children can be supported to develop technological knowledge and design processes in their play with blocks.

For the purpose of the study, technological skills and thinking include investigating materials and needs, designing or planning, making, constructing or building, problem identification and problem solving, appraising, evaluating, and identifying improvements, and demonstrating creativity and innovation.

Contextualising Block Play

Since the first kindergartens developed by Froebel in the early nineteenth century, young children have been provided with wooden blocks for play (Gutek 2005). Creating what he called *Gifts,* Froebel provided young children with a set of beautifully crafted tabletop wooden blocks suitable for building, designed so that, through teacher-directed play, children would come to see the relationships between parts and wholes. Each set of blocks was cut from eight-inch cubes, and became more complex in terms of the number, size and shape of the component blocks in each set or *Gift* (Read 1992). Unlike the teacher-directed manner in which Froebel's blocks were originally to be used, today's play with blocks in most Australian early childhood centres tends to be non-interventionist or what Bruce (1992) has termed *laissez faire,* with children frequently engaging in "free play" with blocks. That is, the tendency is for a block area to be available, but the children are often left to use these blocks without adult intervention, even when other areas of the programme are very clearly planned for and teacher-directed (Siraj-Blatchford and MacLeod-

Brudenell 1999). While the value of block play in supporting young children's learning in areas such as mathematics is recognised, technology is often overlooked or under-utilised. This *laissez faire* approach can be contrasted with teacher-directed and with interactionist methods, such as the purposeful intervention of adults to help children develop strategies that assist their "blockplay to develop with quality" (Bruce 1992, p. 16), or helping them to solve problems when they arise.

Research Approach

In the study we used a multiple case study approach, drawing on sociocultural/cultural-historical theory, particularly the work of Vygotsky (1987, 1997a, b, 1998, 1999). In particular, attention was paid to the manner in which drawings, materials such as wooden blocks and/or *Lego*, together with talk between teachers and children mediate their technological thinking such as identification of needs, creativity, innovation, evaluation, problem solving.

In Australia, block play is an activity that is provided within a preschool programme on a daily basis. However, as Bruce (1992) identified through the studies conducted in England by the *Froebel Blockplay Research Group* (Gura 1992), in most instances, children engage in what is commonly referred to as "free play" in the block area, with little support from their teachers for the development of any mathematical, scientific or technological concepts embedded within this block play. For this study, after an initial period of observation, we decided to use an interactionist approach (purposeful intervention of adults to promote children's technological thinking) to determine the extent to which the children's block play, and in turn their technological thinking, can be enhanced. This approach was chosen as it allowed us to focus on semiotic mediation (or how technological thinking and meaning could be supported). It was believed that a *laissez-faire* approach which would not afford a focus on interactions between children, adults and materials (such as the blocks), while a teacher-directed approach might not permit the development of innovation, creativity, problem solving and other technological thinking on the part of the children.

The study was situated in two preschool centres known to the researchers. Both centres are located in south-east Melbourne, Victoria, Australia. Preschool 1 has two groups of four- to five-year-olds (up to 26 children in both groups) and two groups of three years olds (15 children in each group). The two teachers in the centre, and their assistants, believe that they adopt a play-based approach to learning and plan according to interests. Blocks are readily available during each session, and at times other artefacts (such as toy cars, farm animals and small dolls' furniture) are added to the play space. Generally, there is a dominance of boys playing with the blocks, and while girls do participate it is not for a long period of time. Moreover, some girls only play with blocks if there are other artefacts associated with "home" play in the block corner. When interviewed, the preschool director's perceptions of technology focussed on computers and electronic gadgets, although she later identified problem solving as an important technological skill. Even though the teachers do not explicitly plan for technology education a significant amount of discrete technology (Fleer and Jane 2004) is often employed within the programme, with the children making teacher-directed artefacts. While sometimes lacking authenticity or meaning for the children, these are important in helping them develop physical and process skills for completing tasks. Siraj-Blatchford and Brudenell (1999) note that it is important for children to develop mastery of fine and gross motor skills if they are to work effectively in technology, particularly during the *making* stage. However, they emphasise that it is important for children to work with their own ideas.

Preschool 2 has two groups of up to 24 four- to five-year-old children attending the programme, with two preschool teachers, plus assistants. The director has a philosophy inspired by her visit to Reggio Emilia, and great attention is given to the presentation of learning activities, displays and provocations, with an emphasis on the use of natural or recycled materials. Open-ended technology materials are freely available, including various types of wooden and plastic blocks stored in tubs within the children's reach, though, again, the teacher does not specifically plan for technology within her programme. Children's block building is valued and opportunities for revisiting are supported when possible. When sessions permit, block buildings are left intact and children are encouraged to use signs created by their teachers that read "*Please leave my work*". When constructing with blocks, children are permitted to source supplementary materials from other areas of the preschool, including the storeroom. Both boys and girls play equally with the blocks that are available in two areas, indoors and outdoors under a shelter. Frequently, the director attempts to make connections between events within the community (for example, the building of apartments near the preschool) and the focus of children's block building. When asked to identify aspects of technology within the programme she described children working with computers and stated that she feels she needs to learn more about what constitutes technology.

In both the preschools we only documented the work of those children for whom permission to participate in the study had been gained (though there was no attempt to exclude other children from the block area). Potentially, there were 45 children across the two preschools who were part of the study. Pseudonyms are used when reporting the children's participation. Two fourth year pre-service teachers, with a special interest in technology education, worked with the first two authors to form a research team, while the preschool teachers worked with the children not involved in the study. The study extended across one and a half terms, with the research team visiting each centre once or twice a week to work with the children in the block area. At the start of the study both pre-service teachers interviewed the teachers to identify their perceptions of what constitutes technology and how they plan for and support technology education.

In each centre the pre-service teachers gained initial data through observations, photographs and conversations about the stages of block building demonstrated by the children, and observed the level of support for technological thinking offered by the teachers. In the second part, across several weeks the pre-service teachers and first author interacted with the children prior to and during their block building.

Emphasis was placed on assisting the children to use a design, make and appraise approach (Fleer and Jane 2004) where possible, and to implement planning, evaluating, problem solving and innovation in their block building. During this process, further observations and photographs were taken, and later analysed for evidence of increased technological thinking and development in the stages of building. Frequent conversations occurred with the preschool teachers, focussing on the observational data and photographic evidence, as well as the children's comments and drawings.

Initial analysis of the observations and photographs revealed some interesting issues in relation to contextual factors, especially the teaching approaches utilised, and the mediation of children's technological thinking. In both centres, in relation to block building, a *laissez faire* teaching approach was by far the most predominant style observed being used by the teachers (Robbins et al. 2008), with an occasional interactionist approach (often aimed at promoting cooperation among children) observed in the second centre and a teacher-directed approach sometimes observed in the first centre.

Theoretical Informants

Sociocultural/cultural-historical theory provides a powerful framework for examining social, cultural and historical aspects of development and learning, specifically focussing on relationships between people, contexts, actions, tools and artefacts, meanings, communities and cultural histories. From a cultural-historical perspective, learning and development occur through a process of changing participation in dynamic cultural communities, in which there are active contributions from individuals, their social partners, practices (current and historical), traditions, cultural tools, technologies and artefacts, and values and belief systems (Rogoff 2003). Cultural-historical theory recognises that individuals and their social partners and the activities in which they engage are continually transforming and developing in mutually integrated ways. Likewise, communities or contexts (and the technologies within those contexts) are constantly changing and being changed, which in turn result in changed opportunities for learning and development. Therefore, a cultural-historical approach to research focusses on the dynamic interactions between individuals, social groups and contexts, and the values, practices, artefacts and "ways of doing things" embedded within those contexts. Importantly, rather than studying learning or development *at one point in time*, as much traditional research tends to do, cultural-historical theory allows us to examine how "*children grow into the intellectual life of those around*" (Vygotsky 1978, p. 89; italics in the original). However, this approach is rarely used to inform research in technology education (Fleer 2008).

Siu and Lam (2005) discuss the emphasis that is frequently placed on visible physical outcomes in technology education in early childhood at the expense of promoting creativity, innovation, problem solving and critical thinking (often core objectives of technology education documents). Fleer (2008) in advocating for a cultural-historical perspective to research in design and technology education states that studying "the dynamic processes, as opposed to the 'end product', offers a useful research framework for technology education" (p. 89). She adds that this is particularly useful for researchers interested in how very young children consider and develop their understandings of technological concepts.

Vygotsky's work, which has gained increasing interest in several disciplines over the past 20 years, including education, provides a compelling way of understanding qualitative changes that occur, across time, in children's thinking. This change occurs through several complex, inter-related processes. One important process is *semiotic mediation*, or how thinking and problem solving, through signs (such as language) and cultural tools (such as paper and drawing implements), moves from an intermental level (where thinking occurs between people engaged in joint sociocultural activity) towards an intramental level (where thinking occurs within the individual) as shown in Fig. 13.1. Semiotic mediation is concerned with examining how shared ideas and activities are gradually internalised and transformed by a child, until the child makes them her/his own (Vygotsky 1987, 1997a, b). It is through semiotic mediation that over time children's thinking moves towards complex higher mental functioning (Vygotsky 1999).

Accordingly, what is examined within this study is the manner in which adults structure block play situations, and through conversation and drawing, assist children to *begin* to internalise and transform their ideas about design and technology, and to move towards more conscious awareness of thoughts and actions (Vygotsky 1987). Attention is also paid to examples of creativity, advanced problem solving, innovation and other forms of higher mental functioning (Vygotsky 1998). In addi-

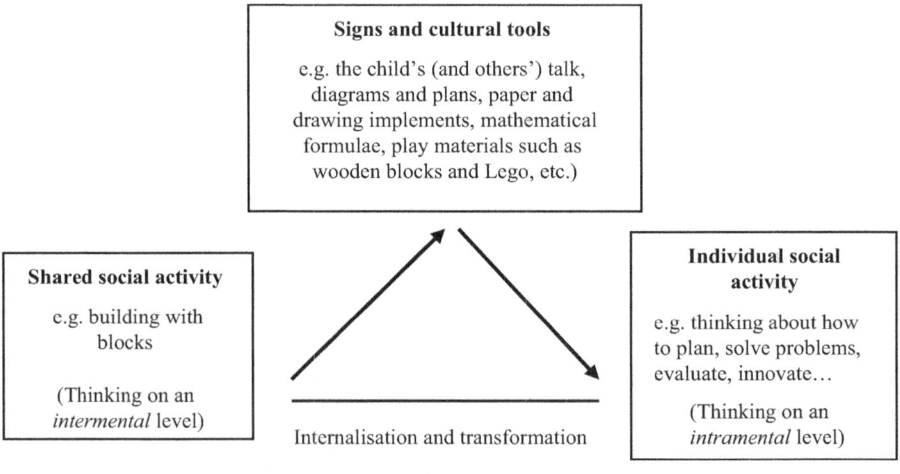

Fig. 13.1 Semiotic mediation. Shared social activity and meaning is internalised through the use of signs and tools, gradually becoming transformed into individual thinking

tion, the way this is interrelated with contextual factors (including teachers' philosophies and pedagogical practices) is considered.

Thus, in relation to technology education, this theoretical framework provides a compelling tool for thinking about what the children are "doing", as well as the inter-relationships and interactional styles, and the contextual factors (availability of materials, teachers' philosophies and pedagogical approaches, and historical "ways of doing things"), that support (or constrain) children's explorations, ability to identify needs, creative thinking and problem solving.

Findings

Within the scope of this chapter we can present and discuss only a limited amount of data that were generated in the study. We have selected two anecdotes from a four- to five-year-old group in each centre for examination and discussion—one relatively short, and the other extending over a number of weeks. The uniting feature of these two anecdotes is that both include an artefact as a stimulus to children's block building and both serve to demonstrate the powerful nature of contexts in children's play and learning.

In Preschool 1, the teacher has placed several photographs of simple block structures on the floor in the block area (see Fig. 13.2), which, in a teacher-directed manner, the children are expected to copy. The rationale for this approach is that, after almost a year in the centre, the children engage in little productive building in the block corner, and the photographs are intended as a stimulus for building.

Other than a suggestion by the assistant teacher that the children "make a building like one of the pictures" there is initially little adult–child interaction. The preservice teacher, Pippa, and the first author, Jill, observe the children's play for ten minutes, noting that, despite there being some four to eight children in the block

Fig. 13.2 A teacher-directed approach, with the teacher providing photographs of simple block buildings for the children to copy

Fig. 13.3 Pippa referring Gabi to the model pictured

area at any one time, little building is done. Some boys take a few blocks from the shelf, but do not use them for building. Instead they play with toy cars, pushing them around the mat, between some simple towers or bridges that two of the girls quickly made. No attention has been given to the photographs. At this stage Pippa, intervenes, showing Gabi one of these photographs. Together they examine the structure pictured, with Pippa drawing Gabi's attention to the upright and then the vertical blocks. Gabi selects some blocks from the shelf and begins building, with Pippa continually referring her to the photograph (see Fig. 13.3). Aidan enters the block area, knocking down Gabi's building. Pippa then talks about the importance of the blocks being firmly balanced, suggesting that she makes a more solid base than that shown in the photograph. Here Pippa is modelling appraisal. As Gabi rebuilds, Pippa demonstrates how to check that the blocks are positioned to ensure that they are stable.

Gabi then announces that she wishes to make her building "higher" (Fig. 13.4a). As the added blocks begin to wobble, Pippa again talks about balance, and how buildings need to be strong. Gabi responds by filling in the spaces between the uprights (Fig. 13.4b) until she eventually has a solid construction. Throughout the process Pippa provides continuous physical and verbal support, mediating Gabi's developing understanding about balance and stability. No other children joined in with the building, although at one stage Pippa does invite them to take part (and Jill attempts to engage other children in focussed block building). Instead some boys lay a few blocks horizontally along the floor for roads, or play with the cars, while one other girl watches Gabi build. No further play develops around Gabi's building, and at the conclusion of the indoor period, the blocks are packed away.

While this activity occurred on one of the last visits to this centre, the teacher continued to express an interest in the study, and sent us some photographs of subsequent block buildings by the children. What was noticeable in these photographs

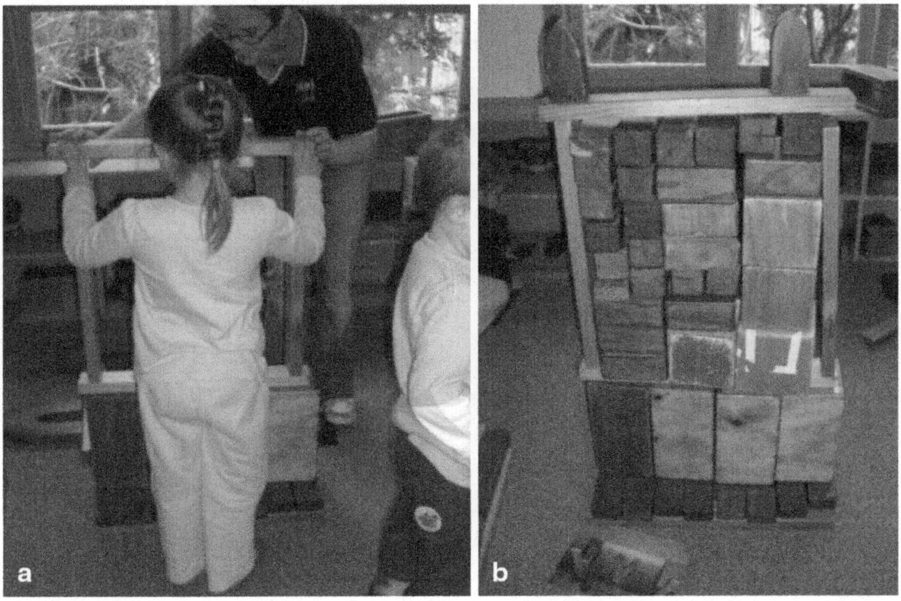

Fig. 13.4 Gabi builds her construction higher (**a**) and with Pippa's prompts concerning balance and stability, makes it stronger (**b**)

and accompanying text was that the teacher placed a dolls' house and furniture, plus vehicles such as trucks, in the block area—as prompts for both girls' and boys' play, rather than supporting what children themselves had created. That is, the block play often appeared to be teacher directed rather than necessarily arising from the children's ideas.

In the following sequence of observations and photographs from Preschool 2, where an interest in lights in block building develops, play is occurring both in a sheltered area outside, as well as inside. Initially, baskets of drawing tools and measuring tapes are provided outdoors, together with a wall display including the preschool floor plan and photographs of previous block buildings. Following the first incident described below, a similarly resourced block area is created inside by the preschool teachers not only to support the ongoing research but also the children's interest in block building. The two block areas become increasingly dynamic, with photographs and design drawings being added to the walls by the children, baskets of different sized blocks added occasionally, and buildings from previous sessions often being left intact with signs saying "Please leave my work" for other groups of children to view.

Having become familiar with the process of designing a building before construction, a strategy previously introduced by the third author, pre-service teacher, Jacinta, Brock has drawn the *Bridge walkthrough* design, that is subsequently labelled by Jacinta (Fig. 13.5). Brock then collaboratively builds this design with Gareth. When it is finished, Gareth begins describing additions he wants to make

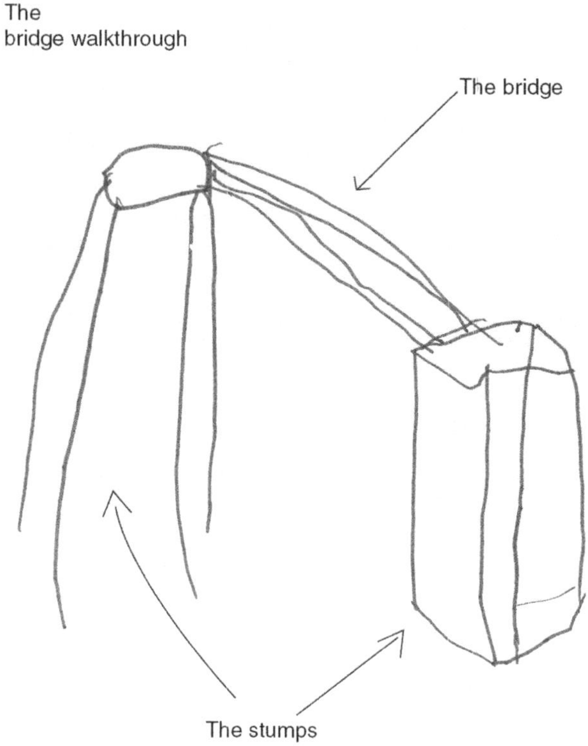

Fig. 13.5 Brock's design of the *Bridge Walkthrough*

(identifying an area for improvement) to Jacinta who is simultaneously assisting Brock to test (appraise) the bridge's strength.

Gareth talks about how he needs some lights:

Gareth: All you need is you just need some pipes…you need some pipes to make lights…We need some lights 'cos we…I know why we need the lights so when the cars drive under then the lights shine…lights turn on and then the cars, and then they can see…

Jacinta: So the cars can see? No, so the people can see?

Gareth: See, the lights shine…

Jacinta is concentrating on the appraisal process with Brock, and later states that her focus on a predetermined "outcome" precludes a simultaneous opportunity to extend the learning possibilities based on Gareth's interest in lights, even though he has identified a problem and a particular need—an important feature of design technology (Victorian Curriculum and Assessment Authority [VCAA] 2007).

During the following visit, Jill begins to mediate further learning, providing an opportunity for Gareth to follow up this need to provide lights for seeing the cars in his building. A group of three children (Sabella, Charlie, Gareth) are making a *Castle for Ponies*. During 35 minutes of shared thinking and collaborative building, they encounter and overcome many problems. Sabella exclaims: "It fell down a hundred times!" Charlie appraises the building, saying that it is "too dark" inside. He finds a bulb-shaped block on the shelf that he names as representing a light, and places it inside the doorway of the building (see Fig. 13.6a). Again, a need has been recognised, an area for improvement identified, and a problem "solved". Meanwhile, Jill asks if the centre has a torch/flashlight or materials to make a circuit. A torch is then located (but not materials for an electrical circuit) and Charlie and Gareth explore the torch together while sharing their knowledge of torches, batteries and switches, with ideas moving between the two children and the researchers. For example, Charlie locates the switch on the torch and states that switches turn a light on and off. As the conversations continue Gareth identifies a problem: the batteries are flat. He asks Jacinta to go to the torch shop and buy some more. At his own initiative, he draws a map of how to get there. Jacinta helps him label and display the map (see Fig. 13.6b). She then uses one of the clipboards provided in the block area to list materials that together they identify for her to bring on the next visit.

At the start of the next visit, Jacinta sets up the block area with torches, and materials for making a circuit and cardboard tubes. Photographs from the previous session taken by Jill are discussed. Gareth, Charlie and Jacinta explore the materials together, again with ideas moving between them. Gareth finds a simple switching device and the following conversation occurs.

Gareth: What does this do? What does this part do?
Jacinta: That's a good question. That's called a switch.
Gareth: Turn on and off…?
Jacinta: Yes that's exactly what it does.
Charlie: Turn it on and off…

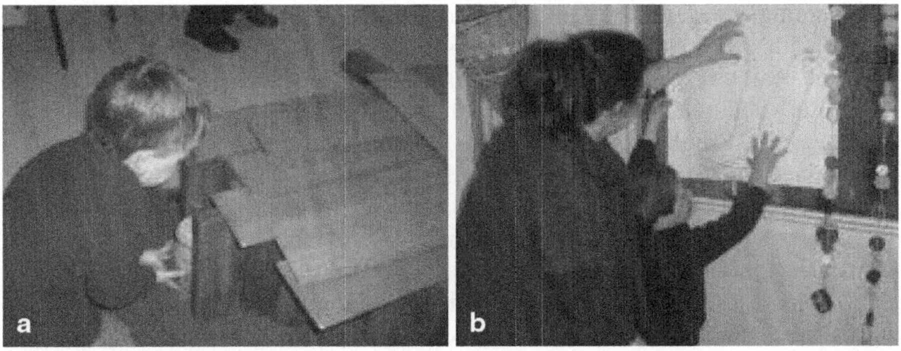

Fig. 13.6 Charlie adding a light to his building (**a**) ["It's dark inside"] and Gareth displaying the map to the torch shop (**b**)

Fig. 13.7 Gareth adds a second battery to the circuit as the globe shines dimly

Charlie announces that the lights at his home also have switches to turn them on and off. Jacinta then assists the two boys to make a circuit by helping, modelling technical language and drawing attention to the finished circuit. Jacinta and Gareth notice the light is not very bright so Jill suggests using two batteries, which Gareth promptly puts in place (see Fig. 13.7).

After he successfully made a circuit with Charlie and Jacinta using a "switcher" and two batteries, Gareth announces that he will now make a block building. He builds collaboratively with Charlie. They investigate the torches and experiment using a torch to light their building, and then Jacinta suggests they might use the circuit they made earlier. Many design challenges arise, and at one stage Jill suggests using tape to stop the globe from falling out. Gareth also uses tape to position the globe in the ceiling of the building. After the circuit is tested, he decides that the switch needs to be placed inside the building, just like real switches. He tries to do this independently before asking Jacinta to help him connect long pieces of wire. He then tests the switch inside the building. After Jill has drawn their attention to the light coming in through the centre's ceiling, more light is allowed into the building with Gareth's addition of a bubble wrap skylight. He had selected the bubble wrap from the nearby recycled construction materials table. Also more light entered due to Charlie's addition of Perspex doors with taped hinges on the front (Fig. 13.8). He said they were like the glass doors he has at home. This collective building resulted in sustained engagement in the task they had identified.

At the end of the session, Gareth disassembles the circuit, placing all materials he requires into an envelope for use the next day. Jacinta records his description of the contents on a label which, at Gareth's request, they laminate together: "The batteries work. Lots of stuff to make the batteries work and 'ect [connect] the wires to the batteries. The batteries go in the holder".

Gareth and Charlie had used a switch in their electrical circuit and had chatted on several occasions about switches at home (everyday concepts). A silence exists here, that no prompting of Gareth occurred, or in fact any of the children, to see if

Fig. 13.8 The finished building with Perspex doors and bubble wrap skylight

they could identify the scientific concept that an electrical current needs to flow in a loop or circuit. It is important not to make assumptions about what children might "discover" from participating in an activity, but that we consciously and purposely ensure we support learning.

Scientific or academic concepts, according to Vygotsky (1987), are deliberately taught and learned. They exist within a system of inter-related logical and hierarchical concepts, are removed from concrete experience, but are used consciously and intentionally by people. They contrast with everyday concepts which are those developed through practical experience of the world, or "borrowed" from others within social communities, and thus generally only contextually relevant. What potentially was lost was the opportunity for the researchers to engage in what Hedegaard and Chaiklin (2005) have termed a "double move" approach to teaching. Here, according to Fleer (2008), the teacher takes into account children's everyday concepts (such as the observable components involved in making a bulb glow—batteries, wires, bulb, switch) (one move), and, at the same time, related academic concepts or subject matter (such as electricity is a form of energy which can be stored in batteries as chemical energy, transferred through wires in a complete circuit and transformed into another form of energy such as light) (the second move). While it was not the intention for the researchers to "teach" children scientific concepts, an incidental opportunity for possible useful and relevant learning was not realised.

Discussion

The observations from these two early childhood centres reveal some interesting issues. At Preschool 1, Pippa was successful in helping one child to think about balance and stability, modelling how to evaluate the building process. With her support, the photographs provided a starting point for building. While there is need for caution in drawing conclusions from one incident, it is useful to contemplate whether this building would have begun without the mediation that occurred (especially as the other children present in the block area did not use the photographs), and whether setting models for children to copy, being teacher-directed in nature, may limit their own creativity. However, it could be argued that as the child decided to make her building taller, there was some originality, but the finished construction was little beyond vertical stacking of the blocks. Once completed, the child was no longer interested in the building, and it was eventually knocked down.

It is also worthwhile to consider this anecdote within the broader context of this preschool centre. As mentioned above, the children in this centre are frequently observed engaging in teacher-directed discrete technology activities. Few observations were made of these children being supported to explore or add materials, to modify designs or to continue working on a construction across several days. Jill and the pre-service teachers had limited success in enhancing technological thinking. Although there are waste and recycled materials available for the children to use, they are somewhat restricted in range, with the time-frame tightly controlled by the teacher. Also, it is useful to reflect on the extent to which having children work with teacher-designed models (either in the block area or elsewhere) may be helpful in promoting technological thinking such as creativity and innovation.

At Preschool 2, the children's building activity also commences with an artefact, but in this case it is a drawing of a building the child wishes to construct, with the idea and design originating from him. The building activity that ensues engages several children over a sustained period of time. At the request of the children, issues that arise during the making of the building (such as the need for lights) are investigated at the time and in subsequent sessions. The researchers' mediations include the addition of torches plus materials to make circuits, together with interactions involving talk, modelling and helping the children solve building problems using materials freely accessible in the centre. Clear connections to everyday applications of the technological and scientific concepts are also mediated, for example, investigating the torches, placing the switches inside the building, making links between the switches in the circuits and the switches for the lights at home and pointing out that the bubble wrap "window panes" were similar to the skylights in the centre. While, as stated above, Gareth appeared to have internalised "how" to make a circuit, it could not be assumed that he understood the underlying scientific concepts.

Within the context of this preschool centre, the children are sometimes asked to participate in construction activities designed by the teacher. However, they are often observed across the indoor and outdoor programme engaging in making and modifying their own designs at collage and waste materials, weaving, sewing and carpentry. Some technological activities are designed to meet specific community needs, for example, making signs to identify seedlings in the vegetable garden. Others are for major events such as an annual art show that features children's creative, artistic and "craft" work. A wide range of construction materials (both in the form of sets such as *Lego* and *Brio*, and non-structured or recycled materials) are provided and opportunity exists for children to engage in sustained creative pursuits, often sharing ideas with their peers. Oral planning of work, as well as some drawn designs, is encouraged. Children also frequently revisit their earlier work, in order to modify or extend. Their attention is regularly drawn to technological phenomena and artefacts within the preschool environment, such as exploring and experimenting with materials to fix a broken step on the playground "fort".

Interest is shown in the apartment construction occurring across the road, which involved many physics concepts related to structures such as stability, symmetry, strength and durability, as well as the aesthetics of the design of the apartment itself. The extensive use of concrete and the various stages in the concreting process provided many opportunities for discussions about technological processes. Moreover, the machines involved in the preparation and delivery stages are motivating for the children who want to know how these machines work. Although the teacher did not capitalise on all the opportunities the building site offered, learning often occurred within the context of local knowledge and happenings, and contextual needs and wants. Of particular importance are the long periods of time allowed for children to construct and revisit. Though the director states she does not know a great deal about technology education, her programme has many potential opportunities for technological learning to occur, particularly if it is supported through careful mediation.

Hedegaard (2007), in her work on children's conceptual development, reminds us that it is teachers' choices of materials and experiences, together with the interactions they provide, that can determine much of the children's learning. Obviously, this choice is likely to reflect what is considered important and worthwhile within a particular social, cultural, institutional and/or historical context. In Preschool 1, what is valued are the end products and physical outcomes, described earlier by Fleer (2008) and Siu and Lam (2005). Here it was less easy to develop creativity and innovation as the children had possibly become "entuned" to the discrete technology experiences provided by the teacher. Even the anecdote from this centre described earlier, where Pippa used the model provided by the teacher as a prompt for helping a child to build, required a considerable amount of guidance on her part. Within this sort of context, we hypothesise that it would be less likely that children would internalise and transform their ideas about technological thinking than in Preschool 2. In this second centre, where the children's ideas are valued, and the director believes that it is important for them to have access to a wide range of materials for making and creating, the children have grown into the intellectual life of those around them (Vygotsky 1978). Creativity, simple designing, problem identification and solving are regularly engaged in by the children, and are mediated by the dynamic interactions between children, adults, ideas and materials, with constantly changing opportunities for learning developing. Within this context we also saw more evidence of the beginnings of conscious awareness and what Vygotsky (1997b) termed higher mental functioning—such as the planning that some of the children engaged in, and the focussed attention on building, with constructions extending over days and weeks, as they are revisited, appraised and modified their ideas.

Implications for Professional Knowledge of Early Childhood Teachers

Rather than focussing on the "end products" and "outcomes" of block play, we examined the way children's technological thinking is constituted with social and cultural/contextual factors. Over the extent of the study, we paid attention to the manner in which drawing and other forms of planning or designing, materials including wooden blocks, together with talk between teachers and children, mediate the children's technological thinking (such as identification of needs, creativity, innovation, evaluation and problem solving).

Cultural-historical theory has allowed us to interpret how talk among children engaged in block play can result in collective planning, joint problem solving, and shared thinking, such as that demonstrated by Gareth, Charlie and Sabella. Talk between adults and children is also very important, but the *form* of talk is most significant in shaping design technology possibilities. On the one hand, a pedagogical approach such as an interactionist style, where there is purposeful intervention by adults, can support children's thinking. On the other hand, a teacher-directed approach, especially when the incentive for the building has come from the teacher rather than the children's ideas or wants, may only provide limited opportunities for engaging the children in sustained or dynamic technological thinking.

Even though children may be involved in playing with blocks, technological thinking does not automatically follow, and thus a *laissez-faire* approach to teaching is perhaps the least effective. Recognition of pedagogy is important, but the significance of teachers' pedagogical content knowledge (PCK) needs also to be emphasised. Where teachers do not hold understanding of what constitutes technology and technology education, children's concepts may remain at the everyday level. For example, if the previously mentioned silence in the discussion of the scientific concepts related to electricity remains, Gareth, while knowing how to construct a circuit, may never develop an understanding of the scientific concepts embedded within this. In relation to working within the early years, this exemplifies the importance of PCK, subject knowledge and pedagogy—that is, understanding scientific concepts, and also knowing how and when to teach them to young children.

Although it has been beyond the scope of this chapter to provide extensive examples of children planning, drawing and designing their buildings, across the study this has occurred. While recognising that drawing designs presents some challenges for young children, we have found that this *is* possible with adult scaffolding (see Robbins et al. 2008). Useful strategies we have implemented in this study (though not discussed due to the word limits) include occasionally modelling the design process, as well as providing examples of different designs and floor plans, such as architects' plans and preschool centre floor plans. For some children, we found that oral planning and identification of materials to be used was an acceptable alternative to drawn designs. In addition, as Fleer and Jane (2004) have previously suggested, it is advantageous to consider the investigating, designing, producing and evaluating process as being flexible. For example, children who construct a building from blocks may be helped to appraise and then draw or plan modifications, or another building. Another strategy is to suggest children draw a representation of a completed building, so that they become familiar with how a three-dimensional structure can be represented in a two-dimensional form.

Furthermore, it is useful to provide drawing materials in or alongside the block area, not only for drawing plans, but also for teachers and children to make lists of materials that may be used in building projects. Although not expanded upon in this chapter, the use of cameras and recorders for later appraisal not only supports children, but also assists teachers to plan and reflect, thus offering more possibilities for developing children's technological thinking. The availability of a range of waste and recyclable materials, such as those accessible to the children at the second preschool, may mediate innovation and creative thinking, such as Gareth's decision to use bubble wrap for skylights in his building and Charlie's use of Perspex for doors.

A crucial factor in promoting technological thinking is affording sufficient time for the processes of investigating, planning and designing, building, evaluating and modifying. Allowing time promotes sensitive teaching interactions that can sustain the building process as well as offering more possibilities for children's technological thinking to develop. In the example from Preschool 1, Pippa's verbal and physical encouragement sustained Gabi's construction for longer than usual in that context. In the case of Preschool 2, Gareth, Charlie, Jacinta and Jill engaged in a sustained, interdependent "dance" of shared thinking for almost a whole indoor session. The ongoing appraisal and design modifications required commitment, co-operation and the continual renegotiation of roles, for example, the lending and borrowing of skills (such as writing or stripping wires) and the use of two pairs of hands when one pair was not enough to hold materials in place.

Conclusion

Ultimately, we aim to develop pedagogical approaches to technology education that will enhance early childhood teachers' understandings of how they can support young children's creative thinking, especially in relation to investigating, designing, producing and evaluating. We are beginning to outline what we believe are useful strategies to meet this aim. The teachers enthusiastically supported our study, and some are beginning to change their practices by providing clipboards and drawing materials in the block area. They are using the words *technology* and *design* during their interactions with children, and adding signs to the block area such as "Please leave my work" and "Watch out. Construction zone!" Furthermore, they are displaying and discussing pictures of local buildings under construction, and identifying opportunities for technological thinking in other aspects of their programme. Hopefully these and other strategies may promote children's on-going engagement with, and sustained interest in, design and technology.

Acknowledgments We wish to sincerely thank the children who are part of this study, and their teachers who have allowed us to work in their programmes. Thanks also to Philippa Hawkins for assistance with data generation. We also acknowledge the assistance of the Faculty of Education at Monash University who have supported the research through a small Faculty Grant, and the Office for Children Research Coordinating Committee, DEECD, who have permitted us to conduct the study in the preschools.

References

Benson, C. (2008). Young children as discriminating users of products: Essential experiences for language development. In H. Middleton & M. Pavlova (Eds.), *Exploring technology education: Solutions to issues in a globalise world* (Vol. 1, pp. 19–26). Surfers Paradise: Griffith Institute for Educational Research.

Bruce, T. (1992). Children, adults and blockplay. In P. Gura (Ed.), *Exploring learning: Young children and block play* (pp. 14–26). London: Paul Chapman Publishing.

Cohen, L., & Uhry, J. (2007). Young children's discourse strategies during block play: A Bakhtinian Approach. *Journal of Research in Childhood Education, 21*(3), 302–315.

Cowie, B., & Moreland, J. (2007). It's a Kiwi: Multimodal approaches to exploring and extending young children's science ideas. *Journal of Australian Research in Early Childhood Education, 14*(2), 37–46.

Department of Education, Employment and Workplace Relations (DEEWR) (2009). *Belonging, being and becoming: The early years learning framework for Australia.* http://www.deewr.gov.au/EarlyChildhood/Policy_Agenda/Quality/Documents/Final%20EYLF%20Framework%20Report%20-%20WEB.pdf. Accessed 30 Oct 2009

Department of Education, Science and Training (DEST). (2006). *Science, engineering and technology (SET) skills audit summary report.* Canberra: Department of Education, Science and Training.

Fleer, M. (2000). Interactive technology: Can children construct their own technological design briefs? *Research in Science Education, 30*(2), 241–253.

Fleer, M. (2008). A cultural-historical perspective on research in design and technology education. In H. E. Middleton (Ed.), *Researching technology education: Methods and techniques* (pp. 89–98). Rotterdam: Sense Publishers.

Fleer, M., & Jane, B. (2004). *Technology for children: Research based approaches* (2nd ed.) Frenchs Forest: Pearson Education.

Gura, P. (Ed.). (1992). *Exploring learning: Young children and block play.* London: Paul Chapman Publishing.

Gutek, G. L. (2005). *Historical and philosophical foundations of education: A biographical introduction.* Upper Saddle River: Pearson/Merill/Prentice Hall.

Hedegaard, M. (2007). The development of children's conceptual relation to the world, with focus on concept formation in preschool children's activity. In H. Daniels, M. Cole, & J. Wertsch (Eds.), *The Cambridge companion to Vygotsky* (pp. 246–275). Cambridge: Cambridge University Press.

Hedegaard, M., & Chaiklin, S. (2005). *Radical-local teaching and learning: A cultural-historical approach.* Aarhus: Aarhus University Press.

Jane, B., & Robbins, J. (2004). Block play as a potential for learning for innovation in technology education in early childhood classrooms. In H. Middleton, M. Pavlova, & D. Roebuck (Eds.), Learning for innovation in technology education (Vol. 2, pp. 60–67). Surfers Paradise: Centre for Learning Research, Griffith University.

Mawson, B. (2005). Where do I start? Technology in early childhood. *ACE papers: The first years, Issue 16* (pp. 104–116). Auckland: Faculty of Education, The University of Auckland.

Moreland, J., Jones, A., & Northover, A. (2001). Enhancing teachers' technological knowledge and assessment practices to enhance student learning in technology: A two-year classroom study. *Research in Science Education, 31*(1), 155–176.
Qualifications and Curriculum Authority (QCA). (1999). *Early Learning Goals*. London: QCA. (QCA/99/436).
Read, J. (1992). A short history of children's building blocks. In P. Gura (Ed.), *Exploring learning: Young children and block play* (pp. 1–13). London: Paul Chapman.
Robbins, J., Jane, B., Bartlett, J., & Hawkins, P. (2008). Developing young children's technological thinking through play with blocks. In H. Middleton & M. Pavlova (Eds.), *Exploring technology education: Solutions to issues in a globalise world* (Vol. 2, pp. 129–136). Surfers Paradise: Griffith Institute for Educational Research.
Rogoff, B. (2003). *The cultural nature of human development*. Oxford: Oxford University Press.
Siraj-Blatchford, J., & MacLeod-Brudenell, I. (1999). *Supporting science, design and technology in the early years*. Buckingham: Open University Press.
Siu, K. W. M., & Lam, M. S. (2005). Early childhood technology education: A sociocultural perspective. *Early Childhood Education Journal, 32*(6), 353–358.
Struss, D. J., & Stremmel, A. J. (2004). A sociocultural investigation of the effects of peer interaction on play. *Journal of Research in Childhood Education, 18*(4), 293–305.
Victorian Curriculum and Assessment Authority (VCAA). (2007). Design, creativity and technology. http://vels.vcaa.vic.edu.au/essential/interdisciplinary/design/index.html. Accessed 21 June 2009.
Vygotsky, L. S. (1978). *Mind in society: The development of higher psychological processes* (M. Cole, V. John-Steiner, S. Scribner & E. Souberman, Eds.). Cambridge: Harvard University Press.
Vygotsky, L. S. (1987). *The collected works of L. S. Vygotsky, Vol. 1: Problems of general psychology* (N. Minick, Trans.; R. W. Rieber & A. S. Carton, Eds.). New York: Plenum.
Vygotsky, L. S. (1997a). *The collected works of L. S. Vygotsky, Vol. 3: Problems of the theory and history of psychology* (R. van der Veer, Trans; R. W. Reiber & J. Wollock, Eds.). New York: Plenum.
Vygotsky, L. S. (1997b). *The collected works of L. S. Vygotsky, Vol. 4: The history of the development of higher mental functions* (M. J. Hall, Trans; R. W. Reiber, Ed.). New York: Plenum.
Vygotsky, L. S. (1998). *The collected works of L. S. Vygotsky, Vol. 5: Child psychology* (M. J. Hall, Trans; R. W. Reiber, Ed.). New York: Plenum.
Vygotsky, L. S. (1999). *The collected works of L. S. Vygotsky, Vol. 6: Scientific Legacy* (M. J. Hall, Trans; R. W. Reiber, Ed.) New York: Plenum.
Webster, A., Campbell, C., & Jane, B. (2006). Enhancing the creative process for learning in primary technology education. *International Journal of Technology and Design Education, 16*(3), 221–235.
Wolfgang, C., Stannard, L. L., & Jones, I. (2001). Block play performance among preschoolers as a predictor of later school achievement in mathematics. *Journal of Research in Childhood Education, 15*(2), 73–230.

Chapter 14
Re-conceptualizing the Teaching of Physics for Non-majors: Learning from Instructor-Driven Reform

Sandy Martinuk, Anthony Clark and Gaalen Erickson

Introduction

Throughout North America the curriculum of introductory physics courses at the post-secondary level is almost universally standardized, with four textbooks dominating 90% of the introductory physics market (Tobias 1992). These texts typically cover material in kinematics, dynamics, momentum, energy, and electricity, topics which are certainly crucial building blocks to future understandings in physics but are substantially less useful for students who are not planning to major in physics. In the fall of 2007 the instructional team at the University of British Columbia (UBC) made some significant curricular and pedagogical changes to their introductory course for non-physics majors aimed at improving the value that this course will have after their students have left the physics classroom. Specifically, the faculty wanted to show students how physics was relevant in the real world and enable students to develop capabilities for applying physics to everyday situations. The themes of energy production and use and climate change were incorporated into the curriculum both to demonstrate the key role of physics in everyday life and to enable students to better understand current social and scientific challenges in these areas.

This paper will present the changes in this introductory course in narrative form, focusing on the perspective of the head instructor (referred to as Ken) and discussing some of the interesting challenges and concomitant learning that he and the other course faculty experienced. Interviews with the faculty are used to explore some of the changes in their attitudes towards teaching and their professional knowledge. The challenges of connecting school science with the everyday are also explored by Rennie (this volume), who offers a much broader perspective.

Sandy was involved intimately in this change process as a member of the team that developed the new course curriculum, a teaching assistant for the course, and a researcher monitoring the students' learning and attitudes. To clearly situate him within the narrative, this paper will use the first person to refer to his involvement.

S. Martinuk (✉)
University of British Columbia, Vancouver, Canada
e-mail: sandy.martinuk@gmail.com

D. Corrigan et al. (eds.), *The Professional Knowledge Base of Science Teaching*,
DOI 10.1007/978-90-481-3927-9_14, © Springer Science+Business Media B.V. 2011

The other two authors provided key advice on both the change process and the paper as critical friends.

Background

Physics 100 at the University of British Columbia is a non calculus-based introductory physics course offered to science students who require a physics credit to complete their degree and who have taken only one year of high school physics. It is a large lecture style course which divides 700–900 students into three sections, each taught by a different faculty member.

Prior to the course changes outlined in this paper, the Physics 100 syllabus and format was similar to many North American introductory physics courses. The course consisted of 3 hours of weekly lectures supplemented with bi-weekly alternating 3-hour laboratory sessions and 2-hour optional tutorial sessions where students worked in groups on practice problems. The course followed a common sequence of topics in mechanics, DC circuits, and geometrical optics. To improve student engagement, the faculty used an electronic response system (also known as clickers) to periodically ask short questions during lectures.

Because Physics 100 is required for many of UBC's Arts and Science programmes the student population is very diverse. Approximately 60% of the students are in the Science program, but the vast majority of them are not intending to major in physics and are required to take only one additional physics course. The remainder of the students are human kinetics, food and nutrition science, forestry, or arts students, and Physics 100 is the only physics course they will take in their undergraduate program.

Motivations for Change

The background for the changes to Physics 100 was the recognition that this population might not be well-served by the traditional course content and format. This course presented physics content in ways that might be useful for further study in physics but few, if any of its students would be pursuing further studies in physics. The overriding question for developing the new course was: other than a fulfilment of an externally set requirement for a degree programme, what use can a physics course be to a student who is not primarily interested in physics?

The idea for changing the content of the course came from Ken, a seasoned tertiary physics instructor who felt that university courses in general were of limited use once students entered the real world. He was particularly concerned with the lack of public understanding of scientific issues such as energy conservation and climate change. These issues are crucial to the success and prosperity of society but his feeling was that the public (including many journalists) may not understand them very well. Worse, it seemed that many people adopted an attitude of either taking scientific statements in the media at face value or writing off all science as "just theories"; taking the presence of scientific debate as an indicator that nothing in these issues was settled or worthy of action.

Ken was also concerned about what he saw as students' poor retention of facts and concepts learned in physics class. It had become clear that the learning strategies employed by students, which often involved last-minute cramming and wholesale memorization of textbooks, resulted in shallow and temporary knowledge (Roediger et al. 2009, Kornell and Bjork 2007). This was particularly true of students in Physics 100 who would take only one or two physics courses, and would have little opportunity to re-learn and deepen their understanding of physics concepts. He was concerned that these students would literally forget everything they learned in this class.

Ken came to believe that a key piece of this problem of lack of retention was the issue of relevance: the Physics 100 students did not see the physics presented as being relevant to anything outside of the physics classroom, and therefore had no motivation to learn it in a deep way. His feeling was "Even the best student will expel [physics] knowledge if they don't see it as being relevant."

Another key goal of the course changes was motivated by survey results that examined the students' attitudes towards physics. In his 2006 Physics 100 course, the Colorado Learning and Attitudes towards Science Survey (CLASS) had been administered as a means of measuring how his students' attitudes and expectations towards physics had changed after taking the course (Adams et al. 2006). The CLASS survey uses 42 statements about physics and learning which the students are asked to agree or disagree with on a 5-point Likert scale. Each statement has a clearly favourable and unfavourable response, validated by physics professors and other expert physicists. Student responses are scored by determining the percentage of items for which the student has given a favourable response, termed the % favourable score. For example, if a physics expert would respond "strongly disagree" to a particular item, both "strongly disagree" and "disagree" are considered favourable responses. The response scale is collapsed in this way to avoid implicitly treating a "strongly disagree" response as worth two "agree" responses, an assumption which is difficult to justify. However, the % favourable score for a particular student can be calculated for both a pre- and post-test, which enables calculation of the shift in a student's scores.

As is often the case in traditional introductory physics courses, the attitudes score of Physics 100 students in 2006 became less favourable over the course of instruction (Perkins et al. 2004). This suggested that the course was actually having a deleterious impact on student attitudes toward physics, reinforcing undesirable attitudes and assumptions about the nature of physics and learning. This negative impact in Physics 100 was especially concerning because Ken felt that these negative opinions would be the principal legacy of the course, and would colour the way his students perceive physicists and scientific information for the rest of their lives.

To address these concerns, Ken wanted to offer his students an education in physics that would meet three major goals. He wanted: to educate his students about socially relevant issues; to enable them to learn how to apply physics to other scientific issues in the public sphere; and to encourage students to see physics as relevant to themselves and to their lives. He hoped that this course would contribute to the students' scientific literacy, which for him meant that they would understand enough about science content and the nature of professional science that they would be willing and able to use their own knowledge to evaluate scientific messages in the media and take action to engage with socioscientific issues. While these goals may seem overambitious to those more familiar with the history of educational research and curricular reform, Ken felt that they were attainable within the introductory physics setting. The changes to the course began with these goals, and with his idea of teaching the physics of energy production and use and climate change as a means to achieve them. He hoped that by teaching subjects that were the topic of significant public discussion, students would start to see connections between the physics classroom and the real world and would start to see physics as relevant to themselves.

Development of the New Physics 100

Because the students in Physics 100 did not feed into the main stream of Physics majors students, Ken was able to begin the process of changing the course curriculum and pedagogies without much formal process. Because there was no need to provide a foundation for higher level physics courses it was not necessary to involve many other professors or administrators in the curricular changes. Consequently, Ken and his co-instructors had an unusual degree of freedom to modify the course.

In May 2007 the team was assembled that would be developing the curriculum and materials for the new course. The team included the three faculty who would be teaching the course as well as two of their colleagues who had some experience teaching the physics of climate change in another introductory course. This team began meeting weekly to develop the syllabus for Physics 100, choose a textbook, and subsequently develop the lecture material, examples, and assessment strategies.

At the time I was a new physics PhD student with some college teaching experience and a declared interest in physics teaching and physics education research. I was invited to participate in the faculty meetings and contribute comments on the materials being developed. My role in the course development was twofold: participating with the team of faculty in developing course goals, curriculum, and materials as well as offering perspectives on physics pedagogy gleaned from readings in the literature and conversations with other science education researchers. Although a novice in the field, I did my best to offer the Physics 100 instructors access to the results of physics education research, and to encourage them to consider new perspectives and pedagogies.

Many of these new ideas were brought to my attention via the Carl Wieman Science Education Initiative (CWSEI). This organization is dedicated to supporting reform in undergraduate science education and was founded in January 2007 when

Dr. Wieman, a 2001 Nobel laureate in Physics, came to UBC. Although the changes for the course had begun before the CWSEI was founded, the arrival of Carl Wieman at UBC and the subsequent development of the CWSEI did have a significant influence on how the course was developed.

Course Goals and Themes

One of the first tasks undertaken by the development team was to begin defining explicit course goals. After attending a workshop on learning goals by Dr Wieman, the Physics 100 faculty agreed to use specific course- and lecture-level goals as the skeleton for the course. These goals were developed concurrently with (and sometimes after) the course materials. Despite this, the faculty still found the process of developing course goals to be worthwhile. The goals served as a way for the faculty to negotiate consensus on what the students should be learning and were useful in guiding assessments throughout the course.

Collaborative development of explicit course goals represents an important shift from the method that faculty had previously used in multi-section courses. Prior to 2007 the faculty would agree on an overall course outline, but each instructor would then proceed to independently develop their teaching materials, usually using the outline as a *de facto* list of the material to be learned. However, this method tended to bias the lessons towards covering content, in which the emphasis is on what is taught rather than what is learned. Working out the course goals for Physics 100 was an important opportunity for the faculty to recognize and develop the other goals they held for their students which had previously remained tacit (Redish 2003), such as goals for their students' problem solving skills and attitudes towards physics. Explicit consideration of these aspects of the student experience helped the faculty to begin the process of developing a deeper understanding of their teaching practices. This process of becoming more sensitive to the students' background and attitudes continued throughout the course development and evolution.

The question of what students would retain from this course was central to framing the faculty members' approach to developing course goals. Research on student retention of concepts from introductory physics courses supported the faculty members' experience that students retained little declarative knowledge from their prior courses (Semb and Ellis 1994). However, their experience was that formal education can still be useful after one has forgotten the facts and formulae. They felt that often it is a way of approaching problems (such as the practice of cutting them down into smaller pieces) or perhaps fragmentary knowledge that might help people interpret new situations. Their approach was instead to highlight things that they felt could be remembered: problem solving approaches and ways of seeing that might allow their students to address real-world problems.

This perspective guided the faculty by emphasizing opportunities for students to learn how to apply physics in real-world situations, rather than to simply amass declarative knowledge of physics content. The intent was not to do away with learning of declarative knowledge altogether, but to emphasize that the key component of learning is recognizing which knowledge is applicable and how to apply this knowledge.

In order to enable students to apply physics knowledge in everyday situations, the faculty felt that it would be crucial to situate the physics in the context of the students' everyday world. The use of real-world examples and contexts throughout the class was intended to demonstrate its relevance to the students and hopefully to improve their engagement and learning. While the emphasis on real-world relevance of course content is not new in the global sphere of education, this commitment on the part of faculty in an introductory university physics course was groundbreaking. However, it became clear later in the year that the faculty's perception of what was "relevant" and "everyday" did not always match the students' perception.

Specific Curricular Changes

In addition to fulfilling his goal of improving the public understanding of climate change and energy consumption, Ken felt that including these topics in Physics 100 would provide clear evidence of the relevance of physics to students' lives. The faculty agreed that climate change was a worthwhile topic to teach, but were concerned about the challenge of interpreting this very complex system for an introductory audience. However, an introductory text that was geared towards Earth and ocean science students offered several perspectives on how simple models of climate change could be used to illustrate important characteristics of the Earth's energy system. These models neglected all dynamics of the atmosphere and treated it simply as a single layer with different absorption for different wavelengths of light. The instructors felt that this model would strike a balance between simplicity and complexity.

One concern was that these models would require students to learn thermal physics and radiation, topics which are typically taught in more advanced courses. However, one of the other professors on the development team had been teaching the physics of home-heating, a thermal equilibrium system which requires many of the same physics concepts as the Earth's thermal equilibrium. The problem of understanding of home-heating seemed like the perfect context to motivate the development of the thermal physics required for climate change.

With the addition of thermal physics to the curriculum, the instructors realized that energy could be expanded from a single unit to a unifying theme for the entire course. The unit on mechanics could include a discussion of fuel efficiency and energy consumption in transportation, and the unit on electricity could be motivated by discussions of household energy consumption. The idea of conservation of energy would be used as a touchstone throughout the course, making connections between the various applications to thermal physics, mechanics, and electricity.

Because of the prevalence of energy in the curriculum, the faculty decided to try to construct a lesson sequence that started with a discussion of energy. This decision was not easy; it was difficult for the instructors to find a new way of introducing this material to their students in a way that deviated from the traditional mode and method of instruction. There was the feeling that observable quantities such as velocity and height would be easier to explain to students, so from this expert perspective it seems sensible to discuss the more abstract concept of energy only after kinematics and dynamics have been thoroughly explored. However, I suggested that energy as a scalar quantity might be mathematically easier to deal with at the beginning of the course than the vector trio of position, momentum, and velocity. In the end we all felt that the course should be about energy, and so the faculty agreed to try introducing energy at the beginning of the course and referring to this concept throughout. It is clear that the faculty believed that learning about energy was relevant to their students, and we set out to convince the students of that.

In order to make room for a discussion of climate change and supporting materials approximately six weeks of the previous course material needed to be cut. Because the faculty saw few everyday situations where geometrical optics might be useful to their students, this material was cut from the syllabus. Vector analysis was another topic that faculty felt was of limited value to this student population. For physics majors, the mathematical and conceptual tools of vector decomposition were essential, but the faculty's experience had been that students in this non-major population tended to regard these aspects of physics as purely formulaic exercises. Hence, vector analysis was cut from the mechanics curriculum along with all discussion of two-dimensional motion such as cars on inclined planes and projectile motion.

Last and most contentious on the chopping block was conservation of momentum. As one of the key conservation laws in physics, it is traditionally taught alongside any treatment of introductory mechanics. However, the team could not find any applications of conservation of momentum in everyday life other than collision analysis, which we felt would have limited appeal for the students. After much debate, the faculty agreed to cut this material based on the fact that we would be greatly expanding our coverage of conservation of energy, one of the other grand principles in physics. This example highlights how the faculty's emphasis on topics they felt had real-world relevance led to a dramatic reconfiguring of the course curriculum away from traditional topics.

The faculty began working to find specific example problems, diagrams, and instructional methods that would present this new course syllabus in terms of the students' everyday experience. This task proved to be quite challenging. Traditional physics instruction is often engaged in the business of reduction: each physical effect is discussed and demonstrated in absence of other effects, and far from the complex reality of everyday situations. Unfortunately, examples that are rooted in the everyday world of the students and cleanly demonstrate the action of a single physics principle are rare: most everyday situations involve a combination of several influences which makes any discussion of the physics quite complicated, especially for novice learners.

In order to tackle the complexity of everyday physics, the Physics 100 faculty adopted several strategies. The first was to include in the course some explicit discussion of the role of modelling and simplification in physics. The process by which an experienced physicist develops a simple model of a complex process is often taught implicitly in more traditional physics courses. We hoped that by discussing these techniques explicitly, we might be able to address some of the students' customary discomfort with making sweeping approximations. This process of making assumptions to simplify everyday situations was discussed in some detail throughout the course and was reinforced in the tutorial sessions which required the students to perform this type of modelling. Another strategy for dealing with this complexity was to search for real-world examples where seemingly extreme simplifications such as "frictionless" or "rigid" could be justified via commonsense reasoning. This task was quite difficult, but we did find a few examples of situations where we believed students would accept such simplifications as being natural. Finally, we did include a few complex problems where several physical effects needed to be integrated in order to understand the physics of the situation. These were typically built up from simpler models by a process of successive approximation, gradually adding complexity and accuracy. For example, the full model of heating costs for a home included both conduction and radiation as loss mechanisms and lighting, heating, and the radiation of its occupants as energy sources. The choice of these real-world examples and the decisions about how complex to make the models were quite challenging, and even now are an area of active improvement in the course.

Teaching the New Physics 100

In the fall of 2007 the new course was offered to the standard cohort of around 750 students split up in the three sections. The faculty continued to have weekly meetings in order to maintain synchronization of their presentation of the material, plan lessons and homework for upcoming weeks, develop learning goals for each week, examine feedback on the course progress, and discuss any difficulties they were having. I continued to participate in these meetings, and also filled several other roles within the course. In collaboration with the faculty, I wrote the weekly tutorial problems, coordinated the course's complement of 20 TAs, and conducted my own tutorial sessions. I also conducted ongoing interviews with students to hear their perspective on the course, and fed back results from these interviews to the faculty.

This feedback was one of a variety of sources of feedback that the faculty accessed to evaluate the students' response to the new course. The faculty solicited input from the TAs during weekly TA meetings and also received regular feedback about the students' understanding through regular clicker questions conducted in lectures. For example, see Fig. 14.1.

Because so much of the course content was being taught for the first time, the faculty made good use of this steady stream of feedback to adjust their teaching practices during the first year of the transformed course. The weekly tutorials and in-class clicker questions provided a window into students' understanding of the material and allowed the instructors to follow up on previous material or, in the

Fig. 14.1 Example clicker question

We have seasons because:

1. In the Winter BC is farther from the Sun than in the Summer
2. In the Winter the Sun is lower in the sky so the angle is between the incoming radiation and ground is small
3. The days are short so there is less solar radiation to heat the ground
4. 1 and 2 are both important
5. 2 and 3 are both important
6. 1 and 3 are both important

case of widespread student difficulties, to circle back and devote more class time to ensure their learning goals were met.

One major challenge faced by the faculty during the first year was interpreting their non-traditional goals for their students' problem solving skills and attitudes towards physics within the context of each of the course's learning activities. In lectures as well as labs and tutorials there was an ongoing tension between new and traditional teaching techniques and assessments, as well as the logistical restrictions of working within such a large course. The faculty needed to learn how to support these goals within their course's diverse learning environments. For example, the discussion on climate change was a significant deviation from the normal curriculum of an introductory course, but the style of presentation of this knowledge and the expectations for student learning followed a familiar pattern. Students were presented with diagrams representing physical phenomena and accompanying formulae, and these formulae were used to perform quantitative calculations.

However, some of the discussion on climate change was conducted in a new way: the class moved away from unproblematic calculations based on simple models and concrete given information, and discussed some of the challenges in the application of mathematical climate models to the Earth's climate. These discussions were largely qualitative, and covered several important issues such as the role of feedback in climate change, the history of Earth's climate, and the evidence for and against humanity's role in inducing climate change. Several common arguments about the nature of climate change and humans' role in it were presented and critiqued by the instructor using the physics in the course. In addition, several common statements about climate change that have been recently debunked in the literature were presented as myths of climate change.

Despite the importance of this material to his goal of social change, Ken presented this material fairly quickly. His intention was not for the students to learn the details of the arguments in climate change, but rather to demonstrate the dynamic and tentative nature of authentic scientific knowledge, and to convince the students that it was appropriate for them to evaluate these arguments themselves using basic physical reasoning. The process of examining and critiquing several arguments was meant to encourage the students to take responsibility for assessing scientific statements they encounter. He also hoped the students would learn some techniques for critiquing and evaluating scientific arguments.

The assessment for this segment of the course was quite different than that for the more content-oriented segments. Because it was focused on very general capabilities rather than specific mathematical tools or models, the faculty felt that typical quantitative problems would be inappropriate. Ken couldn't find any satisfactory ways of testing students' capability of critiquing scientific arguments that would be appropriate in the sequestered setting of an exam, and so this material was not significantly tested on the midterm or the final exam.

The best assessment of these capabilities was through the course's final research project, a new course component which required groups of students to research and present on a topic of their choice related to the course. Student groups were asked to quantitatively model a real-world situation to make a choice between two options or settle a dispute. Following the theme of the course, many of these projects were related to energy conservation, efficiency, or greenhouse gas emissions. For example, one suggested project topic was for students to assess the statement that "in the UK walking to the store damages the planet more than going by car", a claim that is only true under a very particular and unlikely set of conditions. The faculty saw this project both as a way for students to further develop their capability of developing simple models to evaluate scientific statements and also as an assessment of their problem solving and modelling capabilities. This project also offered interesting feedback because it was the first chance for the students to make decisions about what they thought was relevant to themselves.

Reflection on the First Year

After presenting and examining the students' knowledge of the material the impression was that many students had not developed any deep conceptual understanding of the material on climate change. Despite discussion of the conceptual underpinnings of energy equilibrium of a body bathed in radiation, many students were not able to apply these concepts to any contexts other than the climate change model which was explicitly developed in class. A final exam problem which asked students to compute the equilibrium temperature of the moon was passed by only 45% of the students. The faculty saw this exam problem as an example of near transfer of the energy equilibrium model of a spherical body bathed in constant-intensity sunlight, but most of the students were unable to apply the model developed for the Earth in this new context.

However, the 45% pass rate on that question demonstrated that this portion of the class had indeed developed some flexible knowledge about climate modelling, which the head instructor felt was an indicator of the success of the first year of implementation. He felt that the results from the first year of teaching indicated that some of the students were now better informed about and better able to assess important socioscientific issues which represented an important, if partial, success. He took the results of this final exam question as an indicator that the course was on the right track, but that it was necessary to refine the teaching of the material on climate change.

Several reasons were considered that might account for the students' difficulties on this topic. There were some indications that the mathematical issues in working with the Stefan Boltzmann law of blackbody radiation were confusing some students (that is, that it contains Greek letters and a T^4 term), but it was the head instructor's impression that the physics of climate change is no more difficult than any of the other physics presented in the course. Rather, he felt that other factors contributed to students' difficulty with this material.

We also realized that because thermal and radiative physics are not often part of the high school curriculum, it was unlikely that students had any prior formal knowledge of thermal physics. Unlike mechanics, in which the students typically have some familiarity, all of the language and concepts in thermal physics would be new to these students which could necessitate a slower instructional pace to allow students time to develop and integrate a new vocabulary. In addition, the rarity of teaching these topics at an introductory level would mean that there are fewer external resources (for example, tutors, textbooks, or chat forums) that the students could access for help.

In light of these difficulties, the faculty felt that in the first year they may have been trying to do too much on the topic of climate change. After building up the component theories of energy equilibrium and radiative heat transfer, the faculty tried worked through several successive iterations of a model for the Earth's temperature, working up to such sophisticated concepts as the effect of differential absorption of visible and infrared radiation by different species of gas. This idea is reliant on an understanding of the concept of the electromagnetic spectrum, something which is fairly simple for the faculty but much less so for their students. After reviewing the course results we felt that perhaps all of these extra details had left the students without a clear idea of the conceptual interpretation of formulae embedded in the climate change models. Without a deeper understanding of these models, a slight change in context meant that students did not know which principles to apply. This echoes the result that novice solvers of physics problems tend to rely on the surface, physical features of problems to choose their problem solving approach (Chi et al. 1981).

Another of the challenges in teaching a subject as complex as climate change in a first-year course is designing appropriate assessments. While several of the goals of this segment focused on students' ability to make approximations and develop simple models for real-world systems, the assessment on the final exams was still largely focused on performing calculations. This type of question does not seem to support the emphasis on students' critical thinking and ability to develop of physical models in everyday contexts.

The use of the final project as the main assessment for students' skill at applying physics to evaluate a scientific argument was also somewhat problematic. It was not explicitly discussed with the students that they were expected to learn from the faculty's critique of climate arguments and myths, and that they would be required to apply similar techniques in the final project. This, combined with their prior experience with science in high school may have led students to infer that these critiques were "extra stuff" in the course that would not be assessed. Such a perception would certainly affect how students attended to this material, and their subsequent learning of these critical techniques.

Additional Research

To further probe the impact of the course changes, I conducted research which provided another source of feedback on the first year of implementation. I interviewed 16 student volunteers throughout the term with the aim of getting general feedback on the course changes. I also administered the CLASS attitudes survey at both the beginning and end of the term.

The items on the survey are grouped into eight categories, one of which is a collection of four questions around the theme of real-world connections in physics. Despite the introduction of the course changes in 2007, the average pre/post shift in this category was statistically identical to the results from 2006. Interviews conducted with ten students after the 2007 course revealed several possible reasons why the expected improvement did not occur. One common student comment was that several of the major topics in the course used real-world examples that were not directly relevant to their lives. For example, while all of the students were familiar with environments where the physics of home-heating is important, most of them have never paid a heating bill and were therefore not really invested in the details of household heat loss. Other students who do not drive a car felt that examples based on driving were not directly relevant to themselves. In general, these interviews revealed a tendency for students to judge an example as relevant only if its context was directly related to their immediate life or career plans. This finding highlighted the potential for significant differences between the students' perception of the real-world relevance and the faculty members' perception.

Because many of the course changes were geared towards showing students how physics is relevant to themselves, the recognition of this gulf between faculty's and students' perception of relevance was disappointing. The faculty realized that they needed to develop a better understanding of their students' perspective, and to that end I began planning another series of interviews and surveys for the following year. Instead of exclusively using the collaboration of faculty to determine which physics would be relevant to the students, we began trying to learn this from the students themselves.

Evolution of the New Physics 100

The results of these follow-up studies were not available in time to develop the second year of the course, and so the instructors made changes based on the feedback already available. Ken developed changes to the second year of the course using a process that was largely intuitive and informal. He supplemented his judgments honed by decades of physics teaching with the formal results of our research to make small changes in his materials. One strategy was to adjust his presentation slides and notes shortly after teaching a class in order to capture the feedback gleaned from that day's teaching.

The changes made in the second year addressed some of the problems discussed above and also addressed the issue of focus: all three said that they felt the content of the lectures had changed very little, but the emphasis of the lectures had changed significantly. They had seen the difficulties encountered when the students could not identify the key ideas, and therefore spent more class time discussing the fundamental concepts that tied the physics together. Some material was discussed more slowly and carefully, which took up the same amount of time as circling back had in the first year, but the faculty's feeling was that it resulted in increased student understanding. The climate change material was also clarified: only two successive models were presented to eliminate some of the complication. More sophisticated aspects of the models were still mentioned, but were not assessed.

Another important evolution was a shift in the emphasis of real-world examples away from the societal towards the personal. The student interviews had shown students made a distinction between "relevant to me" and "relevant to the world in general". To pursue their goal of engaging and motivating the students, the course instructors tried to use personal-scale examples and to emphasize the former type of connections more explicitly.

After conducting the second year of the course, feedback from the student interviews and surveys has shown some significant improvements in students' attitudes towards physics. While it is impossible to tease out exactly what made the difference, the instructors felt that the feedback from the first year allowed them to better understand their students' beliefs and knowledge, and thereby identify and focus on the essential elements of the physics.

Discussion

The story of the changes to Physics 100 offer several lessons for educators and researchers.

The Key Role of Collaboration

Conducting a comprehensive revision of such a large course required significant effort on the faculty members' part. This commitment was supported by the tightly collaborative nature of the course development team: faculty co-created the course goals, themes, and syllabus, and helped to develop each other's materials. This collaboration also created social and supportive environment that helped to encourage each participant to continue working to bring new ideas and materials into the collective pool.

In addition to motivating participation, this collaboration also helped the instructors to develop new ways of seeing their material and their students. Ken cited the iterative discussion and revision enabled by the course's collaborative development as being crucial to the success of his endeavours to re-conceptualize upper-year material for an introductory audience and to develop examples and contexts that related to the students' lives.

The collaborative feedback process also helped the faculty to enact their goals of making substantive changes in their approach to the course. One of the challenges cited by the Ken in developing the new material was forcing himself to actually deviate from old habits of teaching that had not been very fruitful in the past. When asked to reflect on the course development process, he said that he was often tempted to just make minor changes to how the material is presented, rather than really considering how to improve the fundamental curriculum and pedagogy employed in the course. He cited that feedback and critiques from his colleagues were "an important signal to let you know if you are doing things too much in the old way". In this way their collaboration enhanced the instructors' reflective practice.

Ownership of Change

Another important factor that helped to motivate the faculty in conducting these course changes was the freedom to choose the course material. Because of the unusual freedom to choose the course learning goals, the faculty could teach material they personally felt was relevant. This increased connection with their teaching has made this course evolution much more satisfying for the faculty on a personal level. Ken reports: "It is definitely more pleasant to teach something that they appreciate, but also to teach something you think *should* be taught" (emphasis added). One of the other faculty described how the freedom to make changes led to an increased sense of ownership: "The realization that the more I put into it the more I care about it is in a way a no-brainer, but I found that surprising." This increased sense of ownership translated higher satisfaction and deeper commitment to improving Physics 100.

Development of Instructors' Professional Knowledge

As described above, the changes in both the content and the pedagogical approaches being introduced to the course along with the deliberations on the rationale for making these changes were both time-consuming and potentially risky for the faculty members. They were charting new territory that none of them had seriously explored before and there was very little literature or models available for them to draw upon. Hence, the course development process and subsequent teaching of the new materials and approaches proved to be extremely fertile ground for the development not only of new teaching materials and strategies but also for the emergence of new perspectives on the teaching and learning of physics content.

For example, they became much more attentive to current understandings and interests of their students as they searched for relevant contexts and appropriate physics content. The emergent sensitivity of the distinction between relevance-to-the-faculty and relevance-to-the-students has continued to drive efforts to understand the students better and shape the course to their needs. This new perspective transformed their approach to teaching in other courses as well as Physics 100. One of the other faculty reported, "The Physics 100 experience has also influenced how I teach Physics 101… In particular, I now look more at the learners, what may motivate them."

The Physics 100 faculty also became much more sensitive to the need to support their pedagogical aims with appropriate assessments. After the first year, more attention was paid to developing assessments that would communicate to the students exactly what type of learning the instructors were aiming for.

Another benefit of the changes to Physics 100 is that the experience of successfully collaborating on development and setting explicit course goals will certainly be beneficial to other courses these faculty subsequently teach, even if those courses are not undergoing extensive revision. Thus, the process of curricular revision provided an unanticipated personal and professional development opportunity for the faculty. Although perhaps not at the level of those directly engaged in the Scholarship of Teaching and Learning initiative in other areas of the university, this development was certainly an important and significant outcome: the faculty were researching their own practice.

The changes to Physics 100 were the first in a sequence of course revisions that the UBC Physics Department is undertaking. Starting this sequence with a course where there is significant freedom to change the course content has allowed three faculty members to become more engaged with the change process, and consequently develop their skills and expertise at course revision. The experience of these instructors is now a valuable resource for the department as the sequence of course revisions continues. Their sensitivity to their students and enhanced pedagogical knowledge may be beneficial in updating other courses, even those that have more stringent external constraints for the course goals.

Conclusion

To improve the quality of their students' learning, the faculty of UBC's Physics 100 course chose to significantly alter the content and context of their course. In contrast to many contemporary educational reforms, these changes were made not only to improve students' grades and understanding of physics concepts, but also to offer them a form of physics education the students would find meaningful and relevant beyond the boundaries of the course itself. While there have been some major challenges in implementing these course changes, the strongly collaborative nature of the team of faculty has supported the sustained effort. Considerable time and effort was spent on generating new pedagogical approaches and exemplars in their teaching. These changes also were accompanied by some risks as the faculty members had to depart from their established repertoire of standard physics problems and the extensive use of quantitative methods for solving these problems. In the process of constructing some new approaches for the teaching of novel physics contexts—such as climate change—the faculty were also generating new understandings of their students as well as the application of standard physics content to novel pedagogical contexts. The faculty's sensitivity and response to feedback from in-class research, their TAs, and their students has also helped them to overcome difficulties and make important changes to the course. The difficulties encountered in the first year helped the faculty recognize the challenge of developing material that would be seen as relevant by their students. They became more attentive to the voice of their students, and more interested in accessing research on their students' perspective and other educational research. Engaging is this process of curricular change has enabled the faculty to bring material into their course that they feel is personally and socially relevant, which has helped to sustain their motivation to bring meaningful education to their students. By offering a set of educational experiences that appears to be more meaningful to their students, the act of teaching becomes more meaningful for the faculty as well.

References

Adams, W. K., Perkins, K. K., Podolefsky, N. S., Dubson, M., Finkelstein, N. D., & Wieman, C. E. (2006). New instrument for measuring student beliefs about physics and learning physics: The Colorado Learning Attitudes about Science Survey. *Physical Review Special Topics-Physics Education Research (APS), 2*, 010101.

Chi, M. T. H., Feltovich, P. J., & Glaser, R. (1981). Categorization and representation of physics problems by experts and novices. *Cognitive Science, 5*, 121–152.

Kornell, N., & Bjork, R. A. (2007). The promise and perils of self-regulated study. *Psychonomic Bulletin & Review, 14*, 219–224.

Perkins, K. K., Adams, W. K., Pollock, S. J., Finkelstein, N. D., & Wieman, C. E. (2004). Correlating student beliefs with student learning using the Colorado Learning Attitudes about Science Survey. In J. Marx, P. Heron, & S. Franklin (Eds.), *Physics Education Research-2004* (Edited by AIP Conference Proceedings 2005, No. 790, pp. 61–64). Melville: American Institute of Physics.

Redish, E. F. (2003). *Teaching physics: With the physics suite.* New York: Wiley.

Roediger, H. L., Agarwal, P. K., Kang, S. H. K., & Marsh, E. J. (2009). Benefits of testing memory: Best practices and boundary conditions. In G. M. Davies & D. B. Wright (Eds.), *New frontiers in applied memory* (pp. 1–41). Brighton: Psychology.

Semb, G. B., & Ellis, J. A. (1994). Knowledge taught in school: What is remembered? *Review of Educational Research, 64*, 253–286.

Tobias, S. (1992). *Revitalizing undergraduate science: Why some things work and most don't. An occasional paper on Neglected Problems in Science Education.* Tuscon: Research Corporation.

Chapter 15
Developing the Knowledge Base of Preservice Science Teachers: Starting the Path Towards Expertise Using Slowmation

Stephen Keast and Rebecca Cooper

This chapter is based on the authors' teaching of the science curriculum specialism in the secondary teacher education program at Monash University, Australia. This unit is part of the fourth and final year of the students' double degree program (for example, Bachelor of Science/Bachelor of Education) or the one year Graduate Diploma of Education course. The year consists of core education units, elective education units and teaching specialism units. The teaching specialism units are designed to engage and inform students about teaching in their teaching specialism. As part of this unit, the authors use a Teaching Procedure called Slowmation (Hoban 2005) to create a shared experience for the preservice teachers. This shared experience forms the basis for rich conversations about their teaching and developing ideas of pedagogy that puts them on the path towards understanding what it required to for expertise in science teaching.

An Introduction to Slowmation

Slowmation is a Teaching Procedure where students design and create short movies of their understanding of science concepts. The Slowmation procedure is taught to students by initially having students' complete a storyboard of the main components of the particular science concept (for example, DNA replication) being examined. At this stage students conceptualise their knowledge and understanding of the concept. The teacher collects the storyboards at the end of class, edits and annotates them with comments and returns them to students ready for the next class. Students use the returned storyboard to create models and scenes using plasticine and other materials. They then take digital photographs of the individual scene segments and link the segments together using commonly available software such as Windows Movie Maker or iMovie to produce short animated movies. These can be easily enhanced with the available software by adding labels, a narration and music. At

S. Keast (✉)
Monash University (Clayton campus), Clayton, VIC, Australia
e-mail: stephen.keast@monash.edu

D. Corrigan et al. (eds.), *The Professional Knowledge Base of Science Teaching*, DOI 10.1007/978-90-481-3927-9_15, © Springer Science+Business Media B.V. 2011

this stage students re-present their knowledge of the scientific concept as they understand it. This is similar to clay animation where clay figures are photographed and multiple images are captured to simulate continuous "near life" movement. However, Slowmation is played at two frames per second to produce "slow animations" (hence the term Slowmation). Such movies can be created in little more than an hour, though they tend to be short, lasting between two and five minutes.

In this study, the Slowmation process was introduced to preservice secondary science teachers who then made short movies about particular science concepts. The purpose of using Slowmation in this way with preservice science teachers was not so much as to explore their understanding of the science concepts, but rather provide them with an experience of Slowmation that they could then apply while on their teaching placement. In this way it was envisaged that preservice teachers would see value in using the procedure with their students during placement as one more tool to monitor their students learning.

(More details on creating Slowmation movies along with examples of movies and research can be found at www.Slowmation.com.au.)

Using Slowmation in Teacher Education: How and Why?

An integral part of the General Science Curriculum unit is to encourage preservice teachers to better understand what it means to teach science from a constructivist perspective. Such a perspective is based on the view that "learners interpret and interact with the physical world through their conceptualizations of phenomena" (Scott et al. 1994, p. 201). Drawing on the work of Piaget (1971) constructive research asserts that learners construct meaning rather than receive knowledge. Learners bring their own prior views of science to the classroom based on the ways in which they conceive of particular concepts and ideas (Driver 1994). McCombs (1996) contends that the learner plays an active part in the learning process. Teaching informed by a constructivist perspective relies on the teacher acknowledging and identifying the learners' alternative conceptions and creating experiences and opportunities for them to experience conceptual change. The expectation being that students might then develop deeper understandings of concepts as they move from their informal prior views towards the "accepted school view" of science (Fensham et al. 1994).

Such a view is regarded as cognitive constructivism (Fok and Watkins 2008). An alternative view, social constructivism (Brown et al. 1989), is based on the work of Vygotsky (1978) that asserts that knowledge is mediated through the social setting, the cultural experiences and interactions of the learner (Tobin and Tippins 1993). Viewing learning from this perspective, teachers encourage students to learn within their own social context (Fok and Watkins 2008), e.g. table groups, learning communities and the like. The difference between these two views of constructivism is that cognitive constructivism asserts that the individual develops complex representations of the world by building on their prior knowledge and experiences, while social constructivism asserts that learning occurs through participation with others in some meaningful way (Windschitl 2002).

Another view of knowledge, situated cognition positions learning as integrally situated in the everyday world of human activity (Henze-Rietveld 2006). From this perspective the context of the learning, the surrounds and situations are all important factors of that learning. Combining these three perspectives Clarke suggests that knowledge is personally constructed, socially mediated and inherently situated (Clarke 1995). Building on this view of learning for preservice teachers, the authors contend that it is important for preservice teachers to be given the opportunity to be involved in conversations about planning for teaching and, also after placement, of their teaching. Offering opportunities for rich conversations about teaching where each preservice teacher has had a common experience in the deeply contextual situation of the classroom has proved problematic in the past. Slowmation has been trialled as a vehicle for engaging preservice teachers in such rich conversations about practice. First as a pre-placement activity and second as a post-placement task where student generated Slowmation movies from their placement classrooms become the catalyst for those conversations.

In teaching Slowmation as a teaching procedure, the teacher educators grapple with the purposes of their own teaching. Discerning the differences between conceptual knowledge, procedural knowledge, organisational knowledge and content knowledge is not always easy for preservice teachers, and Slowmation also highlights the overlap between each. This gives rise to different ways through which preservice teachers view the purpose of using Slowmation in their own classrooms. Initially the task is procedural, but (as this paper will demonstrate) on returning to university after teaching Slowmation the discussion of the task highlights the conceptual and thus focuses attention on aspects of that which we would term pedagogical intent. Hence Slowmation in encouraging discussions of pedagogical intent works for the preservice teachers as a pedagogical scaffold.

Purpose of Teacher Education Unit

A common dilemma of teacher educators is meeting the preservice teachers' expectations while still fulfilling their own purposes for teaching. At the beginning of each semester, the authors have a clear purpose for their teaching. The authors hold a view that teaching is a life long journey for teachers where they develop a deeper knowledge about teaching and a greater understanding of students learning. Preservice teachers often enter the course viewing teaching as a set of skills they need to learn to deliver content to students. Such a transmissive view of teaching is based on a skill set best described as "tips and tricks" for teaching-specific content. The teacher educators have the view that university classes should be planned to engage preservice teachers in the discourse and practice of teaching for purposeful learning.

Understanding Knowledge Development of Preservice Teachers

To understand the progress of the preservice teachers, the authors draw on several models that describe preservice teachers' knowledge development. The first draws on the work of Fuller and Brown (1975), Calderhead (1987), Sitter and Lanier (1982), Guillaume and Rudney (1993) and Furlong and Maynard (1995) which all describe stages of development or concerns that preservice teachers experience during their teacher education course and these are strongly bound up in the supervised teaching placement. Katz (1972) identified four developmental stages for teachers in their first five years of teaching: "survival", "consolidation", "renewal" and "maturity". Fuller and Brown (1975) identified three stages of development for preservice teachers and labelled these "survival", "mastery" and either "routine based" or "consequence orientated". After mastery, Fuller and Brown (1975) argued that preservice teachers either became routine based and resistant to change or consequence orientated and concerned about their impact on their students and responsive to feedback of their teaching.

Sitter and Lanier (1982) found that although preservice teachers move through stages of development (Fuller and Brown 1975) they each experienced common concerns about self and identity, survival, teaching tasks, student learning, materials and curriculum development. They noted that while preservice teachers moved through developmental stages these concerns were dealt with all at the same time as ongoing issues during placement. Guillaume and Rudney (1993) identified six categories of concerns in their research: lesson planning and evaluation, discipline, working with pupils, working with supervisors, adjusting to their classrooms, working with others in the profession and transition from student to teacher. Similar to Sitter and Lanier (1982), they found that preservice teachers held such concerns as ongoing issues during placement. They noted that the nature of the preservice teachers concerns shifted as they gained experience. Guillaume and Rudney (1993) assert that preservice teachers' knowledge of teaching occurs as they develop more complex thought patterns as their learning and experience as teachers grows.

Katz (1972) also identified that preservice teachers as part of their learning move through several stages of concerns and identified them as: anxiety about teaching, learning about typical students, responding to student's individual needs, experimenting with their teaching and finally seeking a "more meaningful search for insight, perspective and realism" (p. 53). Furlong and Maynard (1995) noted that many researchers adopted stages of development and/or concerns, but make the point that preservice teachers do not move through these linearly but rather erratically. Incorporating many of the ideas discussed here Furlong and Maynard (1995) developed a five-stage model of knowledge development for preservice teachers: "early idealism", "personal survival", "dealing with difficulties", "hitting a plateau" and "moving on".

The authors have found the Furlong and Maynard (1995) model useful for understanding the knowledge development of their preservice teachers. In stage 1, preservice teachers have an idealistic notion of teaching and their role as teachers particularly before placement. Once they begin teaching they move into stage 2, where they are often overwhelmed by everything that is happening in the classroom and their main motivation is survival. As they become more experienced their sense is that they can survive and they move to stage 3 where they are concerned with the difficulties of teaching. As they learn to solve their difficulties and come to terms with the messy and complex nature of teaching they move into stage 4 where they start to feel and act like teachers. At this stage they often feel ready for teaching, but lack much of the thinking that experienced teachers have before entering the classroom. The final stage, "moving on", signifies a change from thinking about teaching for the task to thinking about teaching for student learning. After the second and final teaching round the authors seek to engage the preservice teachers in conversations that will not undermine their confidence but still make them see that there are things they need to learn and reflect on about their growing understanding of teaching.

The model proposed by Morine-Dershimer and Kent (1999) with its construct of facets of pedagogical knowledge has proved useful for the teacher educators to understand how the preservice teachers develop their understanding of pedagogical knowledge. It has also given the teacher educators a lens through which to observe their preservice teachers growth of pedagogical knowledge and the factors that influence such growth.

One issue that had been causing problems for the authors was that after the second professional placement in semester 2, preservice teachers returned to university in "state of plateau" (Furlong and Maynard 1995), believing that they were now teachers and there were just a few boring weeks of university left. It was often difficult to engage preservice teachers at this stage and they were often focused mostly on the assignments that were due. At this time the authors sought to engage the preservice teachers in rich conversations of their growing understanding of pedagogy that drew on their recent teaching experience and drawing on elements of Clarke's knowledge development (Clarke 1998). It was the authors purpose to have preservice teachers reconsider their "plateau" position (Furlong and Maynard 1995) and start to focus their attention of teaching on student learning in the "moving on stage" (Furlong and Maynard 1995).

The journey from novice to accomplished teacher is dependent on the preservice teacher's personal interactions with other preservice teachers, their teacher education course and those in their school placement. Their learning becomes unique to them and their experiences on placement. It also becomes more complex as they try to incorporate the context of their school with the ideas conveyed at university. Their learning can also be erratic as they move through their own stages of development and their concerns influence their teaching as their confidence shifts. Furlong and Maynard (1995) assert that part of the role of teacher educators in light of this is to help preservice teachers "see" teaching in conceptual terms to build more appropriate and complex understandings of teaching and learning. Copeland (1992) argues that preservice teachers need help to make sense of the classroom and to build practical professional knowledge.

Formulating the Study

The research design in this study draws on the work of Nicaise and Crane (1999) who conducted similar research on the knowledge construction of teachers enrolled in higher degrees at university when using hypermedia authoring tools. Like Nicaise and Crane, this Slowmation research design used a qualitative and descriptive approach drawing together data from multiple sources. The authors collected data by recording classroom presentations, conducting individual interviews of volunteer preservice teachers, and each recorded field notes while observing the other teach their class. In addition, following each lesson, the authors also met regularly to discuss their teaching (guided reflection) which offered more formalised individual data sets through reflections in their journals. The General Science Curriculum unit from which journal data and field notes were drawn comprised 38 preservice teachers in 2007 and 34 in 2008. The tape recordings of the presentations were transcribed and along with the field notes and journals maintained by the authors, the Slowmation movies the preservice teachers presented were analysed by identifying themes, then collating data thematically to identify similarities and differences. All quotations used are drawn from these data sets outlined above and are offered as indicative quotes of given situations.

The following presentation of the data is organised around the themes identified by the analysis of the data collected from classroom conversations and individual preservice teacher interviews. Often there were many similar comments made by different preservice teachers about the same theme. Where there were contrasting comments made these have been included. The authors have chosen indicative comments that illustrate the "typical response" of the preservice teachers understanding of the teaching or pedagogical issue being discussed.

Learning How to Teach Slowmation

In terms of structural outcomes (i.e. lesson organisation, planning and instruction), the preservice teachers reflected on the way Slowmation could be taught within the busy school curriculum they each experienced. They reported that Slowmation required several lessons preferably spread over several weeks and that it worked best when developed over about four, 50-minute periods with several days between Slowmation lessons for work outside of class time to be completed. It was useful to have a double period to take the photographs so that models and scenes did not need to be packed up and set up again. A typical program developed by the preservice teachers consisted of setting the scene in the first lesson, showing some examples, allocating topics and beginning the storyboards; students completed the storyboards between lessons one and two. The second and third lessons were generally used for making the models and taking the photographs. Between lessons three and four, students edited their movies and presented them back to the class in lesson four.

As is the case in secondary schools, the specific time period allocated to subjects imposes certain constraints which led the preservice teachers to consider how Slowmation could be incorporated into their teaching. A typical response from a preservice teacher:

> On reflection I wouldn't do such a big content area for the students' Slowmation movies…I would reduce the amount of content that had to go into their movies if I had to do it again. (PST1 Classroom presentation 2007)

This preservice teacher found that the scientific concept to be studied needed to be manageable or self contained; such as DNA replication or how electricity flows in wires. Concepts that had multiple parts or were overly complex led to difficulties in students completing their Slowmation movies in the allotted time and made it difficult for students to explain their understanding without over simplifying the movies. Several preservice teachers struggled with how to utilise Slowmation with the content to be taught, for example:

> I had Year 9 Forensics and Year 11 Biology, and for the life of me, my supervisor and I, couldn't think of anything to do with forensics that was suited to Slowmation. So I did it for animal behaviour in Year 11. Which didn't particularly suit the Slowmation either but it worked out ok. …
> [Later in the discussion] Yeah, it's like what could the tiny specific topics be that you could tell the kids would be appropriate [in forensics]. (PST2 Classroom presentation 2007)

An additional benefit that preservice teachers found in using this approach was that by covering several related scientific concepts relevant to a topic, Slowmation presentations worked well as a revision session for students. Preservice teachers began to recognise the complexity within concepts they were required to teach. These were not quite as simple and straightforward as they had initially imagined or the textbook had indicated.

> In all the movies we looked at the students showed DNA just as a helix, but we know that in the cell, the DNA doesn't exist as just a helix. They are more like, "squished" together, none of the ones we have looked at have shown this. And in my class when students made models of DNA they made just the helix and I don't think they ever thought how it was connected within the cell. (PST8 responding to a movie Classroom presentation 2008)

> I also had students ask me how do you get such a long thing (DNA helix) inside a tiny cell. (PST3 Classroom presentation 2008)

Over two different presentations the complex nature of teaching DNA arose and different issues about the complexity of this topic and how it was represented overly simplistically were discussed by preservice teachers. Sharing these ideas with fellow preservice teachers began a conversation about what needs to be taught about concepts, how much detail do teachers need to cover for students to understand, and when does more detail help and when does it confuse students?

The authors regard such conversations about how to teach using a teaching procedure, the timing and content of lessons and the complex nature of the content being taught as an example that the preservice teachers were "moving on" (Furlong and Maynard 1995) in their thinking about teaching. These conversations were also assisting the preservice teachers to reflect on their developing understanding of pedagogical knowledge. Using the Morine-Dershimer and Kent (1999) framework the conversations about structural outcomes in the classroom focused the preservice teachers thinking on "Instructional Models and Strategies" and how to best use Slowmation as a teaching procedure to improve their teaching.

The Nature of Learning with Slowmation

Organising Groups for Groupwork

Slowmation leant itself to group work and allowed several students to work together to meet a common goal. Most preservice teachers allowed students to choose their own groups, with various levels of success as this comment demonstrates:

> And for Year 9…their groups didn't work well, but for Year 10 the groups did work well together… (PST1 Classroom presentation 2007)

Another preservice teacher spent a lot time planning her groups which became a powerful learning experience for her.

> I grouped them, there were different ability levels and different personalities, so I grouped them and that really paid off. It got students to work together who didn't normally work together. (PST3 Classroom presentation 2007)

For this preservice teacher arranging the groups herself, considering the students who would work together and those who would not was important for how she came to understand student learning in groups through the Slowmation task. These conversations aided the preservice teachers involved in teasing out the complexities of students working in groups and whether the students should choose the group or the preservice teacher should.

Realising that the effort they put into their teaching was rewarded by improved student learning had a big impact for some of the preservice teachers who realised that extra planning of their teaching improved how the lessons went and learning outcomes for students. In discussing group work and how to best cope with students within their own context the preservice teachers were clarifying and making public their understanding of "classroom management and organisation" as a facet of their general pedagogical knowledge (Morine-Dershimer and Kent 1999).

Learning Science in Groups

With the students working in groups raised other issues. One that generated plenty of discussion was when a preservice teacher recognised students being busy and actually learning science are not necessarily the same thing:

> In terms of group work, I found that everyone could be working hard, but not doing very much about science. So in groups of 3, I found one person tended to be the director, someone would be the media player whiz, who was getting it all going on the computer and someone would be really involved in making the models. But they wouldn't really engage with the science at all. So you couldn't say that they weren't working because they were, but they weren't doing the sort of work that I wanted. So I don't know what the answer to that would be, because to get everyone to do an individual one [Slowmation] is too time consuming and resource consuming. (PST2 Classroom presentation 2007)

PST2 (above) grapples here with the nature of the learning that was taking place within the groups. The emphasis from her university classes was that Slowmation should not just be a fun activity, but be understood as a teaching procedure that could be employed in ways to genuinely benefit the science learning of her students. During her presentation when she raised this issue with her peers, good discussion ensued about whether or not the issue could be "solved".

> Was this much different if the teacher was out the front lecturing? (PST6 Classroom presentation 2007)

As a consequence, the preservice teachers involved in that discussion started to see teaching as messy and complex and that their teaching dilemmas did not necessarily have simple answers; rather they came to see a need to find strategies that worked for them in their own context. Importantly, as these preservice teachers began to see teaching as problematic, they also started to recognise that although they were familiar with Slowmation as a teaching procedure, that it was not something that was then simply applied in the same way all of the time. They had to be responsive to the content, their context, their students learning and be more sensitive to the ways in which their students interacted during learning episodes.

Good Teaching

Preservice teachers often grapple with the expectation placed on them about what is perceived as good teaching. Should they have quiet classrooms with students working in rows? The following comment is one example of a number of such comments that occurred during the presentations:

> They really got into the peer reviewing [of storyboards], and this was the day my supervisor from university came to observe my class. And I was getting really upset that they were so noisy because usually I had a really good feeling in the classroom. And I was getting upset that they were all talking. Then at the end the supervisor said they were all talking about the movies and [were] engaged. (PST4 Classroom presentation 2008)

The conversation opened up between preservice teachers and others added their thoughts and experiences. Does learning science require quiet classrooms? What is the level for appropriate noise? What if your level differs from your supervisor, even if the students are enjoying it and on task? More questions were raised and the conversation continued with one preservice teacher concerned about keeping students on task when some have been absent and the group gets behind.

> Keeping kids on task that is a big thing. And absenteeism, because one group had only 2 kids at the planning stage, 3 kids at the making stage, one kid did the written up piece. I think it changed the dynamics of the groups. (PST3 Classroom presentation 2008)

The conversations moved in this way, directed by the preservice teachers. The authors were engaged in setting up the movies on the laptop, so that the preservice teachers were involved in their own discussions not those directed by the teacher educators. Here the preservice teachers were making public their "personal beliefs" (Morine-Dershimer and Kent 1999) and drawing from their "personal practical experiences" (Morine-Dershimer and Kent 1999) of teaching Slowmation to inform their growing understanding of "personal pedagogical knowledge" (Morine-Dershimer and Kent 1999). Rather than seeing themselves in control of teaching as described by the "plateau stage" in Furlong and Maynard's model (1995) these preservice teachers are seeing that teaching is complex and the answers are not simple. This is representative of the "moving on" stage in Furlong and Maynard's model (1995).

Choosing the Appropriate Task for the Content

The preservice teachers became involved in discussions about how, when and why to use a teaching procedure in their science classes. They started to unpack their practice and began thinking carefully about when it was appropriate to use Slowmation. They felt that there were concepts for which Slowmation was more suitable and concepts for which it was not so applicable.

> I probably wouldn't do it with the one I did like Energy, Heat, Light and Sound one, [content areas are too broad]...When we saw the one in class [university class] the Mitosis/Meiosis ones I thought they were really powerful in picking up things [alternative conceptions], or DNA replication those were very powerful for me anyway so yeah those sorts of types I things [self contained topics]. (PST7 Classroom presentation 2008)

Recognising that there was content more suitable to Slowmation and content that was less suitable assisted preservice teachers to realise that there are pedagogical decisions about choosing to use teaching procedures and when they might be appropriate and for whom. The authors regard such preservice teacher reflection on their general pedagogical knowledge as building context-specific pedagogical knowledge (Morine-Dershimer and Kent 1999).

Identifying Students' Alternative Conceptions

Preservice teachers also came to recognise the pedagogical value of storyboarding—since it requires the maker to deconstruct the scientific concept and think about how the different elements "work together". However, in their classes it rapidly emerged that their students did not see the storyboards as important. Not surprisingly, they wanted to get straight on with making the models, taking photos and making movies. One preservice teacher commented:

> The students didn't want to do the storyboard but they learnt a lot from it and I could see their understanding. (PST1 Classroom presentation 2007)

Persevering with storyboarding was recognised as helpful in identifying students' alternative conceptions and offered an opportunity for the preservice teacher to discuss the science with their students:

> I checked their storyboards, and made sure they had science and commented, you know, have you got arrows to show where the forces are acting? (PST3 Classroom presentation 2007)

The storyboarding gave preservice teachers an insight to their students' understanding of the topic and this then directed their teaching as this comment reveals:

> So during the first class when we were doing the whole planning and everything, using the storyboards, I realised that a lot of the students really had no idea what geotropism, phototropism and that sort of thing actually meant. Which was really surprising because it had actually been covered with their regular teacher. So I had to spend most the class teaching how this worked and referring them to the text book. (PST10 Classroom presentation 2008)

After the topic had just been taught the preservice teacher realised that the students did not understand the work sufficiently to make their movies. From this the preservice teacher became aware of the importance for the teacher to monitor the learning of her students.

Understanding the difficulties of conceptual change in students was realised when students reworked their storyboards but included their alternative conceptions in the movies. This had been a difficult concept for the authors to have preservice teachers accept during university classes but was made clear in their discussion of their teaching.

> Like you [the other preservice teacher] said they seemed to understand it when they did the storyboard and explained it to me, but then they put that to the side and went on to the next activity which they thought was supposed to be fun. Hence their movie didn't reflect the understanding in the storyboard. (PST11 Classroom observation 2008)

The preservice teachers are talking their way to understanding of their developing pedagogical knowledge. Rather than viewing teaching as separate chunks of learning, they are identifying the importance of teaching Slowmation as a process, the purpose of the storyboard and how student learning is affected when they do not refer back to it. The preservice teachers agreed that the process of Slowmation was useful for helping them to identify students' alternative conceptions of a science concept. One preservice teacher commented:

> The major conception they had before they started, was they had the light and dark reactions mixed up [in photosynthesis], and didn't show how the process worked, but this was picked up in the storyboard process and I was able to help with that. I didn't have time to follow this up as it was the second last day [of the placement], so I cannot tell if that is what they really meant or [if] it is just how they represented it. But the Slowmation did reveal

to me areas that students didn't fully understand and where I needed to re-teach the ideas. (PST1 Classroom presentation 2007)

The student's work on their Slowmation movies gave the preservice teachers opportunities in which they could "see" their students' understanding and could monitor their learning.

From their experiences, these preservice teachers were able to identify common areas of alternative conceptions amongst their students and began to consider how to improve their explanation of this area of content in relation to their subsequent teaching.

I think the reason for that was, because they [the students] were basing their thinking [understanding] solely on a diagram they got out of the text book that was very, very, simplistic and they didn't really do much of the reading. And I didn't realise that until after I had seen their movie. …and I thought maybe I should have included that [in my teaching]. (PST3 Classroom presentation 2008)

One valuable aspect of the Slowmation process for the preservice teachers was that it (sometimes) helped them to become more aware of small aspects of individual understanding that needed clarification, this was a valuable reflection for their teaching. Several made similar comments indicating they were reflecting on their teaching given the alternative conceptions they had observed from Slowmation. A typical example follows:

One of the really good things I found out from them doing it [Slowmation] was, when you saw those little green snake things coming in, I hadn't realised, but the students thought that was the sugar-phosphate backbone of the DNA, and they thought it just miraculously appeared and climbed in there, and joined with the rest of the structure. They didn't realise that it was actually already connected to the bases. So that was a really good thing that I learned from them not understanding that I only would have picked up from Slowmation. So I realised if I taught that again I would have to be really clear on that bit, because obviously they didn't understand it from the way I had explained it. (PST4 Classroom presentation 2007)

She realised that teaching it the way she had the students had not learnt the concept as she had expected. In a similar way another commented:

(PST5 2008 commented about another preservice teacher's student's movie). When I taught Mitosis on my previous round I noticed they thought it was a process that went phase 1, phase 2, phase 3, etc. not that it gradually changed. They just think they [cells] divide and that's it, not that it's a cyclic process. [PST6 2008 responds to that with:] That's what we got at the end of our video, in fact we didn't even get two cells. We actually got, they've pulled apart, that's it. That's the end of the process. I think that's one of the things I would address, like with animation, and could show, the connection of the processes rather than just stage 1, stage 2, stage 3. (PST6 Classroom presentation 2008)

The value of presenting the movies to the class is demonstrated in the exchange above, where one preservice teacher recalls something she learnt about teaching this content. It adds to and illuminates the discussion about how students appear to understand this as single process when it should be a cyclic process. In this way sharing experiences of teaching and their developing knowledge of context-specific pedagogical knowledge (Morine-Dershimer and Kent 1999) helps inform each other and further extends that understanding.

Meta-Learning

Preservice teachers realised the importance of having their students talk about their movies as they played them to the class. By having their students explain their movies, the preservice teachers were able to clarify aspects that were unclear and probe their students' understanding:

> ...they had to go out the front and explain it, what was happening, when the movie showed each frame, otherwise it was quite unclear what was happening in some of them. (PST3 Classroom presentation 2007)

The value of classroom communication articulated here by preservice teachers and the accompanying discourse was identified as building preservice teachers general pedagogical knowledge. This way of accessing students' thinking about the science ideas was revealing to these preservice teachers in ways they had not previously recognised:

> So we turned one of the seeds upside down, so in one the root and the shoot were coming out of the top, and on the other the root and the shoot were coming out of the bottom. So they really had to think about what was happening, and what gravity has to do with it. (PST10 Classroom presentation 2008)

Presenting the movies to their peers clarified for the preservice teachers what their students did not understand or held alternative conceptions about, and what these alternative conceptions were. Moirne-Dershimer and Kent (1999) assert that it is this reflection on general pedagogical knowledge that assists the development of one's context-specific pedagogical knowledge. The authors regard engaging preservice teachers in discussion of their experiences in the classroom as encouraging reflection on general pedagogical knowledge, and hence as important for their developing growth towards expert science teachers.

Talking About Teaching

Slowmation became a process through which these preservice teachers could highlight other skills aside from their students' apparent content acquisition. They commented that some students were able to get involved and use skills that gave them new ways of engaging with the content:

> Here is my student who didn't really like biology, was slow to grasp the concepts etc., but in Slowmation he was king of the class, he knew how to do everything. I think that was a real good thing for him, he was achieving in biology for the first time. (PST2 Classroom presentation 2007)

One preservice teacher used her placement Slowmation tasks as part of her students' assessment for the unit she had taught. The issue of assessing group work later generated heated discussion at university, as the preservice teacher defended her rationale:

> I gave them a peer and self assessment sheet, based on their own self assessment. If all members of the group said they wanted the same mark then I gave them the same mark. Some other groups said they didn't think they had worked well together and they deserved individual marks. In this case I would give them individual marks based on how I thought they contributed to the group and what they reported on their self assessment sheet. (PST2 Classroom presentation 2007)

Another student also used Slowmation as part of the students' assessment.

> I did assess them. I used it as an assessable task, I assessed them on their contribution to research, their contribution to model making, their contribution to IT, quality of what they produced; not whether their models were any good, but whether their end movie conveyed the science correctly. (PST1 Classroom presentation 2007)

This preservice teacher wanted her students to realise that she valued all of the components of the process, an important step in her learning about teaching given that at least initially, preservice teachers tend to focus on the "end product" more than the process of learning. By sharing her ideas with her preservice teacher colleagues she raised issues about assessment to monitor the learning of her students and their understanding of scientific phenomenon.

Finding Freedom to Act

A few preservice teachers found it difficult to initiate lessons in Slowmation during their placement. Some of their school supervising teachers were reluctant to allow their preservice teachers to try Slowmation because they did not see that it would fit in with their term/unit plan. These supervisors viewed Slowmation as taking too much time away from teaching and felt that it did not achieve their particular needs. Other supervising teachers recognised Slowmation as a great activity but could not see the benefits of it for student learning or improving teachers' knowledge of their practice.

One preservice teacher had the following experience:

> I wasn't allowed to do Slowmation initially, my supervising teacher said it wasn't good and I wasn't allowed to do it, because they have all their lessons planned out for the semester. But then by the last week, she said, "ok you can do your Slowmation thing now". (PST3 Classroom presentation 2007)

The authors wanted the preservice teachers to explain the purpose and use of Slowmation to their supervisor to emphasise what the purpose was for themselves. Initially, pre-placement the preservice teachers were excited by Slowmation as a fun activity that students would enjoy. Post-placement they were aware of the difference and recognised that while Slowmation was a fun activity for students, it offered so much more to them as teachers.

> The supervisor thought it [Slowmation] was great. (PST10)
> Did he see it as a nice activity or as way to get look at what the students had been learning? (Author 1)

> I think he saw it as a nice activity really. Thought because we were using it as an assessment task he saw it as that way as well (a monitoring of learning tool) but he also saw that it took a lot of time and so maybe it wasn't as good for senior classes where there is a lot of content to cover. (PST10 Classroom presentation 2008)

This preservice teacher was not only able to discuss the difference between activity and teaching procedure, but recognised that her supervisor saw Slowmation as merely an activity where she recognised it as a valuable teaching procedure. She could see the value of the learning Slowmation promoted to make it more than an activity.

Though all preservice teachers managed to use Slowmation on their teaching rounds, for some it required considerable negotiation with their supervising teacher. Overall, these preservice teachers learnt new and interesting things about teaching by experimenting with this teaching procedure in their science classes during their placement.

Discussion: Developing Pedagogical Understanding

The preservice teachers came to see a variety of ways in which Slowmation could be used in the classroom. They came to recognise in rich conversations about their practice (Clarke 1998) how it could assist in the development of their practice by identifying what students knew and where they still had alternative conceptions, but also gave valuable insights into their students' learning. The authors regard the strength of the preservice teachers conversations came from the fact they had personally constructed meaning of using Slowmation, this knowledge was inherently situated in their own school and classroom context and was socially mediated in their post-placement presentations. Slowmation was valuable in encouraging the preservice teachers to reflect on all facets of their developing pedagogical knowledge. They found that Slowmation worked best when they had a series of lessons in which to complete the task and when the lessons were structured in a particular way for example: first lesson for storyboarding, second and third for taking photos and the final lesson for presentations. This demonstrated that the preservice teachers had engaged with classroom management and organisation knowledge that leads to a better understanding of pedagogical knowledge (Morine-Dershimer and Kent 1999). Additional time was required for working on Slowmation as homework (completing storyboards and editing the movies). The preservice teachers' experiences suggest that Slowmation movies were most effective when the concept chosen was small, self contained and easy to chunk and represent. This recognition by the preservice teachers about what Slowmation was suited for demonstrated during the class presentations they had been absorbed in thinking about purposeful teaching for student learning.

Slowmation worked well as revision for a topic; taking the major concepts of the topic and distributing them amongst the groups. These preservice teachers used Slowmation to identify their students' alternative conceptions and were able to offer feedback to their students about their learning in a variety of forms. Seeing their students' alternative conceptions led the preservice teachers to think carefully about the ways in which they had presented science concepts and to consider revising their teaching approach for future classes.

The preservice teachers realised that which Scott et al. (1994) noted, that when teaching from a constructivist perspective, even when the concept has been well taught with opportunities for practical activities and discussion, students may still formulate conceptions that are not in accord with that which the teacher intended or expected. Slowmation gave these preservice teachers the opportunity to see how their students had interpreted what they taught in ways that otherwise might not have been recognised. They were also able to consider how, when and why they would use Slowmation as a Teaching Procedure in the future and in other classes; aside from the one that they had used during their placement.

The many aspects of producing a Slowmation provided some great opportunities for these preservice teachers to vary their modes of assessment. They found that the storyboard, movie, script and presentation to the class could all be assessed as separate entities which meant that a large variety of skills could be assessed, essentially through one task. These preservice teachers also saw that their students were able to further their abilities in many other areas aside from "just the science". These areas included: computer skills; creative writing; group work and research.

Some of these preservice teachers found that they had to do a little convincing in order to persuade their supervising teacher to let them use Slowmation during their placement. In some cases they identified that the supervisor had viewed Slowmation simply as a fun activity that could be used to fill in time—something which the teacher educators had recognised as important to avoid. The teacher educators involved viewed this as a central issue within their pedagogy of teacher education (Loughran 2006) and further reinforces Appleton's (2002) concern about preservice teachers seeing teaching through the lens of "activities that work".

The preservice teachers returned to their teacher education program with new insights into how to use Slowmation as a Teaching Procedure. They incorporated Slowmation into their teaching and came to see benefits for their students' learning as it assisted them in coming to better understand the ways in which their students developed their science knowledge.

The value in Slowmation for the authors was not as a fun activity for their preservice teachers but rather as a shared experience that was very different for each of them due to the contextual nature of teaching and the content being taught. The conversations generated by showing artefacts of student learning via the Slowmation movies, were rich because they each had personally constructed their knowledge of Slowmation, it had been socially mediated during their presentation and situated inherently in their own context (Clarke 1998). Using Morine-Dershimer and Kent's Model (1999) the presentations represented a chance for preservice teachers to discuss classroom organisation (group work), instructional models and strategies (the lesson breakdown for Slowmation), engage in discussions about classroom communication and discourse to build general pedagogical knowledge. During the presentations the preservice teachers made public their personal beliefs through their personal practical experience of using Slowmation furthering their understanding of personal pedagogical knowledge (Morine-Dershimer and Kent 1999). The authors contend that this process coupled with reflection on the general pedagogical knowledge engaged preservice teachers in thinking about their context-specific pedagogical knowledge. For several preservice teachers the discussion of the content in terms of their own class, such as the DNA example above, demonstrated a beginning understanding of pedagogical content knowledge (Shulman 1987).

Conclusion

Slowmation had proved to be highly effective for the authors to focus the preservice teachers after their final professional placement. Not as a busy task, but rather to have them focus on the many facets of teaching and how these impact on student learning. It is through the discussion of Slowmation that the preservice teachers begin to reconstruct their pedagogical knowledge within the framework of Morine-Dershimer and Kent (1999) that the teacher educators have been assisting them to deconstruct for the semester, and form it into a coherent whole. The features of Slowmation that have proved important for informing the preservice teachers' pedagogical knowledge development can be readily identified by the authors. Key to this is the preservice teachers' shared experience. Even though they teach different year levels, different topics across different scientific domains and across different school sectors, the experience is familiar in ways that allow preservice teachers to engage with each other about their students' learning. The value of the shared experience cannot be underestimated here, as Clarke recognised (1998).

Another feature that has proved valuable in the discussion is the movies themselves. Preservice teachers were able to share student work that demonstrated their students' understanding and alternative conceptions of a topic, which has proved to be highly informative for them. Even though the preservice teachers do an alternative conceptions assignment in semester 1, often their strategy to aid in changing students thinking is to "tell" students or explain it better and then they will understand. It is when standing in front of their peers showing their movies that they vocalise their dismay that teaching students by telling does not change students' alternative conceptions and the realisation that conceptual change for students can be difficult to achieve. As a group they discuss the difficulties of conceptual change from a position of shared experience.

The features of teaching Slowmation lay the way for discussion of many of the teaching issues that the teacher educators have been raising with the preservice teachers all semester. Teaching Slowmation includes group work, using ICT in science, model making, representation of abstract concepts, and the movies give insights into the students' thinking of scientific concepts. Discussion of these aspects of pedagogical knowledge assist preservice teacher's development in this area as these discussions cover each feature of the facets of pedagogical knowledge (Morine-Dershimer and Kent 1999).

Slowmation has proved useful for the authors to move the preservice teachers on in their thinking and understanding of teaching from the plateau level (Furlong and Maynard 1995) post-placement. They recognised and acknowledged the problematic nature of teaching and that there were no "simple solutions". During their discussions of problematic nature of teaching the preservice teachers are articulating their knowledge of teaching by talking their way to understanding and how it impacts on student learning. In this way they are making sense of the classroom by seeing teaching not more simply but rather as complex and messy. As Copeland (1992) suggested with support preservice teachers can make sense of classroom as they build practical professional knowledge.

Finally, what has been significant for the teacher educators is not that Slowmation is an innovative and novel activity for preservice teachers to use in classrooms with their students, but that Slowmation has been used in a way that encourages preservice teachers through interpretative discussion of their experience to articulate and synthesise their developing understanding of pedagogical knowledge. Such discussions appear to place them well down the path on their journey to understanding the pedagogical knowledge they need to be an expert science teacher.

References

Appleton, K. (2002). Science activities that work: Perceptions of primary school teachers. *Research in Science Education, 32*(3), 393–410.

Brown, J. S., Collins, A., & Duguid, P. (1989). Situated cognition and the culture of learning. *Educational Researcher, 18*(1), 32–42.

Calderhead, J. (1987). The quality of reflection in student teachers' professional learning. *European Journal of Teacher Education, 10*(3), 269–278.

Clarke, A. (1995). Professional development in practicum settings: Reflective practice under scrutiny. *Teaching and Teacher Education, 11*(3), 243–261.

Clarke, A. (1998). Born of incidents but thematic in nature: Knowledge construction in practicum settings. *Canadian Journal of Education, 23*(1), 47–62.

Copeland, W. (1992). Clinical experiences in the education of teachers. *Journal of Education for Teaching, 7*(1), 3–16.

Driver, R. (1994). *Making sense of secondary science: Research into children's ideas.* London: Routledge.

Fensham, P. J., Gunstone, R. F., & White, R. T. (Eds). (1994). *The content of science: A constructivist approach to its teaching and learning*. London: Falmer.

Fok, A., & Watkins, D. (2008). Does a critical constructivist learning environment encourage a deeper approach to learning? *The Asia-Pacific Education Researcher, 16*(1), 1–10.

Fuller, F., & Brown, O. (1975). Becoming a teacher. In K. Ryan (Ed.), *Teacher education: The seventy-fourth yearbook of the national society for the study of education* (pp. 25–52). Chicago: University of Chicago Press.

Furlong, J., & Maynard, T. (1995). *Mentoring student teachers: the growth of professional knowledge*. London: Routledge.

Guillaume, A., & Rudney, G. (1993). Student teachers' growth towards independence: An analysis of changing concerns. *Teaching and Teacher Education, 9*(1), 65–80.

Henze-Rietveld, F. (2006). *Science teachers' knowledge development in the context of educational innovation*. Unpublished PhD thesis, Leiden University, Leiden.

Hoban, G. (2005). From claymation to slowmation: A teaching procedure to develop students' science understandings. *Teaching Science: Australian Science Teachers Journal, 51*(2), 26–30.

Katz, L. (1972). Developmental stages of pre-school teachers. *The Elementary School Journal, 73*(1), 50–54.

Loughran, J. (2006). *Developing a pedagogy of teacher education: Understanding teaching and learning about teaching*. London: Routledge.

McCombs, B. (1996). Alternative perspectives for motivation. In L. Baker, P. Afflerbach, & D. Reinking (Eds.), *Developing engaged readers in school and home communities* (pp. 66–87). Mahwah: Erlbaum.

Morine-Dershimer, G., & Kent, T. (1999). The complex nature and sources of teachers' pedagogical knowledge. In J. Gess-Newsome & N. Lederman (Eds.), *Examining pedagogical content knowledge: The construct and its implications for science education* (pp. 21–50). Dordrecht: Kluwer.

Nicaise, M., & Crane, M. (1999). Knowledge constructing through hypermedia authoring. *Educational Technology, Research and Development, 47*(1), 29–50.

Piaget, J. (1971). *Science of education and the psychology of the child*. London: Longman.

Scott, P., Asoko, H., Driver, R., & Emberton, J. (1994). Planning and teaching a chemistry topic from a constructivist perspective. In P. J. Fensham, R. F. Gunstone, & R. T. White (Eds.), *The content of science: a constructivist approach to its teaching and learning* (pp. 201–220). London: Falmer.

Shulman, L. (1987). Knowledge and teaching: Foundations of the new reform. *Harvard Educational Review, 57*(1), 1–22.

Sitter, J., & Lanier, P. (1982). *Student teaching: A stage in the development of a teacher or a period of consolidation!* Paper presented at the Meeting of the American Education Research Association, New York, March 19–23.

Tobin, K., & Tippins, D. (1993). Constructivism as a referent for teaching and learning. In K. Tobin (Ed.), *The practice of constructivism in science education* (pp. 3–22). Hillsdale: Lawrence Erlbaum.

Vygotsky, L. (1978). *Mind in society: The development of higher psychological processes*. Cambridge: Harvard University Press.

Windschitl, M. (2002). Framing constructivism in practice as the negotiation of dilemmas: An analysis of the conceptual, pedagogical, cultural, and political challenges facing teachers. *Review of Educational Research, 72*(2), 131–175.

Chapter 16
Teaching Science in Informal Environments: Pedagogical Knowledge for Informal Educators

Lynn Uyen Tran and Heather King

Introduction

In this chapter, we discuss the knowledge required and applied by educators working in environments other than schools. Some such educators may have originally trained as classroom teachers, many others arrive at the role from a range of circuitous routes. Either way, their responsibilities in these non-school environments are distinct and different from their counterparts working in formal sector classrooms. In this discussion, therefore, we use the term "educator" to refer to paid staff working in non-school, or "informal" environments whose primary task it is to provide a range of educational programmes, activities and other experiences for the users or visitors of the particular space.

Rennie (Chap. 2) argues that in terms of teaching outside the classroom, a different interpretation of content knowledge and pedagogical practice is required. Furthermore, she explores a more interdisciplinary view of the curriculum, which can serve both the need for disciplinary knowledge and the need for students to be able to apply their learning outside of school. We consider Rennie's arguments as we explore the nature of knowledge required by educators, for we note that out-of-school settings, and the opportunities they afford, are inherently interdisciplinary. By examining the nature of their knowledge in this way we seek to develop a framework for understanding, guiding and enhancing the practice of educators. Thus, just as it has been argued and demonstrated that the professional and practical knowledge of schoolteachers comprises an important factor in shaping the nature and quality of student learning in schools (Abell 2007), we argue that the study of educators' knowledge is important for understanding and enhancing visitor learning in informal environments.

L. U. Tran (✉)
Lawrence Hall of Science, University of California at Berkeley, Berkeley, CA, USA
e-mail: lynn.tran@berkeley.edu

D. Corrigan et al. (eds.), *The Professional Knowledge Base of Science Teaching*,
DOI 10.1007/978-90-481-3927-9_16,

Science Pedagogy in Informal Environments

Informal science environments that are "intentionally designed for learning about science and the physical and natural world" (National Research Council (NRC) 2009, p. 127) offer visitors of all ages the opportunity to learn about, talk about, and engage in a range of science concepts and issues. These informal environments include museums, science centers, aquariums, zoos, and botanic gardens. Some researchers describe such environments as "free-choice" (Falk and Dierking 1992), noting that their non-evaluative nature promotes and nurtures learning. Others have argued that free-choice environments serve to foster visitors' intrinsic motivation to learn (Csikszentmihalyi and Hermanson 1995).

Educators working in informal environments do different types of work and have many responsibilities (Tran 2008). Their primary role is to act as a mediator between the visiting public and the objects within the environment that represent particular areas of knowledge and culture. As such, educators develop, coordinate, and implement programmes for schoolchildren, families, teachers, and the general public (Bailey 2006; Brüninghaus-Knubel 2006). They are involved in designing and developing exhibits (Bitgood et al. 1994; Roberts 1997), and in some cases, in conducting research on visitors' experiences at those exhibitions and programmes. In addition, they create and nurture relationships with community groups in order to attract visitors and make their institutions accessible, relevant, and inclusive of the people they serve (Henry 2006).

The teaching of science in informal environments is not organized according to a visitors' grade level, age, or ability, but instead by the ways educators interact with their learners. For example, educators engage in casual interactions with visitors at exhibits but also lead more structured and organized programmes for school groups and the general public (Tran 2008). It is important to note that both types of interactions are typically one-off experiences. In general, they last for only a few minutes, although some programmes may extend over several days. Furthermore, such interactions regularly comprise different configurations of learners from single-aged school groups to multi-generational families. In addition, the teaching is not generally compartmentalized into scientific domains, but may be better described as interdisciplinary and situated within real-world examples. For instance, in an aquarium, an educator may teach at an exhibit or in a class about how ocean water circulates (and thus explore aspects of physics), while doing so within the context of other concepts such as trade routes (more usually covered in social studies, history, or economics), the transport of organisms (a feature of biology), the transfer of heat and energy (concepts in meteorology, climatology), and movement of nutrients (a component of both biology and physics). Finally, the teaching is not solely designed to abide by government standards or curricula, but is more likely to be driven by the nature of the objects and resources available, and the interests and expertise of the educators responsible for developing and implementing the programmes. While most institutions align their educational provisions to standards in order to accommodate the needs of schoolteachers, most educators also emphasize the importance of providing experiences that support cultural or affective agendas (Schauble et al. 1996).

Review of Educators' Practice

Of the small number of studies that have previously examined the practice of educators, many have been disparaging in their assessments. For example, Cox-Peterson et al. (2003), in their observation study of the tours conducted by docents (volunteer educators) at a natural history museum in California, concluded that the tours tended to be lecture-oriented, and that the science content was presented in a didactic or narrative style. They also found that the activities were docent-directed, rather than learner-centered. Tal and Morag (2007), in their observation study of educators in four natural history museums in Israel, found that educators frequently used scientific jargon with limited explanation, and posed questions that promoted lower-order thinking only. These authors concluded that in planning and leading their educational programmes the educators essentially followed a knowledge-transmission model of teaching, which resulted in limited opportunities for student learning. They asserted that:

> Unlike the typical science class in school, where the teacher follows a curriculum that considers the students' previous knowledge, and presents new ideas with a suitable hierarchy, pace, cognitive demand, the museum lessons we observed were quite often overwhelming for the amount of content being covered. (Tal and Morag 2007, p. 757)

Some researchers, however, have presented a far less negative description of educators' practice. In the United States, Tran (2003) made detailed observations of educators teaching and in combining her findings with a series of stimulated recall interviews found that educators employed a variety of strategies in order to facilitate visitors' learning, and were highly receptive to visitors' needs. In a further study, Tran (2007) described the ways in which educators demonstrated flexibility in the design of their educational programmes and made adjustments quickly in order to respond to the needs of their learners. Such adjustments were both practical, for example, editing an activity to accommodate late arrival, and intellectual, for example, modifying the depth and details of content matter to be in keeping with the supposed ability of students gauged from responses to initial questions and comments.

Piscitelli and Weier (2002) also noted that educators employed a variety of purposeful techniques in order to enhance children's learning. In describing the manner by which educators facilitated young people's engagement at an art exhibition in Australia, they reported the following:

> Initially, [educators] used casual conversation to develop rapport with young visitors and elicit personal stories...A range of specific interaction strategies then came into play as [educators] endeavored to expand children's repertoires of experience and their knowledge, using the authentic objects and materials available to be explored and manipulated in the exhibitions and studio environment. [Educators] combined non-directive, scaffolding and

> directive teaching-learning behaviors in isolation during episodic dialogues with visitors, or as part of longer, carefully guided learning experiences. (Piscitelli and Weier 2002, p. 139)

King (2009), in her study of educators at a natural history museum in England, identified a variety of strategies used by educators in their verbal interactions with visiting school groups. Educator "moves" comprised a variety of utterances that served to promote responses from students and shape the wider exchange. These included prompts to describe specimens, to make comparisons or to speculate on a specimen's provenance or classification; the technique of repeating a student's comment to highlight its significance to the wider group or, alternatively, to signal that the comment was in some way incorrect; the re-voicing of student's contributions to "tidy" them up or to imbue them with appropriate scientific language; and the technique of emphasizing the epistemological bases of contributions. Other less common moves included modeling the process of enquiry or engagement; creating a shared experience by using "we" statements; and seeking to inspire learners with the breadth and range of objects in the museum. In analyzing the impact of the various strategies, King identified the techniques that were effective in promoting visitor learning—primarily those that served to emphasize the role of evidence—and those which appeared to impede or limit visitor learning, such as the non-selection of visitor contributions and the withholding of answers to visitor questions.

King's, Piscitelli and Weier's, and Tran's investigations have described and categorized a broad range of strategies used by educators. Their findings also point to a body of knowledge applied by educators, although King (2009) notes that few of the educators she studied were consciously aware of the various strategies they used or their relative efficacy. If this finding is relevant to the field as a whole, it suggests that the body of knowledge is neither recognized nor shared by educators working across various institutions and settings. Without a shared knowledge base underpinning practice, it may be argued that the pedagogical support provided by educators in the informal sector is inherently compromised (Tran and King 2007). Furthermore, a lack of an explicitly articulated body of knowledge raises concerns as to whether the field can become a profession (Abbott 1988; Freidson 1994; Larson 1977) and further develop its practices.

A Knowledge Base for an Emerging Profession

A profession is identified by the use of a well-defined body of knowledge and skills necessary to provide a service to society that is not offered elsewhere (Abbott 1988). Individuals within the profession claim exclusive access to this knowledge and skills through qualifying examinations coordinated by professional associations, which are themselves independent of individual institutions where the work takes place. In so doing, jobs are not disparate positions in single organizations created by management for their own purposes, but are opportunities in careers that extend across organizations. Through the educational process, new members are introduced to the technical language and inculcated into the culture of the profession. Moreover, they are equipped with a body of knowledge and skills that are theoretically grounded, thus allowing for a certain degree of abstractness in the description and application of the work. As a result, the profession may be versatile, and adapt to changes over time and to needs in society (Abbott 1988; Freidson 1994; Larson 1977).

Educators in informal environments have been in the process of professionalizing for many decades, but they have yet to make explicit a shared body of knowledge that underlies their work (Tran and King 2007). During the 1970s and 1980s, there were extensive discussions among educators in the United States regarding their professional value and the types of knowledge and skills they should have in order to do their work (Nichols 1992; Nichols et al. 1984). In 2000, Yellis (2000) discussed the field's revelation for the need for practice to be informed by theory and research. Despite these discussions, practitioners remain concerned about their relevance as professionals (Scott 2006), and how to prepare the next generation of leaders in the field (Nolan 2009). The task of identifying knowledge for the field is not straightforward: professions such as engineering and school teaching, continue to study and develop their knowledge base and technical language (Carlsson 1999; Hargreaves 2000; Hiebert et al. 2002), while others, such as journalism, fail to do so (Aldridge and Evetts 2003). In the case of schoolteachers, their knowledge is often held tacitly (Cochran-Smith and Lytle 1999). Furthermore, as Hargreaves (2000) notes, the explicit application of knowledge and research from the fields of sociology, psychology, cognitive science, and education to everyday classroom practice is limited. Fortunately, a considerable school-focused research community exists whose aim it is to study the knowledge base and practice of teachers, and to use evidence-based research to inform and guide future practice. Indeed, a key focus of this research is to make explicit the nature of a teacher's professional knowledge and judgment—the way they make subject matter understandable based on a strong understanding of that subject matter, the processes of learning, and varied methods of teaching.

In contrast, the informal sector lacks an established focus on, and rigorous research exploring, the professional practice of its educators. Furthermore, while many have proposed numerous lists of knowledge and skills that educators should possess, they have not grounded their proposed knowledge base upon research and theory on learning and teaching in informal environments (for examples see Fines 1984; Jackson-Gould 1992; Rutowski 1992). To address this shortfall, and to highlight the existence of the specialist knowledge of educators in informal environments, we reviewed the limited research and theory on learning and teaching in informal science environments, and identified core components from which an educator builds his or her practice (Tran and King 2007). In establishing the theoretical basis of each component, we followed the sociological arguments regarding the nature of professions (Abbott 1988; Freidson 1994; Larson 1977), which contend that it is only by grounding practice in theory that workers are able to operate at a more abstract and non-institution specific manner with the result that the occupation may be better described as a profession, and the workers as professionals. We briefly describe these components below.

Core Knowledge for Practice

A knowledge of *context* pertains to an understanding of the ways in which the physical, social, and temporal dimensions of informal environments interrelate. The physical context of the institution can be novel and awe-inspiring, which can both promote and impede learning (Carson et al. 2001; Falk et al. 1978; Griffin 1994; Orion and Hofstein 1994). The institution is also a social space where members of the community come to engage with science. In order to support learners, educators need to recognize and respect the different ways in which various groups respond to the environment. In addition to acknowledging the particular social dynamics of visiting groups—similarly aged school parties, multi-generational families, couples, and adult peers—educators must also attempt to cater for the learning needs of the individual. In this way, while most interactions with educators generally last for a short period, they must nonetheless create the opportunities for learners to engage with the objects and content for as long as they want by building understanding and inspiration to visit again or visit other related institutions.

A knowledge of *choice and motivation* refers to the need to recognize the way in which learners are free to engage in topics and materials in informal environments that interest them, thus nurturing their intrinsic motivation for learning (Csikszentmihalyi and Hermanson 1995). Research and theory in psychology show that people are more able to attend to and grasp the importance of an intrinsic goal for their learning when they feel free to decide for themselves to learn rather than feeling forced to do so (Deci and Ryan 2000; Vansteenkiste et al. 2004). Thus, while educators may design the layout of the space, decide what and how objects are displayed, and offer pre-planned educational opportunities, they must also be sensitive to the manner by which visitors use the informal environment in their own way and at their own pace, and are driven by their own intrinsic motivation to learn. In addition, educators must also be conscious of potential influences from learners' cultural backgrounds. For example, Iyengar and Lepper (1999) found that children from cultures where members are more interdependent, such as East Asian cultures, are more motivated to engage in activities when choices are made for them by significant others. Their study challenges the generalized assumption of personal choice and motivation that may be based on Anglo-American culture, where individualism is highly valued and engrained. In short, in supporting visitors, it is important that educators understand how learners' motivations shape their experiences. In this regard, it is important that educators are confident in amending their interactions in order to follow and build on the interests of learners.

A knowledge of *objects* recognizes that informal environments are characterized by the special things—the specimens, artifacts, exhibits, and landscapes—that they possess. Such objects are also the reason why visitors go to such environments

(Leinhardt and Crowley 2002). To mediate a visitor's experience of an environment, educators need to unravel the complexities inherent in an institution's collection of objects and help individuals to find points of personal connection. Thus educators must select from a range of interpretations to best suit their understanding of visitors' needs. In so doing, they may address the provenance of the object, its social history or scientific significance, or they may simply encourage visitors to observe an object and appreciate it for its own sake or aesthetic value. In this way, an understanding about the nature of objects is linked with content knowledge.

A knowledge of *content* includes what we know and how we know what we know. With regards to science, this knowledge component requires that educators understand the concepts and ideas of science as well as how we know and generate this scientific knowledge—that is, the nature and process of science. This understanding is necessary as it enables educators to mediate the objects on display and furthermore explain aspects of the specific disciplines such objects represent. In addition, such knowledge allows educators to indicate the salient features of particular objects that may otherwise have gone unnoticed by the visitor. With respect to objects housed in institutions, educators also need to understand the epistemological basis of their selection for display—the reasons why they are significant to the discipline and broader society. In developing a deep knowledge of content, educators will be able to be flexible to visitor's interests and choices and thus further enhance their experience.

A knowledge of *learning* refers to an understanding of how people learn, and is essential for interpreting the actions of learners and for guiding the ways in which one may best mediate the educational content of informal environments. Visitors to informal environments comprise a range of ages, social groupings, economic status, cultural identities, and experiences; grounding their practice in theoretical models of learning will thus allow educators to support these diverse characteristics and needs of their learners. Learning is a contextualized process of making sense of experiences in terms of extant knowledge within social and physical contexts over time (Rennie and Johnston 2004). Hein (1998, p. 179) argued that "visitors make meaning in [informal environments], they learn by constructing their own understandings." This perspective further emphasizes the informal educator's role as the interface between the institution's construction of knowledge and the visitors' ways of knowing (Rice 2000). The challenge, then, is for educators to support visitors' meaning making when they neither know the visitors nor their previous experiences. In such situations, educators need to rely on their previous experiences and build from a body of case examples, from which they may combine a range of common conceptions, age-specific language, and the particular set of skills required to transform knowledge into content for mediation.

The final knowledge component relates to the need of an educator to employ manifold techniques to communicate with visitors. We refer to this as a knowledge of *talk*, but recognize that this knowledge combines both verbal and non-verbal actions. The inclusion of *talk* concurs with the sociocultural perspective on learning whereby meanings are rehearsed and made explicit as a result of verbal interchanges and interactions between people as a process of becoming internalized by the individual. In face-to-face interactions, museum educators use a variety of verbal strategies to structure or scaffold conversations (King 2006); in organized events, when the visitor's voice appears less dominant (Cox-Peterson et al. 2003), the educator's role is to encourage private mental conversations and again he or she will employ a range of strategies to do so (Tran 2007).

Pedagogical Knowledge for Informal Science Educators

We have argued that the six knowledge components are foundational elements in educators' practice. Here we describe the ways in which educators may combine such elements and draw on them in their practice, and in so doing we propose a set of categories which define an educator's pedagogical knowledge. The categories are: (a) orientations towards facilitation, (b) knowledge about the affordances of objects, (c) knowledge of visitors' learning and understanding of science, and (d) knowledge about facilitation strategies.

Educator's Orientation Towards Facilitation

This category for educators' pedagogical knowledge refers to their views on the content and objects of their informal environments, and how these may be mediated. It is driven by their goals for visitors' experiences in informal environments overall, as well as in specific programmes and interactions, and is influenced by the perspectives on learning advocated by the institution and held by the educators.

In reviewing the literature on educators' practice we note that some educators do not regard visitors' conceptual learning of specific pieces of content to be a key aim. For instance, while Castle (2001) and Tran (2003, 2007) affirmed the educative role of informal educators, both authors also noted that visitor acquisition of new knowledge and skills were not the primary foci for the majority of educators. Rather, Castle (2001), in Canada, and Tran (2003, 2007), in the United States, reported that educators cited the need to establish the experiences as memorable incidences for learners as their main priority. In short, the educators sought to promote an affective experience over a conceptual one. Thus Castle noted that for several of the educators in her study, a key aim was to develop a strong rapport with visitors, with the view that "if they like me, they will like the museum" (2001, p. 6). Interestingly, Tran observed (2007) that in prioritizing the affective goal over the conceptual, educators often taught in a way that cut short learners' comments and contributions in order to make time to complete more memorable activities. As an explanation for the educators' focus on the affective, both authors independently suggested that the educators viewed a school trip to an informal environment not as a one-time event, but as part of a continuum of potential informal learning opportunities lasting well beyond school and childhood.

In considering the nature of educators' orientations, we note that the notion of discovery learning wherein the learner "discovers" concepts for him or herself through the phenomenon-based exhibits on display, appears to be predominant across the field (Hein 1998; King 2009). Unfortunately, the notion of discovery learning in informal settings is often interpreted to mean that the actions of an educator are not required (King 2009). For example, Rahm noted that the National Center for Atmospheric Research "shares a similar philosophy to the Exploratorium in that visitors' active interactions and experiences with the exhibits are emphasized. *Museum staff is only supposed to 'interfere' if asked to do so*, or if visitors appear to be in need of help to make the exhibits work for them" (Rahm 2004, pp. 226–227, emphasis added).

In perceiving the tenets of discovery learning in this way, we note that an educator's orientation towards facilitation may in fact limit the efficacy of their practice. Without guidance, a learner, and children in particular, may not be able to make sense of an object or piece of content and potentially leave an interaction with an incomplete or incorrect understanding of the intended concept (Crowley et al. 2001; Grandy 1997; King 2009; Klahr and Nigam 2004). If a discovery learning orientation to facilitation is indeed the most prevalent among educators, it is clear that this aspect of an educator's pedagogical knowledge base would benefit from greater reflection and further professional development.

Educators' Knowledge About the Affordances of Objects

This category for educators' pedagogical knowledge refers to an understanding of the social, cultural, historical, and scientific concepts represented by the objects in the informal environment, and an understanding of how to use the objects as the primary vehicle for communicating the content. As Eberbach and Crowley (2005) and Hohenstein and Tran (2007) have noted, objects encourage different types of learning conversations and provoke affective connections between learners and the knowledge represented (Leinhardt and Crowley 2002; Macdonald 2004). Indeed, it is through objects that teaching in informal environments takes an interdisciplinary approach, though what content is communicated is dependent on who chooses the object.

We (Tran and King 2009) have observed that, in organized interactions, educators are confident in choosing the objects that provoke affective connections and promote conversations about the scientific concepts they intend to teach. For example, in one classroom-based lesson that we observed, the educator was clearly sensitive to the novelty of both the context and the objects for her learners. In discussing her lesson that included many taxidermy specimens the educator explained that she informed the students of what they were about to see:

> I always prepare the children for what they're going to see and hear. From my experience I've found that the children can be a bit disturbed by seeing stuffed animals and also they

> can want to grab at them. And so I do an introduction so that before the children enter, they're prepared.

From this comment, it was clear that the educator was cognizant of the value of these objects for learning, but also understood the potential negative effects of a novel experience and how this may be assuaged by the appropriate form of communication.

In the casual interactions we observed it was usually the learners that chose the objects, and as a result the educators had the harder task in finding relevant connections to explain the scientific concepts. However, it was also clear that many educators were adept at this task. For example, in a gallery exploring a British city's industrial heritage we observed an educator initiating a conversation with a visitor who stood in awe of a full-size working steam engine. The educator asked the man if he knew the historical significance of the engine, a question that subsequently led to an animated discussion about the role of steam-power in the Industrial Revolution. In this instance, the educator used her knowledge of the object, together with the visitor's apparent interest in the object to develop and sustain a learning experience.

Educator's Knowledge of Visitors' Learning and Understanding of Science

This category for educators' pedagogical knowledge refers to the knowledge educators possess about visitors, which helps them to facilitate visitor exploration and learning of the scientific knowledge that is represented in the institution. It comprises two parts: knowledge of the nature of visitors' learning and knowledge of visitors' understanding of science.

Knowledge of the Nature of Visitors' Learning: This subset of knowledge refers to educators knowing how, when, and where learning takes place in informal environments, and how to support those instances when they occur or to create them when and where they are needed. In particular, it requires an understanding of the interrelationships and implications of the physical, social, and temporal contexts of learning in informal environments. For example, Piscitelli and Weier (2002) have described how educators' responsiveness to these contexts allowed them to promote meaningful experiences for visiting children by designing particular programmes. Thus in an organized programme studied by Piscitelli and Weier the educators made both novel and familiar objects available to their learners, and arranged the space to allow for learners to engage individually, in pairs, and in small or large groups. During their interactions with learners, they encouraged children to work with peers and adult chaperones, offered learners the opportunity to lead explorations, and responded to learners and their ideas both by giving learners access to the objects and talking with learners as they guided them through critical analysis of the material on display. If, on the other hand, educators fail to acknowledge the import of the learning context, visitor learning will be hindered. For example, Tal and Morag (2007) noted that in instances whereby educators disregarded the importance of providing students with time to explore freely in social groups and to make sense of their environment, the educators' practice was found to be lecture-oriented, vocabulary laden, and asocial—hardly features associated with effective learning.

A Knowledge of Visitors' Understanding of Science: This aspect of knowledge refers to an acknowledgement and valuing of learners' prior knowledge and experiences, as well as knowing and understanding learners' possible alternative (naïve, pre-, mis-) conceptions. It also requires educators to be aware of the science topics and concepts in their institution that learners often have difficulty learning. This category follows a constructivist approach to learning science in the sense that it recognizes the role educators and experiences in informal environments play in building visitors' knowledge of, and interest in, science. It also broadens the scope of learning outcomes that have been most commonly espoused by educators and informal environments, from predominantly affective (Castle 2001; Tran 2007) to include other aspects of science learning including the use of scientific language and tools, engagement in the design and physical processes of experimentation, and the generation of excitement and interest in the natural world (NRC 2009).

While it may appear nigh on impossible for educators to gauge and then build upon a learner's prior experiences given the short interaction time, Tran (2007) has noted that veteran educators are able to modify their interactions to suit the needs of learners. For instance, because educators teach the same programme to a range of grade levels, they readily amass a collection of comments reflecting varying levels of comprehension, which allows them to compare one learner's comments with comments made by other learners from previous interactions. Tran's educators utilized such comments, whether self-initiated or elicited by educators' questions, to guide the way they responded to subsequent learners' prior experiences and levels of understanding, resulting in adjustments to the depth and detail of the lesson content accordingly. However, as Tran observed, this practice could also result in educators curtailing learners' exploration and processing of content because they assumed from previous learners' comments that the current participants would not or could not understand particular material. Again, this aspect of educators' pedagogical knowledge needs to be further developed to ensure that educators offer the most effective learning support possible.

Educator's Knowledge About Facilitation Strategies

This aspect of pedagogical knowledge addresses the ways in which educators support learning in the context in which they teach. It is reflected in the ways educators use objects and talk with learners about particular science content. Again, the interrelationships between the physical, social, and temporal context of informal environments are important here as they influence the manner in which educators talk with learners. Thus, in supporting the diverse agendas, interests, experiences, and abilities of various learners, educators need to respect a group's social dynamics while also acknowledge the needs of individuals. They must make judgments about what a group of learners are able to understand (by making assumptions about their age, background, or education), and then communicate an appropriate level of content.

In a recent study, we observed educators employing discursive strategies in different ways according to the type of interaction (Tran and King 2009). For instance, in an auditorium show where learners ranged in age from young children to senior citizens, the educator deliberately asked a wide span of questions from low- to high-order thinking, and then selected respondents according to perceived ability levels. In this way, he was clearly supportive of his learners' abilities, and in looking for a willing respondent, was also sensitive to learners' confidence to participate in front of strangers. In contrast, in the gallery interpretation that we observed, the learners typically knew one another, and in cases with families and school groups, were in multi-generational groups. The educator in this instance explained to us that in order to accommodate the disparity in knowledge and ability of her learners in a more intimate interaction, she sometimes had two ongoing conversations simultaneously.

In a separate study on discursive strategies King (2009) found that educators often used dramatic and exclamatory talk in order to persuade visitors of the intrinsic value of an object. For example, in facilitating a learner's engagement with a meteorite, an educator was observed to speak in a very excited manner, stress particular words, and act as if in awe of the specimen. The educator was observed acting in a similar manner on several occasions suggesting that his knowledge of facilitation strategies centered on the use of enthusiasm and predetermined scripts.

In sum, and from the examples above, it can be seen that educators combine the core knowledge components in different ways resulting in a pedagogic knowledge base appropriate to the informal environment. However, as is also clear from the examples, not all educators make exemplary use of this knowledge base when working with the public. Indeed, we note that educators may either support or hinder learning depending on their ability and confidence in applying the knowledge components.

Next Steps: The Need for Further Examples of Educator Practice

In reviewing educators' practice, we have identified categories of pedagogical knowledge that draw upon a body of foundational knowledge. Furthermore, we have found that while educators may use this foundational knowledge in their practice, such knowledge is often tacitly held. We argue that given the growing attention to learning science in informal environments (NRC 2009) and the growing recognition of the role of informal educators (Hooper-Greenhill 1991; Nolan 2009; Scott 2006), practitioners now need to develop, share, and promote an explicitly recognized knowledge base to underpin and justify their pedagogical practice.

As a first step, we suggest that more examples or case studies of educators applying their knowledge in practice—in a range of settings and across a range of activity types—are needed. By studying exemplar cases of practice in this way, we believe that educators will be able to enrich their pedagogical knowledge by gaining a greater understanding of the interrelationships between the components of foundational knowledge. Furthermore, by examining the nature of exemplar cases, educators will have the opportunity to observe instances of highly skilled practice. Moreover, they will be able to reflect on instances of problematic practice in terms of supporting learning and amend their own actions appropriately. In studying real-world examples, educators will, we hope, gain the confidence to enact (or avoid) similar interactions in their own engagement with visitors. In creating a repertoire in this way, the range of practices employed by educators may be shared and vetted by fellow professionals. In turn, this will prevent any idiosyncratic practices of individuals or individual institutions from dominating the field. In short, we suggest that a body of examples of educator practice, discussed and critiqued by both the field and academic colleagues, will serve to inform and illustrate curricula for pre-service and in-service professional development. As a result, informal educators will be able to claim the professional respect they deserve and gain the pedagogical knowledge base they require in order to best serve learners.

References

Abbott, A. (1988). *The system of professions: An essay on the division of expert labor.* Chicago: The University of Chicago Press.

Abell, S. K. (2007). Research on science teacher knowledge. In S. K. Abell & N. G. Lederman (Eds.), *Handbook of research on science education* (pp. 1105–1150). Mahwah: Lawrence Erlbaum.

Aldridge, M., & Evetts, J. (2003). Rethinking the concept of professionalism: The case of journalism. *British Journal of Sociology, 54*(4), 547–554.

Bailey, E. B. (2006). Researching museum educators' perceptions of their roles, identity, and practice. *Journal of Museum Education, 31*(3), 175–198.

Bitgood, S., Serrell, B., & Thompson, D. (1994). The impact of informal education on visitors to museums. In V. Crane, M. Chen, S. Bitgood, B. Serrell, D. Thompson, H. Nicholson, F. Weiss, & P. Campbell (Eds.), *Informal science learning: What the research says about television, science museums, and community-based projects.* Dedham: Research Communications.

Brüninghaus-Knubel, C. (2006). Museum education in the context of museum functions. In P. J. Boylan (Ed.), *Running a museum: A practical handbook* (pp. 119–132). Paris: ICOM—International Council of Museums.

Carlsson, B. (1999). *The knowledge base in the engineering sector.* Paper presented at the OECD/CERI Seminar: Measuring knowledge in learning economies and societies, Washington.

Carson, S., Shih, M., & Langer, E. (2001). Sit still and pay attention? *Journal of Adult Development, 8*(3), 183–188.

Castle, M. C. (2001). *Interpreters, docents, and educators: Ways of knowing, ways of teaching in a history museum, an art gallery, and a nature centre.* Unpublished PhD dissertation, University of Toronto, Toronto.

Cochran-Smith, M., & Lytle, S. L. (1999). Relationships of knowledge and practice: Teacher learning communities. *Review of Research in Education, 24,* 249–305.

Cox-Peterson, A. M., Marsh, D. D., Kisiel, J., & Melber, L. M. (2003). Investigation of guided school tours, student learning, and science reform recommendations at a museum of natural history. *Journal of Research in Science Teaching, 40*(2), 200–218.

Crowley, K., Callanan, M. A., Jipson, J. L., Galco, J., Topping, K., & Shrager, J. (2001). Shared scientific thinking in everyday parent-child activity. *Science Education, 85*(6), 712–732.
Csikszentmihalyi, M., & Hermanson, K. (1995). Intrinsic motivation in museums: What makes visitors want to learn? *Museums News, 1995*(May/June), 34–37 and 59–61.
Deci, E. L., & Ryan, R. M. (2000). The "what" and "why" of goal pursuits: Human needs and the self-determination of behavior. *Psychology Inquiry, 11*(4), 227–268.
Eberbach, C., & Crowley, K. (2005). From living to virtual: Learning from museum objects. *Curator, 48*, 317–338.
Falk, J. H., & Dierking, L. D. (1992). *The museum experience.* Washington: Whalesback Books.
Falk, J. H., Martin, W. W., & Balling, J. D. (1978). The novel field trip phenomenon: Adjustment to novel settings interferes with task learning. *Journal of Research in Science Teaching, 15*(2), 127–134.
Fines, J. (1984). Imagination in teaching: Reflections on my fortnight's work. In S. K. Nichols, M. Alexander, & K. Yellis (Eds.), *Museum education anthology, perspectives from informal learning: A decade of Roundtable Reports 1973–1983* (pp. 54–67). Washington: Museum Education Roundtable.
Freidson, E. (1994). *Professionalism reborn: Theory, prophecy, and policy.* Cambridge: Polity.
Grandy, R. E. (1997). Constructivism and objectivity: Disentangling metaphysics from pedagogy. *Science and Education, 6*(1–2), 43–53.
Griffin, J. (1994). Learning to learn in informal science settings. *Research in Science Education, 24,* 121–128.
Hargreaves, D. H. (2000). The production, mediation and use of professional knowledge among teachers and doctors: A comparative analysis. In J. -M. Saussois & K. Larsen (Eds.), *Knowledge management in the learning society: Education and skills* (pp. 219–238). Paris: Organisation for Economic Co-Operation and Development.
Hein, G. E. (1998). *Learning in the museum.* London: Routledge.
Henry, B. (2006). The educator at the crossroads of institutional change. *Journal of Museum Education, 31*(3), 223–232.
Hiebert, J., Gallimore, R., & Stigler, J. W. (2002). A knowledge base for the teaching profession: What would it look like and how can we get one? *Educational Researcher, 31*(5), 3–15.
Hooper-Greenhill, E. (1991). *Museum and Gallery Education.* Leicester: Leicester University Press.
Iyengar, S. S., & Lepper, M. R. (1999). Rethinking the value of choice: A cultural perspective on intrinsic motivation. *Journal of Personality and Social Psychology, 76*(3), 349–356.
Jackson-Gould, J. S. (1992). Current approaches to interpretation in zoos. In S. K. Nichols (Ed.), *Patterns in practice: Selections from the Journal of Museum Education* (pp. 317–327). Washington: Museum Education Roundtable.
King, H. (2006). *Why are they called explainers when they rarely explain? Researching and developing the role of museum educators in mediating visitors' experiences in a hands-on gallery.* Paper presented at the American Educational Research Association Annual Meeting, April, San Francisco.
King, H. (2009). *Supporting natural history enquiry in an informal setting: A study of museum Explainer practice.* Unpublished Doctoral thesis, King's College London, London.
Klahr, D., & Nigam, M. (2004). The equivalence of learning paths in early science instruction: Effects of direct instruction and discovery learning. *Psychological Science, 15*(10), 661–667.
Larson, M. S. (1977). *The rise of professionalism: A sociological analysis.* Berkeley: University of California Press.
Leinhardt, G., & Crowley, K. (2002). Objects of learning, objects of talk: Changing minds in museums. In S. G. Paris (Ed.), *Perspectives on object-centered learning in museums.* Mahwah: Lawrence Erlbaum.
Macdonald, S. (2004). Exhibitions and public understanding of science paradox. The Pantaneto Forum 13. http://www.pantaneto.co.uk/issue13/macdonald.htm. Accessed 11 April 2007
National Research Council. (NRC). (2009). *Learning science in informal environments: People, places, and pursuits.* Washington: Board on Science Education, Center for Education, Division of Behavioral and Social Sciences and Education.

Nichols, S. K. (Ed.). (1992). *Patterns in practice: Selections from the Journal of Museum Education.*Washington: Museum Education Roundtable.
Nichols, S. K., Alexander, M., & Yellis, K. (Eds.). (1984). *Museum education anthology, perspectives from informal learning: A decade of Roundtable Reports 1973–1983.* Washington: Museum Education Roundtable.
Nolan, T. (2009). Guest editor's introduction: The museum educator crisis. *Journal of Museum Education, 34*(2), 117–121.
Orion, N., & Hofstein, A. (1994). Factors that influence learning during a scientific field trip in a natural environment. *Journal of Research in Science Teaching, 31*(10), 1097–1119.
Piscitelli, B., & Weier, K. (2002). Learning with, through and about art: The role of social interactions. In S. G. Paris (Ed.), *Perspectives on object-based learning in museums* (pp. 121–152). Mahwah: Lawrence Earlbaum.
Rahm, J. (2004). Multiple modes of meaning making in a science center. *Science Education, 88,* 223–247.
Rennie, L. J., & Johnston, D. J. (2004). The nature of learning and its implications for research on learning from museums. *Science Education, 88*(Supplement 1), S4–S16.
Rice, D. (2000). Constructing informed practice. In J. S. Hirsch & L. H. Silverman (Eds.), *Transforming practice: Selections from the Journal of Museum Education 1992–1999* (pp. 222–225). Washington: Museum Education Roundtable.
Roberts, L. C. (1997). *From knowledge to narrative: Educators and the changing museum.* Washington: Smithsonian Institution Press.
Rutowski, P. (1992). Theater techniques in an aquarium or a natural history museum. In S. K. Nichols (Ed.), *Patterns in practice: Selections from the Journal of Museum Education* (pp. 312–316). Washington: Museum Education Roundtable.
Schauble, L., Beane, D. B., Coates, G. D., Martin, L. M. W., & Sterling, P. V. (1996). Outside the classroom wall: Learning in informal environments. In L. Schauble & R. Glaser (Eds.), *Innovations in learning: New environments for education* (pp. 5–24). Mahwah: Lawrence Erlbaum.
Scott, M. M. (2006). From the editor. *Journal of Museum Education, 31*(3), 243–244.
Tal, T., & Morag, O. (2007). School visits to natural history museums: Teaching or enriching? *Journal of Research in Science Teaching, 44*(5), 747–769.
Tran, L. U. (2003). *Examining science teacher-student verbal interactions in informal settings.* Paper presented at the International Organization of Science and Technology Education, 1st North American Symposium, May, Williamsburg.
Tran, L. U. (2007). Teaching science in museums: The pedagogy and goals of museum educators. *Science Education, 91*(2), 278–297.
Tran, L. U. (2008). The work of science museum educators. *Museum Management and Curatorship, 23*(2), 135–153.
Tran, L. U., & King, H. (2007). The professionalization of museum educators: The case in science museums. *Museum Management and Curatorship, 22*(2), 129–147.
Tran, L. U., & King, H. (2009). Shared professional knowledge: Implications for emerging leaders. *Journal of Museum Education, 34*(2), 149–162.
Vansteenkiste, M., Simons, J., Lens, W., Sheldon, K. M., & Deci, E. L. (2004). Motivating learning, performance, and persistence: The synergistic effects of intrinsic goal contents and autonomy-supportive contexts. *Journal of Personality and Social Psychology, 87*(2), 246–260.
Yellis, K. (2000). Paradigms shifted: Comprehending the meaning from without. In J. S. Hirsch & L. H. Silverman (Eds.), *Transforming practice: Selections from the Journal of Museum Education 1992–1999* (pp. 182–186). Washington: Museum Education Roundtable.

Chapter 17
Knowledge to Deal with Challenges to Science Education from Without and Within

Peter J. Fensham

Introduction

Two challenges to science teachers' knowledge have recently emerged. At first sight, they seem so different in origin, that they can be characterised as *challenges from without Science* and *challenges from within Science*. They both, however, stem from common features of contemporary society, namely, its complexity and uncertainty. Both also confront science teachers with teaching situations that contrast markedly with the simplicity and certainty that have been characteristic of most school science education. The purpose of the Part I and Part II of this chapter is to identify these challenges and to set out the new demands they present for science teachers' knowledge and skill.

In Part I, I am also at last addressing issues that have intrigued me since the Science Research Council of Canada (1984) published its version of Science for All. Four purposes were set out for school science in this report. The first two, addressing the science needs of future scientists and of future citizens, were common to all these reports in the early 1980s. Science curricula, ever since, have acknowledged these two student targets, although most countries are still unable to find a curriculum or curricula that will provide good justice to each.

The other two purposes, *provide an appropriate preparation for modern fields of work,* and *stimulate intellectual and moral growth of students,* were not so identified in any other national reports of Science for All. The first of these purposes was identified at the general level of schooling as a whole by the Royal Society for Arts (1986) in England when it urged that school learning should be related more to the world of work. This report expressed these learnings as "capabilities", in an attempt to avoid the mechanical and behavioural overtones that "competencies" had acquired in vocational education. This purpose was not, however, taken up in

P. J. Fensham (✉)
Monash University & Queensland University of Technology, Brisbane, QLD, Australia
e-mail: peter.fensham@monash.edu

D. Corrigan et al. (eds.), *The Professional Knowledge Base of Science Teaching,*
DOI 10.1007/978-90-481-3927-9_17, © Springer Science+Business Media B.V. 2011

any serious way as a priority in any of the national curricula for school science that emerged in the later 1980s and in the 1990s.

The second purpose, particularly its latter component, *the moral growth of students,* was also largely ignored, except by the more radical curricular examples of the Science/Technology/Science (STS) movement in science education, such as the one developed by Aikenhead (1991) for the province of Saskatchewan, Canada.

The remarkable Canadian foresight that set out these two purposes for school science 25 years ago is now urgently revived in the two twenty-first century challenges this chapter addresses.

Part I: Challenge from Without Science— The New World of Work and the Knowledge Society

Since 2000, school curriculum reports and documents in a number of countries have acknowledged the importance of the learning generic competencies. The impetus for the recognition of these generic competencies stems from the studies of the changing world of work in countries like those that are members of the OECD. Work in these societies, it is argued, is changing in a number of irreversible ways—in kind, in the requirements for performance, and in the permanence of one's engagement. (OECD 1996a, b, 2000; Bayliss 1998). The competencies are seen as necessary for employability and social capacity in societal contexts that are now inherently complex in terms of changeability and uncertainty.

Regardless of their success in traditional school learning, many young persons are now seen as lacking other knowledge and skills that are essential for their personal, social and economic life. This has led to a conviction that learnings must be developed that lie outside or go beyond what is provided by the learning areas into which the school curriculum has been traditionally divided. These new learning intentions are explicitly stated as *more* than the acquisition of the established knowledge content of a number of Key Learning Areas (or subjects). They set out what that *more* is, but as yet, little has been suggested about how in fact these extra learnings are to be taught by teachers whose expertise has been honed in terms of traditional subject knowledge.

The changes that are now occurring in the world of work are very largely driven by new forms of information (the digital revolution), and the emergence of knowledge being increasingly recognised as the primary source of economic growth. The globalisation of production, and innovation of new products and processes result from the producing of new knowledge (Gee et al. 1996; Gilbert 2005), and the *Knowledge Society* is the phrase now generally used to characterise these changes. The priorities for learning in the Knowledge Society are so different from those in traditional schooling that they can be said to constitute a new paradigm.

Learning in the Knowledge Society

In the Knowledge Society knowing has the following characteristics:

- Knowledge is now dynamic, like a verb, rather than static like a noun
- Knowledge is about acting and doing to produce new things, rather than being the storing up of bits of established knowledge

In the Knowledge Society value is associated with:

- Knowing how to learn
- Knowing how to keep on learning
- Knowing how to learn with others
- Seeing alternative possibilities for solving problems
- Acquiring important skills—asking questions, communicating to different audiences, etc.

These values contrast, and indeed clash, with what has been valued in traditional schooling, and in science education in particular, namely:

- The individual accumulation of many bits of knowledge
- Knowing the right answer to questions with only one answer

Another project, the Definition and Selection of Competence (acronym, DeSeCo), led to three broad types of competence—*communicative, analytical* and *personal,* for which the above list provides obvious examples (Rychen and Salganik 2003). A number of the participating European countries have since explicitly listed these competencies as intentions for their school curricula.

The demand for learning in students that these new work situations make on education systems varies, but they all introduce dynamic aspects of learning, like *to adapt to change, to generate new knowledge* and *to continue to improve performance* (Fraser and Greenhalgh 2001). In the case of science education, these dynamic aspects extend the active dimension of learning beyond a prescribed set of specific skills that became common in science curricula in the 1990s under headings like *Working Scientifically.*

The Curriculum in Response

The curricular responses in Australia since 2000 reflect the confusion this new paradigm for knowledge is causing. The state of Tasmania (2002) introduced a new Curriculum Framework consisting of five sets of Essential Learnings—*Thinking, Communicating, Personal Futures, Social Responsibility* and *World Futures* (Tasmanian Department of Education 2002). This Framework was to make the curriculum less crowded, to engage learners more deeply and more relevantly. It was prefaced by a statement of Values and Purposes, summarised as *connectedness, resilience, achievement, creativity, integrity, responsibility,* and *equity,* words that were quite foreign to even the rhetoric of the science curriculum let alone it to its list of detailed topics, concepts and principles.

Another state, Victoria, echoed similar values and purposes in the Introduction to its Essential Learning Scheme (VELS) and described the curriculum as three strands of learning that are *intended to be interwoven* (VCAA 2005). One strand is Discipline-based learning with Science as one of six disciplines. Another strand includes *metacognition* and *interpersonal development,* while the third has *thinking and communicating* as sub-headings.

As part of its "educational revolution" Australia's new Labor Government, has asked a new Australian Curriculum, Assessment and Reporting Authority to develop the four subject areas of English, Mathematics, Science and History as the first stage of a national curriculum (hitherto always resisted in Australia). In addition to these priority subjects, the Authority has recognised the importance of generic competencies, listing them as *communicating ideas and information, working with others in a team, solving problems, using technology, collecting, analysing and organising information* and *planning and organising* (*initiative, enterprise and self-management*). They are "*to be dealt within its four initial subject curricula to avoid the risk that they will languish unattended on the assumption that they will be addressed by schools 'across the curriculum'*" (NCB 2009, p. 8) However, none of the Shape Statements for determining details for the four subjects address what such "*addressing them*" could mean in practice.

New Zealand has had more open discussion of the Knowledge Society than Australia, and Gilbert (2005) in her book, *Catching the Knowledge Wave*, has provided a helpful dialogic account of both the Knowledge Society and of its demands on schooling. She recognises the gulf between its paradigm for knowledge and learning and the paradigm of traditional schooling. She expresses personal concern that the former has too great an emphasis on competencies and skill strategies to the detriment of traditional knowledge. Competencies like *thinking, communicating, investigating* and *problem solving*, she argues, require a context, and students need to know that every knowledge discipline offers them these competencies as active strategies. This is not a retreat to, or a defence of the traditional teaching and learning in these disciplines. The traditional paradigm has overemphasised static, established disciplinary knowledge, and given too little attention to how these competencies are employed as strategies in the disciplines.

Not surprisingly, New Zealand in setting out a new curriculum in 2007 listed *Managing self, Relating to others, Participating and contributing, Thinking*, and *Using language, symbols, and texts,* as five key competencies that are important for schooling for life in a modern economy (Ministry of Education (2007). The role of these competencies in the new Science curriculum is however less clear. A strand of learning, labelled *Nature of Science,* is intended to be "*overarching and unifying*" for the disciplinary science strands of knowledge and understanding.

Gilbert (2005) attempts her own rapprochement for the new and traditional skills and knowledge in which the "elements from one knowledge system need to be put together with elements from another, arranging them so that they work in new ways and do new things" (p. 156). This suggestion about "putting together" in what follows will be seen to have operational merit as we address the question of science teachers' knowledge for the Knowledge Society. The metaphor of "putting together" will also have use in Part II, albeit rather different use, for how science teachers need to respond to the challenges from within Science.

Knowledge for Science Teachers in the Knowledge Society

It can, I believe, be argued that the current educational interest in general competencies is largely because science education (and its counterparts in other knowledge disciplines) has, hitherto, failed to give adequate attention to these aspect of what it means to be educated in science. Layton's (1991) plea for science knowledge-in-action and his metaphor for Science as *a quarry* to be raided for use, compared with Science as *a cathedral* to be reverenced, come to mind.

The "*putting together*" of the Knowledge Society's competencies with the recent interest in the nature of science in school science education immediately spells out that science teachers need to have some knowledge of:

- the ways in which thinking occurs in science;
- how science is communicated within and beyond the scientific community;
- strategies students can use in learning science;
- how adapting to change is an integral feature of science;
- the variety of ways scientific questions are investigated;
- what constitutes a solution to a scientific problem and
- how learning with others can become a reality in science classrooms.

Each of these aspects of knowledge will now be briefly discussed, and it is interesting that the sources are often from non-science educators who have chosen to study science as a special form of human enterprise. The traditional paradigm of science education in schools has emphasised the conceptual knowledge of science to the detriment of the nature of science itself as a human enterprise. This has left the majority of students ignorant of this aspect of science, which was left to hopefully emerge in those few students who go to become practising scientists after university studies. Recent attempts to include the nature of science in schooling are a healthy corrective to this situation. What has been eventually and implicitly learnt by the few in practice, needs now to be explicitly taught to all students.

Ways of Thinking

The extent to which thinking has been neglected in school science is evident by its absence from a list of 61 metacognitive and metalinguistic verbs used in official curriculum documents (Wilson 1999). This may because "thinking in science" covers a variety of purposes, some of which were in the list. Ohlsson (1995), a cognitive scientist, claimed the following are epistemic for science—*describing, explain-*

ing, predicting, defining defining, arguing, critiquing and *explicating*. Each of these has a distinctive character and importance in science. Together they constitute most of what Science has to contribute as a Way of Thinking.

The basic knowledge that teachers need to equip themselves about these epistemic activities is from studies in the history and philosophy of science—knowledge that is now very rarely included in most science teachers' background preparation, despite the strenuous efforts of Matthews and others to revive it (Gauld et al. 2005).

In addition to personally acquiring these ways of scientific thinking as basic knowledge for themselves, science teachers need to know how they can be taught to students for their active use. Fortunately, science education researchers are now attending to some of these activities, so that there is now a significant corpus of pedagogical know-how to pass on to teachers.

Science teachers and science curricula have hitherto concentrated on these as static rather than as active processes by expecting students to learn to repeat a specific and previously discussed description, explanation or definition, rather than on mastering these scientific processes for active use in a scientific investigation. Zuzovsky and Tamir (1999) reported, from an analysis of students' explanatory responses that many of them were simplistic, in which the use of scientific terms was rare, and *describing* was confused with causality. The role of explanation in science has been extended through the work of Ogburn et al. (1996) who elaborated how the need for *explaining* can be generated in classrooms, the variety of explanations and explanatory processes that are involved, and how these can be shared actively with students. The role of models in science education and the reasoning associated with them, such as *predicting* and *explaining*, has now been quite thoroughly studied (e.g. Gilbert et al. 1998) and is again ripe for sharing with teachers.

Defining has been a common feature in science classrooms, but usually without what Gardner (1975) refers to as the intensive meaning of a definition or the extensive range of its application.

Arguing in science has attracted considerable recent research attention (Kuhn 1991; Aufschnaiter et al. 2007), and there is much from their studies that is now accessible to science teachers wishing to develop this aspect of science in school classrooms (see also Chap. 11).

The nature of *critiquing* in science and of *explicating* (as distinct from *describing*) a scientific situation or a scientific phenomenon, have as yet received little research attention, although both are included in the PISA definition of scientific literacy (OECD 2006), and that test has some assessment items for these scientific competencies.

Communicating

Science educators have been interested for many years in the use of specific words in science (Gardner 1972; Sutton 1992) and in the general importance of language in socio-cultural approaches to learning (Mortimer and Scott 2003). Socio-linguists have been more precise in their analysis of the language used in science classrooms and in science itself. Lemke (1990), in his studies of science classrooms, introduced the notion of the students trying to learn to approximate "*the science way of talking*" that is presented by the teacher—a rather daunting responsibility for the teachers whose own education has underplayed science as communication. Kress et al. (2001) have extended language to include the multimodal forms of communication that all participants, teachers and students, need to learn if they are to function optimally as active transformers of meaning in multimodal classrooms and in an even more multimodal society.

Halliday and Martin (1993) identified the particular grammatical constructions that are used in the written communication of science, and Unsworth (2001) has listed as types of practice in written science—procedural recounts, explanations, descriptive reports, taxonomic reports, expositions and discussions. Only as teachers build up knowledge of this variety will they be able to assist their students to achieve the communication skills involved in receiving science information from different sources, and discussing it with diverse target audiences. These practices are a far cry from the undifferentiated prose in most current science textbooks, and the box ticking communication expected from students when multiple choice items are the sole determinants for successful science learning. Ritchie et al. (2008) have shown that a well-defined procedure can be used in classrooms by teachers to engage the children, as young as 8- or 9-year olds, in writing science mystery stories. As well as positive affective and cognitive gains with respect to science, the students acquired communication skills as subtle as recognising the differences between the narrative genre and the science genre.

Managing Self (Knowing How to Learn)

Science educators have given considerable attention to meta-learning, and see metacognition as an important tool for science students to have (Thomas 2006). Baird (1990) defined the metacognition students need to engage in purposeful scientific inquiry and Gunstone (1994) provided evidence for the positive role that specific science content plays in developing this awareness in students.

Some powerful research tools have been described by White and Gunstone (1992) and by Mintzes et al. (2005) for probing students' understanding of science concepts and scientific phenomena. The Project to Enhance Effective Learning (PEEL) in Australia, Sweden and Denmark has introduced a number of these to teachers as ways of assisting them and their students to become aware of themselves as learners, and hence more effective learners (Mitchell 2005). Larkin (2006) and Georghiades (2006) have shown how this awareness can be encouraged in the early years of schooling.

Adapting to Change: An Integral Feature of Science

Scientific knowledge progresses because of its ability to adapt to change that is monitored by well-defined criteria. Indeed, the temporary nature of scientific knowledge is listed as a key feature of science. Although Posner et al. (1982) published a paper on the conditions for conceptual change in science and in science education that was influential among researchers, very little of this feature of science or of these conditions is shared with students in traditional science education. Since the history of scientific ideas and theories has commanded so little attention in the education of science teachers, there is an urgent need for teachers to become knowledgeable about some specific examples of change in science, if they are to assist their students to develop the capacity to adapt to change. A whole genre of popular books have recently provided just such examples from the history of physics, biology, earth sciences and chemistry, e.g. *Five Equations that Changed the World*, by Guillen (1999); *The Double Helix* by Watson (1968); *Rosalind Franklin: The Dark Lady of DNA* by Maddox (2002); and *The Map that Changed the World* by Winchester (2002).

Scientific Inquiry: A Variety of Ways to Solve Scientific Problems

Scientific inquiry has a long history in science education and its importance is uniformly recognised (National Research Council 2000). At times, inquiry in the form of set laboratory investigations, has been advocated as a means for students to learn science concepts and principles, but if the assessment priority for learning them is merely their repetition or simplest applications in a paper-and-pencil test, then this use of inquiry is exaggerated and unwarranted. There has been a tendency, especially in the primary years, to reduce scientific inquiry to *a fair test*. Setting up a *fair test* is just one stage that can be useful in some scientific investigations. Inquiry needs to be recognised as beginning with the observation of a phenomenon sufficiently thoroughly to lead to a question that is scientifically investigable, and then figuring ways to seek enough data from which a solution can be inferred and/or an explanation suggested. When the richness of the variety of skills and understandings that can be involved in scientific inquiry are appreciated by students, the time and effort put into their teaching become worthwhile (Bybee 2004). It is just this flexible learning outcome that is being called for when problem solving is listed as a priority in learning for the world of work. Flick and Lederman (2004) have brought together a number of studies that report positive experiences in developing the learning of inquiry (problem-solving) skills in science teachers and in students.

Learning with Others (Participating and Contributing)

At first sight the use of the laboratory as a common part of school science seems to be an ideal contribution to learning with others. Science teachers, indeed, usually differentiate their classrooms into individual learning and small group learning. The former is associated with what is often described as "theoretical" science content and the latter with the carrying out of "experiments" in a science laboratory. Studies of the dynamics of these small groups carrying out routine or recipe-type experiments suggest that little shared learning occurs. Too often, a dominant student carries out the physical handling of equipment leaving others passive as spectators or recorders.

Pedagogies that do encourage cooperative learning are ones that engage students working together on socio-scientific issues (SSIs) or less familiar problems. The openness of these situations, together with their multi-dimensional character, cannot be fully undertaken by one student or one small group. They thus encourage more students to feel able to contribute and that their contributions are recognised as needed for the problem or situation to be solved. France (2007), Eijkelhof (1986), and Rennie (2007) have provided a wide range of examples, in which teachers have acquired the appropriate knowledge and confidence needed for such teaching.

Part II: Complexity of Issues and Uncertainty in Science

If the complexity and uncertainty of the Knowledge Society demand new understandings and contributions from science teachers, these are certainly matched by the demands that are posed by the role of complexity and uncertainty in science itself.

Since 2000 a number of scientific bodies have been identifying key science-based issues for attention. For example, the American Association for the Advancement of Science used the theme of Grand Challenges and Opportunities to frame the 125th Anniversary issue of *Science* on "What Don't We Know" (AAAS 2006). Omenn (2006) in his presidential address relates these key issues very directly to the achievement of national and international social goals. He refers to grand challenges in environmental sciences, including topics such as climate variability, hydrologic forecasting, land use and recycling (National Research Council 2001), the 14 grand challenges the Gates Foundation identified for global health (Varmus et al. 2003), and several other lists of such grand socio-scientific challenges.

These science-based issues have been increasingly taking the centre stage of societal attention, and although the global economic recession in 2008/2009 may have temporarily overshadowed them, they will re-emerge with even greater urgency.

Their importance for society is underpinned by the urgency they are accorded by the scientific communities, and the priority the public media give to them because of their societal impact. Duggan and Gott (2002), soon after this public attention surfaced, argued that schooling that ignores these SSIs can be accused of selling short its current students as future citizens (Duggan and Gott 2002).

The 2007 World Conference on Science and Technology Education, (ICASE and ASTA with support from UNESCO) brought some of these issues to the school science education community. Lord Robert Winston (biomedical issues), Graham Pearman (global warming issues), Howard Gardner (issues involving multiple intelligences) and Ian Lowe (energy and other conservation issues), were the keynote speakers. Each described a set of SSIs of great significance for society, and hence for science teachers to take seriously with their students. Their examples were all multi-faceted in nature, involving several scientific disciplines, and extending beyond these sciences into economics, social philosophy and ethics.

These examples, and those in the lists of Grand Challenges, enable a number of common features to be noted about the science of these priority issues. In Table 17.1 these features are contrasted with those that are associated with traditional science education.

The striking contrast in Table 17.1 is quite similar to the one presented by Aikenhead (2005) between traditional school science and what he calls humanistic science, an issue he takes up in his chapter in this book (see Chap. 7).

The different senses of knowledge discussed in Part I of this chapter confront science teachers with a different paradigm of educational outcomes—the challenges from without science. The differences in the features of Science that are presented in Table 17.1 are so great that science teachers also face a new paradigm for science itself if they are to teach these important issues in school science classrooms—the challenges from within science. The great majority of science teachers will not, themselves, have been educated about such a paradigm for science. Rather their own education will have been a paradigm for science that is reflected in the science they have hitherto been teaching.

The contrasts in Table 17.1 are so considerable that they need to be teased out more fully. The *Cynefin Framework* from Complexity Theory is a useful way to

Table 17.1 Features of the science in priority SSIs and in traditional school science education

Priority SSIs	Traditional school science
• The science involved is so inter-disciplinary in character that teams of scientists from different scientific disciplines must be involved. • The science that is involved has a degree of uncertainty. • Possible solutions are the goal, not a single solution. • The uncertainty introduces the ideas of "risk" and "probability" as basic features of the scientific response and of any tentative solutions or conclusions. • The issues in their reality involve non-science aspects that make them "multi-disciplinary". Expertise from other disciplines is needed for understanding the issues and for sensible responses in political terms. • Science perspectives alone suggest solutions that distort the reality of the issues.	• Science is taught in discrete disciplinary subjects or strands—*biology, chemistry, earth science, physics*. • The science knowledge is firmly established and authoritative. • Its application is restricted to idealised or contrived situations (*ideal gases, frictionless surfaces, conservation of energy, states of matter, pure substances and isolated biological contexts*). • Learning involves reproduction of static knowledge from these discrete science disciplines. • Problem solving needs science knowledge/information only to achieve the "correct" answer. • Assessment of learning involves answering questions with only one correct answer. • Non-science aspects are used only for motivation and not as integral or essential learning. • Scientific reasoning does not include "risk" or " probability".

make clear to teachers the key contrasts between the science knowledge in the Grand SSI challenges and that which has been the basis of traditional science education. These differentiations, in turn, can then become the grounds for discussing the old and the new paradigm and hence the new teacher knowledge that these challenges demand.

The Cynefin Framework

Kurtz and Snowden (2003), important figures in the development of Complexity Theory, invented the "*Cynefin*" *Framework* to help people make sense of situations for which familiar assumptions, like *order, rational choice* and *clear, singular intent* do not apply. It takes the form of a 2 by 2 matrix as shown in Fig. 17.1 with some medical cases included as examples that are discussed below. The left column is associated with phenomena for which well-established laws (and *order, rational choice* and *singular intent*) hold. The right column is for phenomena where a degree of uncertainty (and *incomplete order, choice not merely rational and lack of agreed intent*) holds.

The introduction of vertical distinctions due to degree of *complication* and horizontal ones due to degree of uncertainty, as characteristics of phenomena enable the concept of risk (of varying significant degrees) to be introduced as a key feature to be considered.

Medical Examples: In the *simple case* of *a broken arm*, it is fully understood why bones break and how to set them so that they will restore themselves. Likewise, in

Established Laws hold	**Uncertainty holds**
simple cases e.g. a broken arm ***risk*** ***zero or very low***	*complex cases* AIDS ***risk*** ***high to very high***
complicated cases e.g. heart bypass surgery ***risk*** ***low to medium***	**CHAOS** ***risk*** ***out of control***

Fig. 17.1 A basic form of the Cynefin Framework (Cynefin is a Welsh word for multiple locations)

the *complicated case* of an expensive, extended open heart bypass operation that was impossible 50 years ago, the multi-staged procedure is quite long and involves the combined efforts of medical personnel with different skills. However, for such skilled professionals, it is now quite routine and the risk associated is low.

The *complex case* of *AIDS* is still not understood or curable after more than 20 years of intensive study. Some progress in controlling its rate of onset and its progression have been made; but these have involved big changes in social behaviour as well as the regular application of costly drug regimes. In a number of countries, these controls have been established too late, or have not been possible, and the illness has become chaotically pandemic.

Science Examples: For scientific phenomena the column under *Established Laws Hold* allows for a differentiation between *simple cases,* involving one science principle or perhaps a short sequence of principles, and *complicated cases,* involving a mix of different principles are involved and where sequencing may have options. In the column under *Uncertainty Holds* scientific phenomena in the top left area are designated as *complex* because of their uncertain or not completely understood character. This uncertain character leaves open extreme phenomena that then fall into the *CHAOS* area in the lower right.

Science in Science Education

The science traditionally involved in most school science curricula, as suggested by its characteristics above, locates itself in the upper left area of the Cynefin matrix (see Fig. 17.2). These curricula present the knowledge of science in mono-disciplinary forms. Only firmly established knowledge is included, and its applications are idealised and contrived (and hence *simple*). They exclude the recognition of non-science knowledge. The established status of this scientific knowledge and its limited applications means it is a "topic" for teaching and learning, rather than, in any sense, an "issue" to be shared and discussed.

In the later 1980s the STS movement for school science education provided strong arguments for, and many interesting examples of real-world topics that required both disciplinary and interdisciplinary science (albeit still usually well established). The application of this science in real-world situations of science and technology (S&T), meant that other fields of knowledge were inevitably involved (Aikenhead 1991; Solomon and Aikenhead 1994). The *complicated* character of the interplay of interdisciplinary science and these other knowledge fields, and the degree of uncertainty this introduced locates STS science in the lower left area of the matrix in Fig. 17.2.

Despite the widespread interest in the STS movement, the new curricula for school science that emerged in the 1990s, for a variety of reasons, retained mono-disciplinary strands, albeit now attenuated across all the years of schooling. School science's location in the upper left of the matrix was retained, with very little atten-

Natural Laws hold	**Uncertainty holds**
Simple cases **Single science disciplinary topics** (traditional school science)	*Complex cases* **Multi-disciplinary socio-scientific issues (SSIs)** (Grand Challenges)
Complicated cases **Interdisciplinary science topics** (STS topics)	***CHAOS***

Fig. 17.2 The disciplinarity of the science in traditional school science education, in STS science education and in school science education for the issues of the Grand Challenges

tion given to the reality of S&T phenomena and situations that were increasingly of personal, social and national interest.

The characteristics of the science involved in the SSIs of the current Grand Challenges were listed above. The uncertainty in some of the science associated with these issues is heightened by the *complexity* of the now not to be ignored non-science aspects. This science thus locates in the upper right area in Fig. 17.2. This location also means that *risk* and *probability* become features of the discussion of this science and its applications. "Water" as a science topic and as an SSI illustrates the difference in these science locations. In traditional school science "water", as a topic, clearly locates in the *simple cases* area, whereas "water" as an " issue" belongs in the *complex cases* area.

Water as a Science Topic

Indicative of the simple case treatment of "water" in traditional school science are the number of pages devoted to it in three commonly used senior secondary textbooks (300+ pages) in Australia for Chemistry, Physics and Biology. Water is described in the Biology text as "*the water of life*", but this significant role is specifically dealt with on only one page. Water was not dealt with at all in Physics. Chemistry provided the most extensive treatment of "water" and the aspects covered are listed in the left-hand column of Fig. 17.3. These aspects make little connection with the key features and concepts that make up the science components of the "Water Issue" in Australia, where it is certainly a grand challenge. Some of these features are listed in the right-hand column of Fig. 17.3.

"Water" (as a Chemistry topic)	**"The Water Issue"** (an interdisciplinary and multi-disciplinary issue in SSI science education)
Properties: B.Pt, M.Pt, surface tension	*Importance*: all living systems. population growth
Composition: preparation, decomposition	*City water*: sources, storage and distribution - cost
Phases: solid/liquid/gas	*Rural water*: sources, storage and distribution - cost
Structure: 108°, polarisation, hydrogen bonding	*Ground water*: sources, ownership, and access - cost
Solvent Properties: polarisation, B.Pt elevation, M.Pt. depression	*Properties/Use*: potability, purification, recycling
	Uses: domestic, industrial, rural
	Sewage/grey water: usefulness, ownership, etc.
	Alternatives for supply: desalination (energy cost)

Fig. 17.3 "Water" as a topic and as an "issue" in traditional and in SSI school science

The Water Issue—A Grand Challenge: An adequate supply of potable water is currently a contentious issue in a number of countries and in a number of key urban and rural areas of Australia. Its character as an issue cannot be appreciated or understood without knowing a considerable amount of the relevant disciplinary and interdisciplinary sciences and their application in relevant technologies. It is an issue that also involves the knowledge from a number of other disciplines, such as economics, social psychology, law, politics and sociology.

The recognition that the SSI of water is a *complex case* in the Cynefin Framework means that there is never one single scientific answer to an adequate supply of potable water for a society, and that attempts to find the most probable answer from within science alone is likely to fail. Political decisions for dealing with the issue cannot wait for certain answers. Rather, what is needed is an approach by teams of experts in the various fields that impinge on the issues, together with a "reins person" who ensures one group of experts does not get too far towards their "solution", before it is checked with the other groups for feasibility and compatibility with their aspects. The mutual goal will be to reach a sufficient understanding of the dimensions of the issue that will result in a first-step solution, that is the most likely to reduce the complexity of the issue away from the *CHAOS* of serious recurring water shortages. Then attention can be focussed on further such steps to the provision of a stable supply for the projected population.

SSIs and the Cynefin Framework

Not all SSIs are Grand Challenges. They vary a great deal in character, and over time, as they are further studied, they can, like the medical case of by-pass surgery in Fig. 17.1, move from being impossibly *complex* to being *complicated* and even *simple*. Others, like smoking, which seemed to be *simply* an optional and popular technology with few serious social or scientific consequences, can be found scientifically to have other societal consequences that now places them in the complicated or *complex* sector with its border on personal, social and national *Chaos*. Conversely, further scientific study can so extend our understanding and control over the source of these issues like smoking and so they shift to the *complicated* or *simple*.

Paradigms for Science Education

In parallel with the location of the different conceptions of science in Fig. 17.2, the science education that promotes their teaching and learning locates in the matrix as shown in Fig. 17.4. In this use, the *risk* feature in the Cynefin Framework reflects as the *certainty or uncertainty of correctness* that is associated with the answers to the questions that are posed for assessing learning in these different types of school science education.

Since the science for both the primary and secondary years is all largely established conceptual and procedural knowledge, the traditional science curriculum requires transmissive teaching that engages the learners through constructivist and heuristic strategies. There is little scope for learners' opinions, and their learning is evident by their ability to answer questions that have a quite certain correct answer. These curricular characteristics locate this science education almost entirely in the top left area of the Cynefin matrix of Fig. 17.4.

By its mono-disciplinary knowledge and its assessment practices, this type of science education avoids the possibility of being fully relevant to the lives of its students. Thus, its very nature poorly prepares students for the grand challenges that are increasingly publicised as what the future holds for them.

The more developed examples of STS science education (see Solomon and Aikenhead 1994) can be located in the bottom left-hand area of Fig. 17.4. They have as their starting point real-world S&T contexts that require science knowledge that was not neatly disciplinary, and which had obvious links to other dimensions of knowledge that were economic, social and political. How these contexts impacted on the learners' lives required input from them as well as from the teacher, and discussion became an essential feature of the teaching. The different views about these social/scientific interactions needed to be reflected in assessments that include questions with more openness to what could be correct.

Science education for SSIs of the Grand Challenge type needs to distinguish the established science and the uncertain science, and to acknowledge the importance of the interactive aspects that require knowledge from other subject areas. Together these features mean a complexity that locates this education in the top right are of Fig. 17.4. *Risk* and *probability* become integral components of both the science

Natural Laws of Science Hold	**Uncertainty Holds**
simple cases **traditional science education** **content:** established disciplinary science knowledge **teaching:** transmission (little scope for student opinion) **learning assessment:** questions have one correct answer ***No uncertainty***	*complex cases* **Grand Challenge SSI science education** **content:** disciplinary and interdisciplinary science plus societal and moral aspects, (uncertainty and probability) **teaching:** transmissive, interpretive and personally involving) **learning assessment:** a mix of certain and probable cognitive understandings plus personal and social commitment possibilities ***Inherent uncertainty and certainty***
complicated cases **STS science education** **content:** disciplinary and interdisciplinary science knowledge, plus recognition of other knowledge dimensions **teaching:** transmissive and discursive (student inputs essential) **learning assessment:** questions with a mix of correct and possible answers. ***Certainty and some uncertainty recognised***	**CHAOS**

Fig. 17.4 Paradigms for the nature of school science education for traditional science, STS science and Grand Challenge SSI

content and its interactions, and they introduce subjective perspectives with moral implications at the personal and social levels. Cognitive assessment is a mix of certain and probable answers together with evidence of the range of possible informed personal and societal responses.

Knowledge for Science Teachers: Teaching Complex SSIs

If science teachers are to address Grand Challenge SSIs in their classrooms, they will need knowledge about complex SSIs, about uncertain science and about *risk* and *probability*. More of the research behind this knowledge has been concerned with the public understanding of science by adult citizens than with students in schooling.

New Conceptions of Knowledge Content

Pioneering studies by Layton et al. (1993) of citizens in a range of situations involving science highlighted the "fragility of much of the available science and its inability to provide unambiguous answers to the questions being asked" (p. 118). These studies also pointed to the importance for teachers of knowing themselves about *trust* with respect to scientific information. This aspect of scientific knowledge was echoed in nine studies by Irwin and Wynne (1996). As a result of these and other similar studies, Jenkins (2000) argued that the "world proves to be much more complicated, uncertain and risky than school science encourages students to believe" (p. 211), heralding the paradigm shift in science education (for teachers and for their students) that this chapter now is urgently demanding. Bingle and Gaskell (1994) also have reported on the role that trust plays in people's views of a scientific issue. The "*trustworthiness*" of scientific information is a knowledge aspect that twenty-first century science teachers need to master for their teaching of SSIs, and for the more general purpose of using the internet to access scientific information of all sorts in schooling.

Ryder (2001) analysed a large number of these studies of public understanding involving both established science and contested science. He argued that six categories of knowledge are necessary for effective participation. These are *science subject matter, collecting and evaluating data, interpreting data, modelling in science, uncertainty in science and science communication.* This list constitutes a useful knowledge set for science teachers wishing to teach science in the new paradigms of *complicated* and *complex* SSIs. "Uncertainty in science" involves an appreciation of the qualitative and quantitative ways that *risk* is defined (Freudenburg 1988; Renn 1992), the role of *the precautionary principle* in making decisions about scientific issues involving *risk* (Harremoes et al. 2002; COMEST 2005), and the distinction between *causal relations* and *correlational ones.*The publicity given to various models for predicting the rise in temperature over the next 50 years, and their likelihood for *tipping point* consequences provides a ready-made example of the *precautionary principle.*

Eijkelhof (1986) as part of the PLON project in the Netherlands developed a Physics teaching unit on dealing with *risks* of ionising radiation. In its evaluation he found the senior secondary students demonstrated the capacity to make risk judgements which did match actual risk statistics. Solomon (2003) suggested that the discussion of *risk* in the classroom is a means of re-engaging some students with science, and Kolstø (2006) reported how central *risk* information became for students in making decisions about the safety of high-voltage power lines.

Science teachers will need to know how to differentiate between the variety of SSIs they may wish, or be required to include in their teaching. The Cynefin Framework and its way of locating the degree of complexity in SSIs is useful here. In the *complicated cases* area, a science teacher needs to use established knowledge from the several sciences involved in the issue, together with the appropriate interdisciplinary concepts, to lay out the alternative solutions.

For SSIs that locate in the *complex cases sector* the first message to learn is that science teachers should be wary of believing they can teach them alone. This may come as a relief to many science teachers who have, in practice, been reluctant to extend their teaching beyond the simple cases of disciplinary ideality to include the *complicated cases* that real-world situations present.

Few science teachers are equipped to do justice to the non-science aspects of the issue. To attempt to do so is likely to lead the students to see the issue as essentially scientific or technical, the solution for which is essentially in the hands of the scientists. At the very least, science teachers must acknowledge the importance of the ethical, the social, the economic, etc. aspects, and indicate their role is providing deep appreciation of the scientific dimensions.

The practice of dealing with these complex issues involving the multi-pronged strategies of different experts has is analogous what happens in "educational events", like field trips or school camps. Teachers from different subject fields plan together to introduce their perspectives on the event so that all students get a broad basic preparation. During the event they develop these various perspectives in some detail, usually in small groups. Finally, the students feed the alternative dimensions into the whole class to explore what coherence about the issue can be reached and what possibilities for resolution can be proposed.

Such an organisational alternative to the usual secondary school pattern of relatively isolated lesson periods devoted to different subjects can do justice to complex SSIs. The educational event over a day or several days becomes a multi-disciplinary teaching and learning experience about the chosen issue. Primary teachers are much more accustomed to include an integrated approach in their teaching, and hence should find such "educational events" relatively simple extensions of what they are doing already.

New Pedagogies

Equipped with this novel range of knowledge, science teachers will then need to develop pedagogies that are consonant with the nature of uncertain science and with *risk* and *trust* as characteristic knowledge features of this new science education paradigm (Ryder 2002).

The old transmissive pedagogy, even with constructivist strategies, that seemed consonant with the authoritative science knowledge to be transferred from teacher to students will need to give way to much more discursive ones in which ambiguity and uncertainty are encouraged and tolerated. Socio-cultural strategies such as debating and discussing that allow students' voices and opinions to be aired, challenged and changed, will be new procedures for many science teachers, but their use is now well supported in the research literature and, indeed, their absence in traditional science education has been specifically identified by students as a major ground for their dislike and disinterest in science (Lyons 2006). Van Rooy (1994), Ratcliffe (1997), Oulton et al. (2004) and Sadler et al. (2004), all report success with these pedagogies in engaging students to be personally challenged with SSIs.

New Approaches to Assessment

The PISA 2006 Science project of the OECD placed a strong emphasis in its assessment design on the use of real-world S&T contexts (OECD 2006)—opening it to the possibility of including complicated and *complex* cases. It also endeavoured to include the interaction of knowledge of, and about, science with these contexts as much as possible. Nentwig et al. (2009) have produced an interesting measure of the extent to which this contextual interaction was, in fact, included in the 2006 test items.

Sadler and Zeidler (2009), impressed by the degree to which the PISA 2006 Science Framework recognises the *complexity* of SSIs, then analysed for this aspect in the 2006 assessment items. They concluded that science assessment in the PISA project has made significant progress towards recognising the *complicated cases* area of the matrix, but the movement into the *complex cases* area is small. For a large-scale paper-and-pencil test involving many diverse countries, this should be regarded as a positive outcome that has not yet been achieved, nationally or more locally, in the modes of assessment being used for science learning. Sadler and Zeidler go on to suggest what seem to be the most promising ways of shifting assessment of science learning so that it is authentic with the intentions of science education for *complex* SSIs.

Conclusion

Modern societies are very significantly defined by science and technology. It is in these societies that the lives of current and future citizens are enacted. The findings from the studies of public understanding of science on both sides of the Atlantic confirm that citizens with more knowledge of science are more in control of their lives (Evans and Durant 1995; Miller 1998). That is, they do not simply reject or uncritically support possible S&T developments. Rather, they show a critical discernment about these issues, supporting some and rejecting others.

Science teachers are privileged in being able to contribute to this personal and social empowerment by the way they teach science. There is strong evidence, however, in many of the more developed countries that too often this teaching is perceived by their students as academic, irrelevant to their lives and hence of no interest (Lyons 2006). This type of teaching is, sadly, a reflection on the way many science teachers were themselves educated in science.

Roberts (2007) set out a case for science teaching that is inspired by what he calls two visions of scientific literacy. For Vision I the source of the science and the goal of this literacy lies in science itself. For Vision II the source of the science and the goal of the literacy lies in the encounters citizens have with science and technology in society. Over the last decade progress has been made through the movement towards Context-Based Science Teaching (Millar 2006) that enables science teachers to gain the knowledge and skills to shift the goal and style of their teaching towards Vision II, which in this paper is represented as the shift from the upper top left area in Fig. 17.2 to the lower left area. This degree of shift is not about a change in the nature of the science knowledge, but about the priority given to particular science topics and their practical applications.

Science through its applications as technologies, has been, and will continue to "simplify" the lives of citizens by providing them with means of enhancing the quality of their living and working. It is now clear that societies are becoming more "complex" in the two ways this chapter has outlined—the uncertainties in the working lives of citizens that are now inherent in the Knowledge Society and the uncertainties in the science of so many pressing SSIs. These increases in "complexity" urgently require science teachers to acquire yet further new knowledge and skills, if they are to fulfil the privileged role they have in relation to the education of young persons.

In Part I of this chapter it has been argued that a more significant emphasis in science teaching on the nature of science as "ways of thinking" would be science education's best means of assisting with the first set of social uncertainties. In Part II a case is argued for teachers to introduce the notion that science itself can be uncertain, and hence that "risk" and "probability" are important concepts to learn about in science education because of their implications for public and private action.

Future teachers will need help, through their preservice education in both science and science education, to gain the knowledge for these new emphases, and existing science teachers must be offered the same help through progammes of professional development. It is in these ways that science teachers will be enabled to make the contributions their privileged role affords.

References

AAAS. (2006). *Annual Report*. Washington: AAAS.

Aikenhead, G. S. (1991). *Logical reasoning in science and technology.* Toronto: Wiley.

Aikenhead, G. S. (2005). *Science for everyday life: Evidence-based practice*. New York: Teachers College Press.

Aufschnaiter, C., Erduran, S., Osborne, J., & Simon, S. (2007). *Arguing to learn and learning to argue: Case studies of how students' argumentation relates to their scientific knowledge.* Dordrecht: Springer.

Baird, J. R. (1990). Metacognition, purposeful enquiry and conceptual change. In E. Hegarty-Hazel (Ed.), *The student laboratory and the science curriculum* (pp. 183–200). London: Routledge.

Bayliss, V. (1998). *Redefining work: An RSA initiative*. London: RSA.

Bingle, W. H., & Gaskell, P. J. (1994). Scientific literacy for decision making and the social construction of scientific knowledge. *Science Education, 78*(2), 185–201.

Bybee, R. W. (2004). Scientific inquiry and science teaching. In L. B. Flick & N. G. Lederman (Eds.), *Scientific inquiry and the nature of science: Implications for teaching, learning and teacher education* (pp. 1–14). Dordrecht: Springer.

COMEST. (2005). *The precautionary principle*. Paris: UNESCO.

Duggan, S., & Gott, R. (2002). What sort of science education do we really need? *International Journal of Science Education, 24*(7), 661–679.

Eijkelhof, H. (1986). Dealing with acceptable risk in science education: The case of ionizing radiation. In M. J. Fraser & A. Kornhauser (Ed.), *Ethics and social responsibility in science education* (pp. 189–200). Oxford: Pergamon Press.

Evans, G., & Durant. J. (1995). The relation between knowledge and attitude in public understanding of science in Britain. *Public Understanding of Science, 4*(1), 57–74.

Flick, L. B., & Lederman, N. G. (Eds.) (2004). *Scientific inquiry and the nature of science: Implications for teaching, learning and teacher education.* Dordrecht: Springer.

France, B. (2007). Location, location, location: Positioning biotechnology education for the 21st century. *Studies in Science Education, 43,* 88–122.

Fraser, S. W., & Greenhalgh, T. (2001). Coping with complexity: Educating for capability. *British Medical Journal, 323,* 799–803.

Freudenburg, W. (1988). Perceived risk, real risk: Social science and the art of probabilistic risk assessment. *Science, 242,* 44–49.

Gardner, P. L. (1972). *Words in science.* ERIC/SMEAC: Ohio State University.

Gardner, P. L. (1975). Science and the structure of knowledge. In P. L. Gardner (Ed.), *The structure of science education* (pp. 1–40). Hawthorn: Longman.

Gauld, C., Matthews, M., & Stinner, A. (2005). *The pendulum.* Germany: Springer.

Gee, J. P., Hull, G., & Lankshear, C. (1996). *The new work order: Behind the language of new capitalism.* Sydney: Allen and Unwin.

Georghiades, P. (2006). The role of metacognitive activities in the contextual use of primary pupils' conceptions of science. *Research in Science Education, 36*(1/2), 29–49.

Gilbert, J. (2005). *Catching the knowledge wave: The knowledge society and the future of education.* Wellington: New Zealand Council for Educational Research.

Gilbert, J. K., Boulter, C., & Rutherford, M. (1998). Models in explanations: Horses for courses? *International Journal of Science Education, 20*(1), 83–97.

Guillen, M. (1999). *Five equations that changed the world.* London: Abacus.

Gunstone, R. F. (1994). The importance of specific science content in the enhancement of metacognition. In P. J. Fensham, R. F. Gunstone, & R. T. White (Eds.), *The content of science* (pp. 131–146). London: Falmer.

Halliday, M. A. K., & Martin, J. R. (1993). *Writing science.* London: Falmer.

Harremoes, P., Gee, D., MacGarvin, M., Stirling, A., Keys, J., Wynne, B., Guedes Vaz, S. (Eds.) (2002). *The precautionary principle in the 20th century.* London: Earthscan.

Irwin, A., & Wynne, B. (1996). *Misunderstanding science? The public reconstruction of science and technology.* Cambridge: Cambridge University Press.

Jenkins, E. (2000). 'Science for all': Time for a paradigm shift? In L. R. Millar, J. Leach, & Osborne, J. (Eds.), *Improving science education: The contribution of research* (pp. 207–226). Buckingham: Open University Press.

Kolstø, S. (2006). Patterns in students' argumentation confronted with a risk-focused socio-scientific issue. *Science Education, 28*(14), 1689–1716.

Kress, G., Jewett, C., Ogburn, J., & Tsatsarelis, C. (2001). *Multimodal teaching and learning: The rhetorics of the science classroom.* London: Continuum.

Kuhn, D. (1991). *The skills of argument.* New York: Oxford University Press.

Kurtz, C. F., & Snowden, D. J. (2003). The new dynamics of strategy: Sense-making in a complex and complicated world. *IBM Systems Journal, 42*(3), 462–483.

Larkin, S. (2006). Collaborative group work and individual development of metacognition in the early years. *Research in Science Education, 36*(1/2), 7–27.

Layton, D. (1991). Science education and praxis: The relationship of school science to practical action. *Studies in Science Education, 19*, 43–79.

Layton, D., Jenkins, E., Macgill, S., & Davey, A. (1993). *Inarticulate science? Perspectives on the public understandings of science and some implications for science education* (1st ed.). Driffield: Studies in Education.

Lemke, J. (1990). *Talking science: Language, learning and values.* Norwood: Ablex.

Lyons, T. (2006). Different countries, same science classes: Students' experiences of school science in their own words. *International Journal of Science Education, 28*(6), 591–613.

Maddox, B. (2002). *Rosalind Franklin: The dark lady of DNA.* London: Harper Collins.

Millar, R. (2006). Twenty first century science: Insights from the design and implementation of a scientific literacy approach to school science. *International Journal of Science Education, 28*(13), 1499–1521.

Miller, J. D. (1998). The measurement of civic scientific literacy. *Public Understanding of Science, 7,* 203–223.
Ministry of Education. (2007). *The New Zealand Curriculum*. Wellington: Ministry of Education.
Mintzes, J. J., Wandersee, J. H., & Novak, J. D. (2005). *Assessing science understanding: A human constructivist view.* San Diego: Elsevier.
Mitchell, I. (Ed.) (2005). *Teaching for effective learning: The complete book of PEEL teaching procedures*. Melbourne: PEEL.
Mortimer, E. M., & Scott, P. H. (2003). *Making meaning in secondary science classrooms.* Maidenhead: Open University Press.
National Research Council. (2000). *Inquiry and the national science education standards: A guide to teaching.* Washington: The National Academies Press.
National Research Council. (2001). *Grand Challenges in Environmental Sciences*. Washington: NRC.
NCB. (2009). *Inquiry and the national science education standards*. Washington: National Academies Press.
Nentwig, P., Roennebeck, S., Schoeps, K., Rumann, S., & Carstensen, C. (2009). Performance and levels of contextualisation in a selected number of OECD countries in PISA 2006. *Journal of Research in Science Teaching, 46*(8), 897–908.
OECD. (1996a). *The knowledge-based economy*. Paris: OECD
OECD. (1996b). *Employment and growth in a knowledge based economy*. Washington: OECD Publications and Information Centre.
OECD. (2000). *From initial education to working life- Making transitions to work*. Paris: OECD.
OECD. (2006). *Assessing scientific, reading and mathematical literacy: A framework for PISA 2006.* Paris: OECD.
Ogburn, J., Kress, G., Martins, I., & McGillicuddy, K. (1996). *Explaining Science in the Classroom*. Buckingham: Open University Press.
Ohlsson, S. (1995) Learning to do and learning to understand: A lesson and challenge for cognitive modelling. In P. Reimann & H. Sparta. (Eds.), *Learning in humans and human machines: Towards an interdisciplinary science* (pp. 37–62). New York: Pergamon.
Omenn, G. S. (2006). Grand challenges and great opportunities in science, technology, and public policy. *Science, 414*(5806), 1696–1704.
Oulton, C., Dillon, J., & Grace, M. (2004). Reconceptualizing the teaching of controversial issues. *International Journal of Science Education, 26*(4), 411–423.
Posner, G. J., Strike, K. A., Hewson, P. W., & Gertzog, W. A. (1982) Accommodation of a scientist conception: Towards a theory of conceptual change. *Science Education, 66*(2), 211–227.
Ratcliffe, M. (1997). Pupil decison-making about socio-scientific issues within the science curriculum. *International Journal of Science Education, 19*(2), 167–182.
Renn, O. (1992). Concepts of risk: A classification. In S. Krimsky & D. Golding. (Ed.), *Social theories of risk* (1st ed., pp. 53–82). Westport: Praeger.
Rennie, L. J. (2007). Learning science outside school. In S. K. Abell, & N. G. Lederman (Eds.), *Handbook of research on science education* (pp. 125–171). Mawah: Lawrence Erlbaum.
Ritchie, S. M., Rigano, D. L., & Duane, A. (2008). Writing an ecological mystery in class: Merging genres in learning science. *International Journal of Science Education, 30*(2), 143–166.
Roberts, D. (2007). Scientific literacy/science literacy. In S. K. Abel & N. G. Lederman. (Eds.), *Handbook of Research on Science Education* (pp.125–171). Mahwah, NJ: Lawrence Erlbaum Associates.
Royal Society for Arts. (1986). *Education for capability*. Windsor: NFER-Nelson.
Rychen, D. S., & Salganik, L. H. (Eds.) (2003). *Key competencies for a successful life and well functioning society.* Gottingen: Hagrefen & Huber.
Ryder, J. (2001). Identifying science understanding for functional scientific literacy. *Studies in Science Education, 36*, 1–44.
Ryder, J. (2002). School science education for citizenship: Strategies for teaching about the epistemology of science. *Journal of Curriculum Studies, 34*(6), 637–658.

Sadler, T. D., & Zeidler, D. L. (2009). Scientific literacy, PISA and socioscientific discourse: Assessment for progressive aims of science education. *Journal of Research in Science Teaching, 46*(8), 909–921.
Sadler, T. D., Chambers, F., & Zeidler, D. (2004). Student conceptualizations of the nature of science in response to a socioscientific issue. *International Journal of Science Education, 26*(4), 387–409.
Science Research Council of Canada. (1984). *Science for every student: Educating Canadians for tomorrow's world.* Ottawa: Supply and Service.
Solomon, J. (2003). Risk: Why don't they listen to us? *Studies in Science Education, 39*, 125–141.
Solomon, J., & Aikenhead, G. (Eds.) (1994). *STS education: International perspectives on reform.* New York: Teachers College Press.
Sutton, C. (1992). *Words, science and learning.* Milton Keynes, Berks.: Open University Press.
Tasmania: Department of Education. (2002). *Essential learnings: Framework 1.* Hobart: Department of Education.
Thomas, G. P. (2006). Editorial—Metacognition and science education: Pushing forward from a solid foundation. *Research in Science Education, 36*(1/2), 1–6.
Unsworth, L. (2001). *Teaching multiliteracies across the curriculum: Changing contexts of text and image in classroom practice.* Buckingham: Open University Press.
Van Rooy, W. (1994). Teaching science using controversial issues: Some guidelines to enhance student learning and motivation. *Australian Science Teachers Journal, 40*(1), 24–27.
Varmus, H., Klauser, R., Zerhouni, E., Acharaya, T., Daar, A. S., Singer, P. A. (2003). Grand callenges in global health. *Science, 302,* 398.
VCAA. (2005). *Victorian essential learning standards.* Melbourne: Victorian Curriculum and Assessment Authority.
Watson, J. D. (1968). *The Double Helix.* London: Weedenfeld and Nicolson.
White R. T., & Gunstone, R. F. (1992). *Probing Understanding.* London: Falmer Press.
Wilson, J. M. (1999). Using words about thinking. *International Journal of Science Education, 21*(10), 1067–1084.
Winchester, S. (2002). *The map that changed the world.* London: Penguin.
Zuzovsky, R., & Tamir, P. (1999). Growth patterns in students' ability to supply scientific explanations: Findings from the Third International mathematics and science study in Israel. *International Journal of Science Education, 21*(10), 1101–1121.

Author Index

D. Corrigan et al. (eds.), *The Professional Knowledge Base of Science Teaching,*
DOI 10.1007/978-90-481-3927-9, © Springer Science+Business Media B.V. 2011

Subject Index

D. Corrigan et al. (eds.), *The Professional Knowledge Base of Science Teaching*,
DOI 10.1007/978-90-481-3927-9, © Springer Science+Business Media B.V. 2011

The manufacturer's authorised representative in the EU is Springer Nature Customer Service Centre GmbH, Europaplatz 3, 69115 Heidelberg, Germany. If you have any concerns regarding our products, please contact ProductSafety@springernature.com

Printed and bound by CPI Group (UK) Ltd, Croydon, CR0 4YY
30/09/2026
02225510-0012